CONTENTS

FIFTH EDITION

MATH FOR THE
AUTOMOTIVE TRADE

0.25
5400.0
0.000001
1.5707963
32.0
4.0
0.000279
144.0
39.2

FIFTH EDITION
MATH FOR THE
AUTOMOTIVE TRADE

John C. Peterson
Professor of Mathematics (retired)
Chattanooga State Technical Community College
Chattanooga, Tennessee

William J. deKryger
Professor Emeritus of Automotive Technology
Central Michigan University
Mount Pleasant, Michigan

DELMAR
CENGAGE Learning

Australia • Brazil • Japan • Korea • Mexico • Singapore • Spain • United Kingdom • United States

DELMAR
CENGAGE Learning

Math for the Automotive Trade, 5e
John C. Peterson, William J. deKryger

Vice President, Editorial: Dave Garza

Director of Learning Solutions: Sandy Clark

Acquisitions Editor: Stacy Masucci

Managing Editor: Larry Main

Product Manager: Mary Clyne

Editorial Assistant: Andrea Timpano

Vice President, Marketing: Jennifer Baker

Marketing Director: Deborah Yarnell

Associate Marketing Manager: Shanna Gibbs

Production Director: Wendy Troeger

Production Manager: Mark Bernard

Senior Content Project Manager: Angela Sheehan

Art Director: David Arsenault

Technology Project Manager: Joe Pliss

For product information and technology assistance, contact us at
Cengage Learning Customer & Sales Support, 1-800-354-9706
For permission to use material from this text or product,
submit all requests online at **www.cengage.com/permissions**
Further permissions questions can be e-mailed to
permissionrequest@cengage.com

Library of Congress Control Number: 2010937937

ISBN-13: 978-1-111-31823-9

ISBN-10: 1-111-31823-9

Delmar
5 Maxwell Drive
Clifton Park, NY 12065-2919
USA

Cengage Learning is a leading provider of customized learning solutions with office locations around the globe, including Singapore, the United Kingdom, Australia, Mexico, Brazil, and Japan. Locate your local office at:
international.cengage.com/region

Cengage Learning products are represented in Canada by
Nelson Education, Ltd.

To learn more about Delmar, visit **www.cengage.com/delmar**

Purchase any of our products at your local college store or at our preferred online store **www.cengagebrain.com**

Notice to the Reader

Publisher does not warrant or guarantee any of the products described herein or perform any independent analysis in connection with any of the product information contained herein. Publisher does not assume, and expressly disclaims, any obligation to obtain and include information other than that provided to it by the manufacturer. The reader is expressly warned to consider and adopt all safety precautions that might be indicated by the activities described herein and to avoid all potential hazards. By following the instructions contained herein, the reader willingly assumes all risks in connection with such instructions. The publisher makes no representations or warranties of any kind, including but not limited to, the warranties of fitness for particular purpose or merchantability, nor are any such representations implied with respect to the material set forth herein, and the publisher takes no responsibility with respect to such material. The publisher shall not be liable for any special, consequential, or exemplary damages resulting, in whole or part, from the readers' use of, or reliance upon, this material.

Printed in the United States of America
1 2 3 4 5 6 7 15 14 13 12 11

PREFACE

MATH FOR THE AUTOMOTIVE TRADES, 5th Edition, is a text-workbook that provides the automotive technology student with examples and problems encountered in this occupation. This edition has been revised to reflect the technological requirements of workers in automotive technology. The text has three parts. The first part, Chapters 1–7, reviews basic mathematics, measurement, and statistics skills and provides an introduction to small business practices. The second part, Chapters 8–14, applies the basic mathematics skills to specific automotive situations. This part begins with a chapter titled "Completing Repair Orders." Repairs orders are used throughout the remaining chapters of this part. The third part, Chapter 15, examines measurement aspects of both analog and digital measurement tools used in the automotive trade.

Changes specific to this edition include the following:

- Applications and examples have been updated to reflect changes in automotive technology. In particular, some of the information about analog meters has been eliminated.

- The chapter "Ratios, Proportions, and Percentages" has been divided into two chapters as follows:
  (a) Chapter 6: Ratios, Proportions, and Percentages
  (b) Chapter 7: Business and Statistics

- The flat rate pages for the Corvette in Appendix A have been replaced with pages for the Ford Fusion, Milan (2006–2009), Zephyr (2006), MKZ (2007–2009), and Milan Hybrid (2010).

- New repair orders have been created.

- Many questions have been added to reflect current concerns with hybrid, electric, and other new technologies.

- Gas prices, tax data, and so on have been updated to reflect current prices.

Ancillary Materials

This list of ancillary materials is comparable to material available for Peterson/deKryger's *Math for the Automotive Trade,* 5e. Much of it has been developed by the authors.

- *Student Solutions Manual* for solutions to all odd-numbered exercises and problems.

- *Instructor's Guide* consisting of solutions and answers to all exercises and problems.

- An e-resource containing

  – Computerized test bank (for both Windows and Macintosh)

  – PowerPoint presentation slides

  – Image library, including

* All textbook figures

* All automotive spec sheets in the text

* All repair orders

* All flat rate pages in the text

The authors wish to thank the following people for their contributions and guidance:

Reviewers for the 5th Edition:

Warren Farnell
Northampton Community College
Bethlehem, PA

Michael J. Peth
Ohio Technical College
Cleveland, OH

About the Authors

John C. Peterson is a retired professor of mathematics at Chattanooga State Technical Community College (CSTCC). Peterson received the Teaching Excellence Award from CSTCC. He received a Ph.D. in Mathematics Education from The Ohio State University. His professional affiliations include American Mathematics Association of Two-Year Colleges (AMATYC) Southeast Vice President 1996–2000 and Production Manager of *The AMATYC Review*, 2004–2008; National Council of Teachers of Mathematics; Tennessee Mathematics Teachers Association; Mathematical Association of America; and American Mathematical Society. He has over 90 professional publications. Peterson is the author of *Technical Mathematics* and *Introductory Technical Mathematics* as well as this text. He is also a consulting editor for Delmar's applied math list.

William J. deKryger is a retired professor of automotive technology in the Department of Engineering and Technology at Central Michigan University. He has been a certified ASE Master Technician for over 35 years and is a member of the Society of Automotive Engineers. During his automotive career working as a professional technician, he was employed by both automotive dealerships and independent garages in Grand Rapids, Michigan; Hollywood and Culver City, California; and Honolulu, Hawaii. Other professional services have included acting as an expert witness for automotive cases; automotive arbitration for the National Center for Dispute Settlement; annual study abroad group leader to Vladimir State University, Vladimir, Russia; and annual Visiting Professor of Automotive Technology to Instituta Technbologico Y de Superiores de Monterrey, Campus Toluca, Mexico.

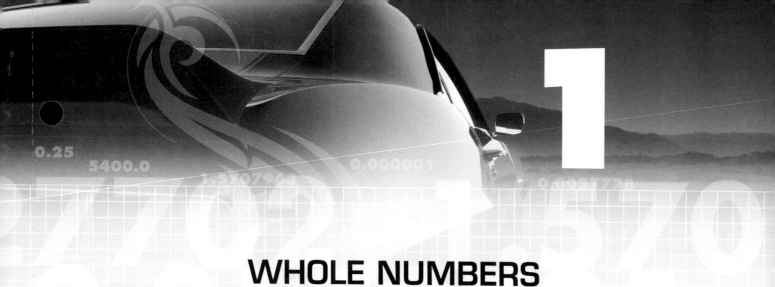

WHOLE NUMBERS

Objectives: After studying this chapter, you should be able to:

- Add, subtract, multiply, and divide whole numbers.
- Apply the basic mathematics operations of whole numbers to solve practical problems.
- Determine total costs related to repair and maintenance.
- Calculate trip distance, fuel consumption, and fuel economy.
- Determine costs related to buying, selling, and insuring a car.
- Determine component inventory volume and production requirements.
- Determine weights of automobiles and automotive components.

Whole numbers are numbers such as 14, 57, and 832. Each digit has a different value depending on its location in the whole number. In the whole numbers 14, 57, and 832, the 4, 7, and 2 are in the ones' place of each number. In a decimal, such as 23.47, the number just to the left of the decimal point is in the ones' place. You could write each whole number with a decimal point. Therefore, you could have written 14., 57., and 832. for the above numbers.

The position just to the left of the ones' place is the tens' place. In the numbers 14, 57, and 832, the 1, 5, and 3 are all in the tens' place.

The third position left of the decimal place is the hundreds' place. The 8 in the number 832 is in the hundreds' place.

Each digit in a whole number has a place value. The chart below shows some of the names of the places for a whole number with seven digits.

millons	hundred thousands	ten thousands	thousands	hundreds	tens	ones
2,	9	8	4,	3	6	5

Large numbers may be written either with a space or a comma separating every third numeral. Hence, the number above could have been written as 2 984 365 or as 2,984,365. Large numbers are usually written with spaces instead of commas when the metric system of measurement is used.

Adding Whole Numbers

In order to answer a question like the following, you must be able to add whole numbers.

A new automobile has a basic price of $14,674. For power sunroof, add $732; for California emission control, add $93; for a cellular phone, add $434. Destination charges are $525. What is the total price of the new car before taxes?

When adding whole numbers, first write the numbers so that the numbers in the ones' place are all in the same column, the numbers in the tens' place are in the same column, and so on. Then add the numbers in the ones' column, then the numbers in the tens' column, and so on. The answer to an addition problem is called the *sum*.

1

■ EXAMPLE 1–1

Find the sum of 214 + 2 + 53.

Solution

ones' place
↓
```
  2 1 4
      2
+   5 3
```

Step 1   Write numbers with digits in the ones' place in the same column.

```
  2 1 4
      2
+   5 3
      9
```

Step 2   Add the digits in the ones' place.

```
  2 1 4
      2
+   5 3
    6 9
```

Step 3   Add the digits in the tens' place.

```
  2 1 4
      2
+   5 3
  2 6 9
```

Step 4   Add the digits in the hundreds' place. The sum is 269.

If the sum of the digits in any column is more than 9, you will have to carry part of the sum to the next column.

■ EXAMPLE 1–2

Add 45 + 28.

Solution

```
  4 5
+ 2 8
```

Step 1

```
    1
  4 5
+ 2 8
    3
```

Step 2   Since 8 + 5 = 13, carry the 1 to the tens' column.

```
    1
  4 5
+ 2 8
  7 3
```

Step 3   Add the digits in the tens' column. Thus, the sum of 45 and 28 is 73.

■ EXAMPLE 1–3

Add 4568 + 927.

Solution

```
      1
  4 5 6 8
+   9 2 7
        5
```

Step 1   8 + 7 = 15. Carry the 1 to the tens' column.

```
      1
  4 5 6 8
+   9 2 7
      9 5
```

Step 2   1 + 6 + 2 = 9, so no number needs to be carried.

```
    1   1
  4 5 6 8
+   9 2 7
    4 9 5
```

Step 3   5 + 9 = 14. Carry the 1 to the thousands' column.

```
    1   1
  4 5 6 8
+   9 2 7
  5 4 9 5
```

Step 4   Add the digits in the thousands' column. Therefore, the sum of 4568 and 927 is 5495.

A four-digit number such as 5495 can be written without a comma or a space. Thus, 5,495, 5 495, and 5495 are all correct.

■ EXAMPLE 1–4

Find the sum of 7805 + 2041 + 926.

Solution

```
      1
  7 8 0 5
  2 0 4 1
+   9 2 6
        2
```

Step 1   The sum of the digits in the ones' column is 12. Carry the 1 to the tens' column.

```
      1
   7 8 0 5
   2 0 4 1
 +   9 2 6
 ─────────
         7 2
```

Step 2 $1 + 0 + 4 + 2 = 7$. No number needs to be carried.

```
   1   1
   7 8 0 5
   2 0 4 1
 +   9 2 6
 ─────────
       7 7 2
```

Step 3 $8 + 0 + 9 = 17$. Carry the 1 to the thousands' column.

```
   1   1
   7 8 0 5
   2 0 4 1
 +   9 2 6
 ─────────
 1 0 7 7 2
```

Thus, the sum is 10,772.

■ EXAMPLE 1–5

Find the sum of $437 + 28 + 741 + 2835$.

Solution

```
         2
     4 3 7
       2 8
     7 4 1
 + 2 8 3 5
 ─────────
         1
```

Step 1   The sum of the digits in the ones' column is 21. Carry the 2 to the tens' column.

```
     1 2
     4 3 7
       2 8
     7 4 1
 + 2 8 3 5
 ─────────
       4 1
```

Step 2   The sum of the digits in the tens' column is 14. Carry the 1 to the hundreds' column.

```
   2 1 2
     4 3 7
       2 8
     7 4 1
 + 2 8 3 5
 ─────────
     0 4 1
```

Step 3   The sum of the digits in the hundreds' column is 20. Carry the 2 to the thousands' column.

```
   2 1 2
     4 3 7
       2 8
     7 4 1
 + 2 8 3 5
 ─────────
   4 0 4 1
```

The sum is 4041.

■ EXAMPLE 1–6

Find the sum of $25\ 354 + 48 + 9784 + 129 + 97$.

Solution

Remember, 25 354 means the same as 25,354. Here we will write 9784 as 9 784.

```
         3
   2 5 3 5 4
         4 8
     9 7 8 4
       1 2 9
 +       9 7
 ───────────
           2
```

Step 1   The sum of the digits in the ones' column is 32. Carry the 3 to the tens' column.

```
       3 3
   2 5 3 5 4
         4 8
     9 7 8 4
       1 2 9
 +       9 7
 ───────────
         1 2
```

Step 2   The sum of the digits in the tens' column is 31. Carry the 3 to the hundreds' column.

```
     1 3 3
   2 5 3 5 4
         4 8
     9 7 8 4
       1 2 9
 +       9 7
 ───────────
       4 1 2
```

Step 3   The sum of the digits in the hundreds' column is 14. Carry the 1 to the thousands' column.

```
   1 1 3 3
   2 5 3 5 4
         4 8
     9 7 8 4
       1 2 9
 +       9 7
 ───────────
     5 4 1 2
```

Step 4   The sum of the digits in the thousands' column is 15. Carry the 1 to the ten-thousands' column.

$$
\begin{array}{r}
{\scriptstyle 1\ 1\ 3\ 3} \\
2\,5\,3\,5\,4 \\
4\,8 \\
9\,7\,8\,4 \\
1\,2\,9 \\
+\quad 9\,7 \\
\hline
3\,5\,4\,1\,2
\end{array}
$$

Thus, the sum is 35 412.

■ EXAMPLE 1–7

At the beginning of this section on adding whole numbers, the following question was asked. Answer the question.

A new automobile has a basic price of $14,674. For power sunroof, add $732; for California emission control, add $93; for a cellular phone, add $434. Destination charges are $525. What is the total price of the new car before taxes?

Solution

To find the total price of this new car before taxes, we need to add 14,674 + 732 + 93 + 434 + 525.

$$
\begin{array}{r}
{\scriptstyle 1} \\
1\,4\,6\,7\,4 \\
7\,3\,2 \\
9\,3 \\
4\,3\,4 \\
+\quad 5\,2\,5 \\
\hline
8
\end{array}
$$

Step 1 The sum of the digits in the ones' column is 18. Carry the 1 to the tens' column.

$$
\begin{array}{r}
{\scriptstyle 2\ 1} \\
1\,4\,6\,7\,4 \\
7\,3\,2 \\
9\,3 \\
4\,3\,4 \\
+\quad 5\,2\,5 \\
\hline
5\,8
\end{array}
$$

Step 2 The sum of the digits in the tens' column is 25. Carry the 2 to the hundreds' column.

$$
\begin{array}{r}
{\scriptstyle 2\ 2\ 1} \\
1\,4\,6\,7\,4 \\
7\,3\,2 \\
9\,3 \\
4\,3\,4 \\
+\quad 5\,2\,5 \\
\hline
4\,5\,8
\end{array}
$$

Step 3 The sum of the digits in the hundreds' column is 24. Carry the 2 to the thousands' column.

$$
\begin{array}{r}
{\scriptstyle 2\ 2\ 1} \\
1\,4\,6\,7\,4 \\
7\,3\,2 \\
9\,3 \\
4\,3\,4 \\
+\quad 5\,2\,5 \\
\hline
1\,6\,4\,5\,8
\end{array}
$$

Step 4 Add the digits in the thousands' and ten-thousands' columns. The final sum is 16,458, so the total price of this car with the indicated options is $16,458.

NAME _____ DATE _____ SCORE _____

1.   34
   + 52

2.   16
   + 41

3.   23
   + 45

4.   19
   + 8

5.   135
   + 42

6.   37
   + 261

7.   435
   + 241

8.   7605
   + 231

9.   37
   + 49

10.   26
   + 38

11. 236
  + 47

12. 254
  + 329

13. 246
  + 365

14. 196
  + 328

15.   97
  + 386

16. 486
  + 39

17. 2014
   191
  + 342

18. 472
  581
  + 406

19. 3462
  341
  2358
  + 902

20.   82
    76
   245
  2351
 + 650

21. $14 + 8 + 19 =$ _____

22. $27 + 15 + 34 =$ _____

23. $62 + 4 + 915 =$ _____

24. $402 + 63 + 9 + 78 =$ _____

25. An engine contains 5 quarts (qt) of oil and 9 qt of coolant. How many quarts of liquid does the engine contain?

25. _____

26. An engine contains 21 liters (L) of oil and 37 L of coolant. How many liters of liquid does the engine contain?

26. _____

27. A summer vacation began and ended in Fremont, Michigan. The first destination was Cleveland, Ohio, with a mileage of 331 miles, then another 564 miles to Hartford, Connecticut, followed by 347 miles to Washington DC, then 475 miles to Sevierville, Tennessee, 285 miles to Cincinnati, Ohio, and finally 419 miles back to Fremont, Michigan. What was the total mileage of the trip?

27. _____

28. A vehicle was driven 1389 kilometers (km) to Knoxville, Tennessee, 2016 km to Bangor, Maine, 576 km to Hartford, Connecticut, and 1467 km to Mt. Pleasant, Michigan. What was the total number of kilometers traveled?

28. _____

29. An automobile owner paid $68 for a tune-up, $85 for a wheel alignment, and $105 for one new tire. What was the total cost of the service to the car?

29. _____

30. An empty automobile weighs 3243 pounds (lb). The driver weighs 173 lb, the front passenger weighs 140 lb, one child weighs 74 lb, and the other child weighs 48 lb. What is the total weight of the vehicle and the passengers?

30. _____

31. An empty automobile weighs 1453 kg. The driver weighs 87 kg and the front passenger weighs 59 kg. If the two children weigh 35 kg and 27 kg, what is the total weight of the automobile and passengers?

31. _____

32. A new basic automobile costs $21,389. For air conditioning add $843, for tinted glass add $88, for larger tires and custom wheels add $385, and for the power convenience package add $1382. What is the total cost of the vehicle before taxes?

32. _____

33. A new hybrid vehicle has a basic price of $23,367. For air conditioning add $911. A power package adds $1894, a premium sound system costs $726, and the destination charge is $543. What is the total cost of this vehicle before taxes?

33. _____

34. A new truck is listed for a base price of $29,875. The power package is listed at $795, the super performance package is listed at $4287, the air deflector and fog lamps are $115, and the destination charge is $585. What is the total price of this truck before taxes?

34. _____

For exercises 35 and 36, use the information in the following table. The "$ List Price" refers to the original base retail price of the vehicle only. The "Avg. Trd-in" is the average price that you can expect to receive from a dealer if you trade in your vehicle. The "Avg. Retail" is the average retail price for which your car would sell. Along the bottom of the table are a list of optional items. This list begins "Add: CD/Radio $100/125." This means that if your car has a CD/radio, you should add $100 to the average trade-in price and $125 to the average retail price for your car. The symbol "w/o" means without.

Model	$ List Price	Avg. Trd-In	Avg. Retail
Sedan 4 dr. . . .	27,135	17,910	23,605

Add: CD/Radio $100/125; Pwr Windows $150/200; Pwr Dr Locks $100/125; Pwr Seat $150/200; Cruise Ctrl $150/200; Tilt Strg Whl $100/125; Custom Whls/Covers $200/250; Convenience package $350/450; V-6 Eng $450/575.

Deduct: w/o Auto Trans $550/550; w/o Air Cond $600/600.

35. Mary is buying another car to replace her four-door sedan, which is like the one described in the above table. It is equipped with a CD player, power windows, air conditioning, and an automatic transmission. How much can she expect to receive for her old car as a trade-in?

35. _____

36. Luis wants to sell his used car as described in the above table. It is equipped with a CD player, custom wheels, a V-6 engine, a convenience package, air conditioning, and a standard transmission. How much can he expect to receive if he sells his car?

36. _____

37. A turbocharged engine has a combustion chamber volume of 72 cubic centimeters (cm^3), the dished piston volume is 11 cm^3, and the head gasket thickness volume is 5 cm^3. What is the combustion volume of this cylinder?

37. _____

To balance a crankshaft, the weights of both the rotating and reciprocating components must be determined.

38. If the piston weighs 416 grams (g), the piston pin weighs 85 g, the two piston pin retainers each weigh 9 g, the two compression rings each weigh 20 g, the oil ring weighs 25 g, and the reciprocating small end of the connecting rod weighs 168 g, what is the total reciprocating weight?

38. _____

39. If the rotating big end of a rod weighs 418 grams, each of the two connecting rod bearing inserts weighs 30 grams, and the crankshaft contains 4 grams of oil, what is the total rotating weight?

39. _____

40. A trip is being planned from New York City to Los Angeles. Two routes are being considered: a northern route and a southern route. Listed below are mileages between selected cities on each route.

Northern Route

New York to Youngstown, OH	416
Youngstown to Chicago, IL	417
Chicago to Des Moines, IA	345
Des Moines to North Platte, NE	424
North Platte to Denver, CO	262
Denver to Grand Junction, CO	347
Grand Junction to Las Vegas, NV	407
Las Vegas to Los Angeles, CA	275

Southern Route

New York to Washington, DC	203
Washington, DC to Asheville, NC	487
Asheville to Memphis, TN	543
Memphis to Oklahoma City, OK	471
Oklahoma City to Tucumcari, NM	372
Tucumcari to Flagstaff, AZ	502
Flagstaff to Barstow, CA	355
Barstow to Los Angeles, CA	120

Determine which route is shorter.

(a) How long (in miles) is the northern route?

40a. _____

(b) How long (in miles) is the southern route?

40b. _____

(c) Which route is shorter?

40c. _____

41. The following parts are required for an engine rebuild. The price of each part is listed.

Camshaft $103
Valve lifters 48
Forged pistons 296
Connecting rod bearings 27
Crankshaft main bearings 35
Oil pump 31
Overhaul gasket set 78
Plug set................................... 12
Bronze valve guide liners 40
Valve springs 55
Reground crankshaft 96
Timing set 28
Camshaft bearings 14

What is the total cost of the parts needed for this rebuild?

41. _____

42. An engine dynamometer using a water brake power absorption unit needed the following amounts of water during one week of engine testing.

Monday .... 2173 gallons
Tuesday 4002
Wednesday 3749
Thursday 73
Friday 5987

What is the total amount of water usage during this one week of engine testing?

42. _____

Subtracting Whole Numbers

In order to answer a question like the following, you need to be able to subtract whole numbers.

At the beginning of a trip, an odometer reads 67,397 miles. At the end of the trip, the odometer reads 71,235 miles. How long (in miles) was the trip?

To subtract two whole numbers, arrange the numbers in columns with the larger number on top, and subtract the ones' digits, the tens' digits, the hundreds' digits, and so on.

■ EXAMPLE 1–8

537 − 124.

Solution

$$537$$
$$- 124$$

Step 1 Arrange the numbers in columns.

$$537$$
$$- 124$$
$$\overline{413}$$

Step 2 Subtract the ones, tens, and hundreds.

Sometimes it is necessary to rename, or borrow, in order to subtract.

■ EXAMPLE 1–9

653 − 407.

Solution

$$653$$
$$- 407$$

Step 1 Write the smaller number under the larger number.

$$\begin{array}{c} {}^{4\ 13} \\ 6\,\cancel{5}\,\cancel{3} \\ -\ 4\,0\,7 \end{array}$$

Step 2 You need more ones. Rename to get 10 more ones and 1 less ten. There are now 13 ones and 4 tens.

$$\begin{array}{c} {}^{4\ 13} \\ 6\,\cancel{5}\,\cancel{3} \\ -\ 4\,0\,7 \\ \hline 2\,4\,6 \end{array}$$

Step 3 Subtract the ones, tens, and hundreds.

It may be necessary to rename, or borrow, more than once.

■ EXAMPLE 1–10

1539 − 857.

Solution

$$1539$$
$$-\ 857$$

Step 1 Write the smaller number under the larger number.

$$1539$$
$$-\ 857$$
$$\overline{2}$$

Step 2 There are enough ones. Subtract the ones.

$$\begin{array}{c} {}^{4\ 13} \\ 1\,\cancel{5}\,\cancel{3}\,9 \\ -\ 8\,5\,7 \\ \hline 2 \end{array}$$

Step 3 You need more tens. Rename to get 10 more tens and 1 less hundred. There are now 13 tens and 4 hundreds.

$$\begin{array}{c} {}^{4\ 13} \\ 1\,\cancel{5}\,\cancel{3}\,9 \\ -\ 8\,5\,7 \\ \hline 8\,2 \end{array}$$

Step 4 Subtract the tens.

$$\begin{array}{c} {}^{14} \\ {}^{0\ \ 4\ 13} \\ 1\,\cancel{5}\,\cancel{3}\,9 \\ -\ 8\,5\,7 \\ \hline 8\,2 \end{array}$$

Step 5 You need more hundreds. Rename to get 10 more hundreds and 1 less thousand. There are now 14 hundreds and no thousands.

$$\begin{array}{c} {}^{14} \\ {}^{0\ \ 4\ 13} \\ 1\,\cancel{5}\,\cancel{3}\,9 \\ -\ 8\,5\,7 \\ \hline 6\,8\,2 \end{array}$$

Step 6 Subtract the hundreds and the thousands. Notice that there are no thousands left to subtract. So, 1539 − 857 = 682.

■ EXAMPLE 1–11

25,246 − 1819

Solution

$$
\begin{array}{r}
25246 \\
-1819 \\
\end{array}
$$

Step 1 Write the smaller number under the larger number.

$$
\begin{array}{r}
{}^{3}{}^{16} \\
25\,2\,4\,6 \\
-1\,8\,1\,9 \\
\hline
\end{array}
$$

Step 2 You need more ones. Rename to get 10 more ones and 1 less ten. There are now 16 ones and 3 tens.

$$
\begin{array}{r}
{}^{3}{}^{16} \\
25\,2\,4\,6 \\
-1\,8\,1\,9 \\
\hline
2\,7 \\
\end{array}
$$

Step 3 Subtract the ones and tens.

$$
\begin{array}{r}
{}^{4}{}^{12}{}^{3}{}^{16} \\
2\,5\,2\,4\,6 \\
-1\,8\,1\,9 \\
\hline
2\,7 \\
\end{array}
$$

Step 4 You need more hundreds. Rename to get 10 more hundreds and 1 less thousand. There are now 12 hundreds and 4 thousands.

$$
\begin{array}{r}
{}^{4}{}^{12}{}^{3}{}^{16} \\
2\,5\,2\,4\,6 \\
-1\,8\,1\,9 \\
\hline
2\,3\,4\,2\,7 \\
\end{array}
$$

Step 5 Subtract the hundreds, thousands, and ten thousands.

■ EXAMPLE 1–12

At the beginning of this section on subtracting whole numbers, the following question was asked. Solve the problem.

At the beginning of a trip, an odometer reads 67,397 miles. At the end of the trip, the odometer reads 71,235 miles. How long (in miles) was the trip?

Solution

To find the length of the trip in miles, we need to subtract 67,397 from 71,235.

$$
\begin{array}{r}
7\,1\,2\,3\,5 \\
-\,6\,7\,3\,9\,7 \\
\hline
\end{array}
$$

Step 1 Write the smaller number under the larger number.

$$
\begin{array}{r}
{}^{2}{}^{15} \\
7\,1\,2\,3\,5 \\
-\,6\,7\,3\,9\,7 \\
\hline
8 \\
\end{array}
$$

Step 2 You need more ones. Rename to get 10 more ones and 1 less ten. There are now 15 ones and 2 tens. Subtract the ones.

$$
\begin{array}{r}
{}^{12} \\
{}^{1}{}^{2}{}^{15} \\
7\,1\,2\,3\,5 \\
-\,6\,7\,3\,9\,7 \\
\hline
3\,8 \\
\end{array}
$$

Step 3 You need more tens. Rename to get 10 more tens and 1 less hundred. There are now 12 tens and 1 hundred. Subtract the tens.

$$
\begin{array}{r}
{}^{11}{}^{12} \\
{}^{0}{}^{1}{}^{2}{}^{15} \\
7\,1\,2\,3\,5 \\
-\,6\,7\,3\,9\,7 \\
\hline
8\,3\,8 \\
\end{array}
$$

Step 4 You need more hundreds. Rename to get 10 more hundreds and 1 less thousand. There are now 11 hundreds and no thousands. Subtract the hundreds.

$$
\begin{array}{r}
{}^{10}{}^{11}{}^{12} \\
{}^{6}{}^{0}{}^{1}{}^{2}{}^{15} \\
7\,1\,2\,3\,5 \\
-\,6\,7\,3\,9\,7 \\
\hline
3\,8\,3\,8 \\
\end{array}
$$

Step 5 You need more thousands. Rename to get 10 more thousands and 1 less ten-thousand. There are now 10 thousands and 6 ten-thousands. Subtract the thousands.

The difference is 3838, so the trip was 3838 miles long.

NAME _____ DATE _____ SCORE _____

1.　　　47　　　　　　2.　　　96　　　　　　3.　　196　　　　　4.　　467
　　　− 1 5　　　　　　　　− 3 4　　　　　　　− 4 5　　　　　　− 2 3 1

5.　4 3 5 6　　　　　6.　2 3 9 7　　　　　7.　5 8 4 2　　　　8.　7 3 8 7
　　− 2 1 4　　　　　　　− 1 8 4　　　　　　− 3 2 1 4　　　　　− 　7 6

9.　　　5 3　　　　　10.　　9 5　　　　　11.　2 4 3　　　　12.　4 8 2
　　　− 3 8　　　　　　　− 6 7　　　　　　　− 1 9　　　　　　− 6 8

13.　6 4 3 2　　　　　14.　3 4 2 7　　　　15.　2 4 9 1　　　16.　3 4 2 9
　　− 2 5 1　　　　　　　− 2 6 2　　　　　　−1 5 3 7　　　　　−1 5 6 2

17.　2 4 7 0　　　　　18.　7 6 8 0　　　　19.　2 4 0 5　　　20.　8 7 0 4
　　− 3 4 6　　　　　　　− 2 6 3　　　　　　−1 3 7 6　　　　　−4 9 1 5

21. $148 - 37 =$ _____　　　　　**22.** $4835 - 213 =$ _____

23. $263 - 85 =$ _____　　　　　**24.** $2347 - 259 =$ _____

25. A vehicle has an odometer that reads 13,854 miles. After a trip, it shows 15,965 miles. How long was the trip?

25. _____

26. An automobile owner has purchased 5 quarts of oil, a can of engine oil supplement, an oil filter, and an air filter for $28. A coupon gives a $5 discount.

   (a) How much did the owner have to pay?

26a. _____

   (b) If the customer paid with two $20 bills, how much change did she receive?

26b. _____

27. Shaft *A* has a diameter of 84 mm and shaft *B* has a diameter of 67 mm. What is the difference in diameter between the two shafts?

27. _____

28. One vehicle weighs 1088 kg and another weighs 1132 kg. What is the difference in weight between the two vehicles?

28. _____

29. It takes 76 minutes to drive 65 miles when you are traveling on the freeway. When you are using surface streets, it takes 2 hours and 3 minutes to drive the same distance. How much time can you save by using the freeway?

29. _____

30. At the beginning of a trip, the odometer reads 14 272 km. When the trip is over the odometer reads 15 189 km. How many kilometers long was the trip?

30. _____

31. You have agreed upon a final price of $23,849 for a new car. The dealer will allow $7371 for the car that you want to trade in. How much will you have to pay the dealer after the trade-in allowance?

31. _____

32. A new car will cost you $18,253. You sell your old car for $4726. If you use this as a down payment on the new car, how much do you still owe on your new car?

32. _____

For exercises 33–36, use the information in the following table. The "$ List Price" refers to the original base retail price of the vehicle only. The "Avg. Trd-In" is the average price that you can expect to receive from a dealer if you trade in your vehicle. The "Avg. Retail" is the average retail price for which your car would sell. An entry in the list at the bottom of the table such as "Pwr Windows (Std Conv)" means that power windows are standard on a convertible. Do not add this value if a convertible has power windows. Subtract this amount if it does not. The notation "w/o" means without.

Model	$ List Price	Avg. Trd-In	Avg. Retail
Sedan 2 dr.	13422	7750	9550
HB 2 dr.	14207	7875	9700
Conv 2 dr.	16899	9950	12550

   Add: Sunroof $250/325; CD/Radio $100/125; Pwr Windows (Std Conv) $150/200; Pwr Dr Locks (Std Conv) $100/125; Pwr Seat $150/200; Cruise Ctrl $150/200; Leather Seats $250/325; Custom Whls/Covers $200/250; 2-Tone Paint $75/100.

   Deduct: w/o Auto Trans $550/550; w/o Air Cond $650/650.

33. George wants to trade in his used car. His car is a two-door convertible like the one described in the table above. It has a CD/radio player, power windows, power door locks, leather seats, standard transmission, and air conditioning. How much can he expect to receive from a dealer if he trades in his car?

33. _____

34. George's used car was described in exercise 33. He decides to sell his car himself rather than trade it in.

   (a) How much should he try to sell his car for?

34a. _____

   (b) How much more than the trade-in allowance can he expect to get if he sells the car himself?

34b. _____

35. Elaine wants to sell her used car. Her car is a two-door hatchback like the one described in the table above. It has an automatic transmission, cruise control, power door locks, leather seats, and two-tone paint. How much can she expect to sell her car for?

35. _____

36. Elaine's used car was described in exercise 35. She decides to trade in her car for a new car that will cost her $19,732 for the options she wants.

   (a) How much can she expect to get on the trade-in?

36a. _____

   (b) How much will she owe on her new car after the trade-in?

36b. _____

   (c) If she can sell her car for what she expected to get in exercise 35 and she uses that money as a down payment on the new car, how much will she still owe on her new car?

36c. _____

37. What is the difference in miles between a northern route of 2938 miles from New York, NY, to Los Angeles, CA and a southern route of 3189 miles?

37. _____

38. One year 1,022,759 hybrid vehicles were sold in the United States. The next year 1,711,164 were sold. How many more were sold the second year?

38. _____

39. One engine is rated at 425 horsepower and 337 pound-feet (lb-ft) of torque. Another engine has a rated output of 385 horsepower and 297 lb-ft of torque. What are the differences in both horsepower and torque between the two engines?

39. (HP) _____

39. (T) _____

40. Some racing vehicles use high-performance engines with aluminum components in an effort to save weight and thereby improve handling and acceleration performance. Listed below are the weights of several iron and aluminum engine parts.

	Cast Iron	Aluminum
Cylinder block.	162 lb	83 lb
Cylinder head.	42	21
Intake manifold.	38	16
Coolant pump.	15	9

   (a) How much is the weight of the cast-iron engine parts?

40a. _____

   (b) How much is the weight of the aluminum engine parts?

40b. _____

   (c) How much weight can be saved on a V-8 engine by using aluminum parts rather than cast-iron parts?

40c. _____

41. A tax credit is a specific dollar amount that is deducted directly off the amount of federal income tax that you owe at the end of the year, lessening your tax obligation. Dan has a tax liability of $13,857 at the end of the year; however, during the year, he purchased a clean diesel–powered car that had a $1375 tax credit. What is the final amount of tax that Dan must pay for that year?

41. _____

42. Fred wants to purchase a new car that will provide excellent fuel economy and hold its value over the years. He went online and checked the used-car value for a popular hybrid and a popular clean diesel. He found two cars that were each four years old and had been driven 185,000 miles. The hybrid was valued at $4650 and the clean diesel had an estimated value of $6275. What is the difference in value between the two cars?

42. _____

Multiplying Whole Numbers

Multiplying whole numbers is used to solve the following problem.

> Each engine requires 14 head bolts. A manufacturer needs to order enough head bolts to assemble 12,782 engines. How many head bolts should be ordered?

The answer to a multiplication problems is called the *product.* The number that does the multiplying is the *multiplier;* the number multiplied by is called the *multiplicand.*

■ EXAMPLE 1–13

$$
\begin{array}{r}
1\,4 \leftarrow \text{multiplicand} \\
\times\ 3 \leftarrow \text{multiplier} \\
\hline
4\,2 \leftarrow \text{product}
\end{array}
$$

In a multiplication problem, each digit in the multiplier is multiplied by the multiplicand.

■ EXAMPLE 1–14

$$
\begin{array}{r}
4\,2 \\
\times\ 3 \\
\hline
1\,2\,6
\end{array}
$$
$\leftarrow 3 \times 2 = 6$
$ 3 \times 4 = 12$

■ EXAMPLE 1–15

$$
\begin{array}{r}
8\,2 \\
\times\ 4 \\
\hline
3\,2\,8
\end{array}
$$
$\leftarrow 4 \times 2 = 8$
$ 4 \times 8 = 32$

Sometimes it is necessary to rename, or carry, during multiplication.

■ EXAMPLE 1–16

$$
\begin{array}{r}
6\,7 \\
\times\ 8 \\
\hline
5\,3\,6
\end{array}
$$
$\leftarrow 8 \times 7 = 56.\ \text{Carry 5.}$
$ 8 \times 6 = 48.\ 48 + 5 = 53.$

■ EXAMPLE 1–17

$$
\begin{array}{r}
8\,9 \\
\times\ 4 \\
\hline
3\,5\,6
\end{array}
$$
$\leftarrow 4 \times 9 = 36.\ \text{Carry 3.}$
$ 4 \times 8 = 32.\ 32 + 3 = 35.$

Remember that when 0 is multiplied by a number, the product is 0.

■ EXAMPLE 1–18

$$
\begin{array}{r}
4\,2 \\
\times\ 3\,0 \\
\hline
1\,2\,6\,0
\end{array}
$$

Write a 0 in the ones' place. Multiply 3×42.

■ EXAMPLE 1–19

$$
\begin{array}{r}
4\,1\,2\,5 \\
\times\ 3\,0\,0 \\
\hline
1\,2\,3\,7\,5\,0\,0
\end{array}
$$

Write 0s in the ones' and tens' places. Multiply 3×4125.

When the multiplier has more than one digit, the multiplicand is multiplied by each digit in the multiplier.

■ EXAMPLE 1–20

$$
\begin{array}{r}
7\,2 \\
\times\ 3\,4 \\
\hline
2\,8\,8 \\
2\,1\,6\,0 \\
\hline
2\,4\,4\,8
\end{array}
$$
$\leftarrow$ (a)
$\leftarrow$ (b)
$\leftarrow$ (c)

(a) 4×72

(b) 30×72. Write a 0 in the ones' place and multiply 3×72.

(c) Add.

The product is 2448.

■ EXAMPLE 1–21

$$
\begin{array}{r}
9\,3 \\
\times\ 2\,5 \\
\hline
4\,6\,5 \\
1\,8\,6\,0 \\
\hline
2\,3\,2\,5
\end{array}
$$
$\leftarrow$ (a)
$\leftarrow$ (b)
$\leftarrow$ (c)

(a) 5×93

(b) 20×93. Write a 0 in the ones' place and multiply 2×93.

(c) Add.

The answer is 2325.

■ **EXAMPLE 1–22**

$$
\begin{array}{r}
3\,2\,7 \\
\times\,4\,1\,8 \\
\hline
2\,6\,1\,6 \quad \leftarrow (a) \\
3\,2\,7\,0 \quad \leftarrow (b) \\
1\,3\,0\,8\,0\,0 \quad \leftarrow (c) \\
\hline
1\,3\,6\,6\,8\,6 \quad \leftarrow (d)
\end{array}
$$

(a) 8×327

(b) Write a 0 in the ones' place and multiply 1×327.

(c) Write 0s in the ones' and tens' places and multiply 4×327.

(d) Add.

The product is 136,686.

■ **EXAMPLE 1–23**

At the beginning of this section on multiplying whole numbers, the following problem was given. Solve the problem.

Each engine requires 14 head bolts. A manufacturer needs to order enough head bolts to assemble 12,782 engines. How many head bolts should be ordered?

Solution

Since each engine requires 14 head bolts, the total number of head bolts needed for the 12,782 engines is the product of $12{,}782 \times 14$.

$$
\begin{array}{r}
1\,2\,7\,8\,2 \\
\times\,1\,4 \\
\hline
5\,1\,1\,2\,8 \quad \leftarrow (a) \\
1\,2\,7\,8\,2\,0 \quad \leftarrow (b) \\
\hline
1\,7\,8\,9\,4\,8 \quad \leftarrow (c)
\end{array}
$$

(a) 4×12782

(b) Write a 0 in the ones' place and multiply 1×12782.

(c) Add.

A total of 178,948 head bolts is needed for these engines.

NAME _____ DATE _____ SCORE _____

1. 8 2 × 4	**2.** 7 4 × 2	**3.** 7 1 × 5	**4.** 6 3 × 3
5. 3 2 × 9	**6.** 4 6 × 7	**7.** 9 6 × 4	**8.** 2 7 × 5
9. 2 7 5 × 3	**10.** 8 3 4 × 6	**11.** 4 0 7 × 8	**12.** 3 0 9 × 7
13. 6 3 × 7 0	**14.** 5 6 × 4 0	**15.** 8 4 × 1 2	**16.** 7 3 × 3 2
17. 4 6 × 2 7	**18.** 3 7 × 5 2	**19.** 6 4 × 5 3	**20.** 8 8 × 2 5
21. 3 4 2 × 3 0 0	**22.** 4 3 1 × 4 0 0	**23.** 9 0 5 × 6 0 0	**24.** 8 0 7 × 5 0 0
25. 3 4 1 × 2 4 0	**26.** 4 7 2 × 3 6 0	**27.** 6 8 8 × 4 0 7	**28.** 7 9 9 × 6 0 8
29. 4 2 1 × 3 2 3	**30.** 5 3 2 × 2 3 2	**31.** 1 4 2 5 × 3 2 7	**32.** 6 2 1 8 × 3 2 4

33. A worker drives 36 miles each day to and from work. How many miles will he travel in 5 days?

33. _____

34. If a gallon of gasoline weighs 7 pounds, how much will 18 gallons weigh?

34. _____

35. A technician spends 7 hours working on an engine. If the labor rate is $68 per hour, what will be the cost of the 7 hours of work?

35. _____

36. A driver travels 17 km to get to work each day. How many kilometers will she drive to and from work in a 5-day week?

36. _____

37. There are 252 working days in a year. A driver travels 17 km to get to work each day. How many kilometers are driven to and from work in one year?

37. _____

38. Your automobile gets 18 km/L. If the fuel tank holds 57 L, how far can you travel on a full tank?

38. _____

39. There are 252 working days in a year.

   (a) If a worker drives 13 miles each day for the round trip to and from work, what will the total mileage be for the working year?

39a. _____

   (b) If it costs $0.53 per mile to drive a car, what will the cost be for a 5-day week?

39b. _____

   (c) What will be the cost for a month (21 working days)?

39c. _____

   (d) What will be the total mileage cost for a year (252 working days)?

39d. _____

40. A car and driver weigh 3550 lb when the car is driven into the gasoline station. The driver puts 13 gallons of gasoline in the fuel tank. If each gallon of gasoline weighs 7 lb, what was the weight of the car and driver when they left the station?

40. _____

41. A worker has two jobs. The worker drives 27 miles (one way) to one job each Monday through Friday. On Monday, Tuesday, and Wednesday, the worker returns home by the same route. On Thursday and Friday, the worker drives 6 miles from work to the second job. It is 25 miles from the location of the second job back to the worker's house. What is the total number of miles the worker drives each week going to and from work?

41. _____

42. A driver wants to improve the appearance of a vehicle and plans to replace all four tires and rims. The new rims each cost $189 and the tires each cost $129. How much will it cost to replace all four tires and rims on this vehicle?

42. _____

43. A car is washed every week at a cost of $12. It also costs $1 to vacuum out the interior.

43. _____

   (a) How much will it cost for these services for one year (52 weeks)?

43a. _____

   (b) If the car owner keeps the vehicle for 5 years, what will be the total cost for these services over that time period?

43b. _____

44. Monthly automobile insurance costs are as listed for the following coverages:

Bodily injury	$19
Uninsured motorist	2
Property damage	3
Personal injury	15
Personal property	2
Standard collision	69
Road service	7
Special assessment	11

Determine the total yearly cost for the insurance coverage listed above.

44. _____

45. The engine oil is changed every other month and the filter is changed every other oil change. An oil change requires 4 qt of oil if the filter is not replaced and 5 qt of oil if it is. If synthetic oil blend costs $3 per quart and the oil filter costs $5, how much will it cost to service the engine for one year?

45. _____

46. An automobile manufacturer is planning to assemble 118,690 automobiles per year. Each automobile requires the following lengths of wire:

18 gauge	53 feet
16	41
14	24
12	19
10	13
6	11
4	5

(a) Determine the total length of wire for a single automobile.

46a. _____

(b) Determine the total length of wire required for one year's production.

46b. _____

Dividing Whole Numbers

In order to answer the following question, you must be able to divide whole numbers.

A certain car traveled 448 miles and used 14 gallons of fuel. How many miles per gallon is this?

Two different ways to write the same division problem are $714 \div 42$ and $42\overline{)714}$. The answer to a division problem is called the *quotient*.

The number you are dividing by is called the *divisor*. In this problem the divisor is 42.

$$714 \div 42 = 17 \qquad 42\overline{)714}$$

quotient ———
divisor

Estimating

Find the number of digits in a quotient before you divide.

■ EXAMPLE 1–24

$$6\overline{)84}$$

Think: Is the quotient 10 or more?
Yes, because $10 \times 6 = 60$ and 60 is not as large as 84. So, the quotient will have two digits.

■ EXAMPLE 1–25

$$8\overline{)576}$$

Think: $10 \times 8 = 80$
$100 \times 8 = 800$
576 is between 80 and 800, so the quotient is between 10 and 100. The quotient will have two digits.

■ EXAMPLE 1–26

$$23\overline{)5673}$$

Think: $10 \times 23 = 230$
$100 \times 23 = 2300$
$1{,}000 \times 23 = 23{,}000$
5673 is between 2300 and 23,000, so the quotient is between 100 and 1000. The quotient will have three digits.

Dividing

Once you know how many digits are in the quotient, start dividing to find the quotient.

■ EXAMPLE 1–27

$$6\overline{)8\ 4}$$

Step 1 There are two digits in the quotient.

$$\begin{array}{r} 1\ - \\ 6\overline{)8\ 4} \\ 6\ \downarrow \\ \hline 2\ 4 \end{array}$$

Step 2 Find the tens in the quotient.

Divide: $6\overline{)8}$ is about 1
Multiply: $1 \times 6 = 6$
Subtract: $8 - 6 = 2$
Bring down the 4.

$$\begin{array}{r} 1\ 4 \\ 6\overline{)8\ 4} \\ 6 \\ \hline 2\ 4 \\ 2\ 4 \\ \hline 0 \end{array}$$

Step 3 Find the ones in the quotient.

Divide: $6\overline{)24}$ is 4
Multiply: $4 \times 6 = 24$
Subtract: $24 - 24 = 0$

So, $84 \div 6 = 14$.

■ EXAMPLE 1–28

$$\begin{array}{r} -\ -\ - \\ 2\ 3\overline{)5\ 7\ 9\ 6} \end{array}$$

Step 1 There are three digits in the quotient.

$$\begin{array}{r} 2\ -\ - \\ 2\ 3\overline{)5\ 7\ 9\ 6} \\ 4\ 6\ \downarrow \\ \hline 1\ 1\ 9 \end{array}$$

Step 2 Find the hundreds in the quotient.

Divide: $23\overline{)57}$ is about 2
Multiply: $2 \times 23 = 46$
Subtract: $57 - 46 = 11$
Bring down the 9.

$$\begin{array}{r} 2\ 5\ - \\ 2\ 3\overline{)5\ 7\ 9\ 6} \\ 4\ 6 \\ \hline 1\ 1\ 9 \\ 1\ 1\ 5\ \downarrow \\ \hline 4\ 6 \end{array}$$

Step 3 Find the tens in the quotient.

Divide: $23\overline{)119}$ is about 5
Multiply: $5 \times 23 = 115$
Subtract: $119 - 115 = 4$
Bring down the 6.

$$
\begin{array}{r}
2\ 5\ 2 \\
2\ 3\overline{)5\ 7\ 9\ 6} \\
4\ 6 \\
\overline{1\ 1\ 9} \\
1\ 1\ 5 \\
\overline{4\ 6} \\
4\ 6 \\
\overline{0}
\end{array}
$$

Step 4   Find the ones in the quotient.

Divide: $23\overline{)46}$ is 2
Multiply: $2 \times 23 = 46$
Subtract: $46 - 46 = 0$

So, $5796 \div 23 = 252$.

Remainders

When the division is not exact, the answer has a *remainder*.

■ EXAMPLE 1–29

$$
2\ 7\overline{)9\ 2}^{\;-}
$$

Step 1 $1 \times 27 = 27$
$10 \times 27 = 270$
Since 92 is between 27 and 270, the quotient has one digit.

$$
\begin{array}{r}
3 \\
2\ 7\overline{)9\ 2} \\
8\ 1 \\
\overline{1\ 1}
\end{array}
$$

Step 2   Find the ones in the quotient.

Divide: $27\overline{)92}$ is about 3
Multiply: $3 \times 27 = 81$
Subtract: $92 - 81 = 11$
Since 11 is less than the divisor, 27, the remainder is 11.

So, $92 \div 27 = 3$ R 11.

■ EXAMPLE 1–30

$$
4\ 3\overline{)3\ 9\ 2\ 1}^{\;-\ -}
$$

Step 1 $10 \times 43 = 430$
$100 \times 43 = 4300$
Since 3921 is between 430 and 4300, the quotient has two digits.

$$
\begin{array}{r}
9\ - \\
4\ 3\overline{)3\ 9\ 2\ 1} \\
3\ 8\ 7\downarrow \\
\overline{5\ 1}
\end{array}
$$

Step 2   Find the tens.

Divide: $43\overline{)392}$ is about 9
Multiply: $9 \times 43 = 387$
Subtract: $392 - 387 = 5$
Bring down the 1.

$$
\begin{array}{r}
9\ 1 \\
4\ 3\overline{)3\ 9\ 2\ 1} \\
3\ 8\ 7 \\
\overline{5\ 1} \\
4\ 3 \\
\overline{8}
\end{array}
$$

Step 3   Find the ones.

Divide: $43\overline{)51}$ is about 1
Multiply: $1 \times 43 = 43$
Subtract: $51 - 43 = 8$
Since 8 is less than the divisor, 43, the remainder is 8.

So, $3921 \div 43 = 91$ R 8.

■ EXAMPLE 1–31

$$
5\ 7\overline{)1\ 7\ 3\ 5\ 6}^{\;-\ -\ -}
$$

Step 1 $100 \times 57 = 5700$
$1000 \times 57 = 57{,}000$
17,356 is between 5700 and 57,000, so the quotient has three digits.

$$
\begin{array}{r}
3\ -\ - \\
5\ 7\overline{)1\ 7\ 3\ 5\ 6} \\
1\ 7\ 1\downarrow \\
\overline{2\ 5}
\end{array}
$$

Step 2   Find the hundreds in the quotient.

Divide: $57\overline{)173}$ is about 3
Multiply: $3 \times 57 = 171$
Subtract: $173 - 171 = 2$
Bring down the 5.

```
        3 0 -
  5 7)1 7 3 5 6
      1 7 1
        2 5
        0 ↓
        2 5 6
```

Step 3 Find the tens.

Divide: $57\overline{)25}$ is 0
Multiply: $0 \times 57 = 0$
Subtract: $25 - 0 = 25$
Bring down the 6.

```
        3 0 4
  5 7)1 7 3 5 6
      1 7 1
        2 5
        0
        2 5 6
        2 2 8
        2 8
```

Step 4 Find the ones.

Divide: $57\overline{)256}$ is about 4
Multiply: $4 \times 57 = 228$
Subtract: $256 - 228 = 28$
Since 28 is less than the divisor, 57, the remainder is 28.

So, $17,356 \div 57$ is 304 R 28.

■ EXAMPLE 1–32

The following question was asked at the beginning of this section on dividing whole numbers. Solve the question.

A certain car traveled 448 miles and used 14 gallons of fuel. How many miles per gallon is this?

Solution

To answer this question we need to divide the number of miles traveled by the amount of fuel used.

$$1\ 4\overline{)4\ 4\ 8}$$

Step 1 $10 \times 14 = 140$
$100 \times 14 = 1400$
448 is between 140 and 1400, so the quotient has two digits.

```
        3 -
  1 4)4 4 8
      4 2 ↓
        2 8
```

Step 2 Find the tens in the quotient.

Divide: $14\overline{)44}$ is about 3
Multiply: $3 \times 14 = 42$
Subtract: $44 - 42 = 2$
Bring down the 8.

```
        3 2
  1 4)4 4 8
      4 2
        2 8
        2 8
        0
```

Step 3 Find the ones in the quotient.

Divide: $14\overline{)28}$ is 2
Multiply: $2 \times 14 = 28$
Subtract: $28 - 28 = 0$

So, $448 \div 14 = 32$. This car got 32 miles per gallon during this period.

NAME _____ DATE _____ SCORE _____

1. $4\overline{)6\ 8}$

2. $7\overline{)8\ 4}$

3. $9\overline{)7\ 4\ 7}$

4. $6\overline{)2\ 1\ 6}$

5. $19\overline{)6\ 0\ 8}$

6. $28\overline{)6\ 7\ 2}$

7. $42\overline{)3\ 1\ 9\ 2}$

8. $97\overline{)3\ 2\ 9\ 8}$

9. $38\overline{)5\ 8\ 1\ 4}$

10. $68\overline{)3\ 5\ 1\ 5\ 6}$

11. $7\overline{)2\ 6}$

12. $4\overline{)3\ 5}$

13. $7\overline{)2\ 3\ 4}$

14. $3\overline{)4\ 2\ 5}$

15. $12\overline{)7\ 0\ 5}$

16. $24\overline{)6\ 2\ 7}$

17. $32\overline{)8\ 2\ 1}$

18. $43\overline{)9\ 2\ 5}$

19. $47\overline{)3\ 2\ 4\ 8}$

20. $56\overline{)7\ 1\ 8\ 3}$

21. $5\overline{)4\ 5\ 2\ 9}$

22. $6\overline{)4\ 8\ 5\ 8}$

23. $49\overline{)9\ 9\ 6\ 0}$

24. $43\overline{)8\ 9\ 7\ 4}$

25. An automobile was driven 368 miles and used 16 gallons of fuel. How many miles per gallon is this?

25. _____

26. The labor charge for 6 hours of work was $252. How much is the labor rate in dollars per hour?

26. _____

27. Six piston ring sets cost $84. How much did each set cost?

27. _____

28. A manufacturer has purchased 12,112,996 head bolts that are to be used for engine assembly. Each engine uses 14 head bolts. How many engines can be assembled from the supply of head bolts?

28. _____

29. A driver has just traveled 350 miles in 7 hours. What was the average speed?

29. _____

30. An automobile traveled 935 km and used 55 L of fuel. How many kilometers was the auto able to travel on each liter?

30. _____

31. A driver has just purchased four new tires for $468. How much did each tire cost?

31. _____

32. A manufacturer has purchased 56,500 feet of window weather stripping. Each car requires 32 feet of weather stripping. How many cars can be outfitted with windows using this supply of weather stripping?

32. _____

33. An empty pickup truck weighs 5760 pounds. It is loaded with a cargo of 15 identical boxes. With the boxes, the truck now weighs 7605 pounds. How much does each box weigh?

33. _____

34. A manufacturer purchased 7,756,658 identical bolts. Each engine assembly uses 8 of these bolts and each transmission needs 9 bolts. How many engine and transmission combinations can be assembled using this supply of bolts?

34. _____

35. A set of forged pistons for a V-8 engine costs $536. What is the cost of each individual piston?

35. _____

36. A fleet of company vehicles logged 120,552,886 miles over a one-year period of time. During that time they consumed 4,464,918 gallons of fuel. Determine the average fuel mileage for this fleet of vehicles.

36. _____

37. An owner has purchased a new sports car and has just discovered that the annual insurance bill amounts to $2676. In an effort to make the payments more manageable, a request was made to make monthly payments rather than semi-annual payments. The insurance company replied that monthly payments could be made but that an $8 service charge would be added to the payment each month.

   (a) What is the amount of each monthly payment?

37a. _____

   (b) What is the annual total cost for making monthly payments?

37b. _____

   (c) How much extra is the owner paying over one year for the convenience of making monthly payments?

37c. _____

38. A stock clerk has 736 spark plugs. There are 8 plugs in a box. How many boxes are in stock?

38. _____

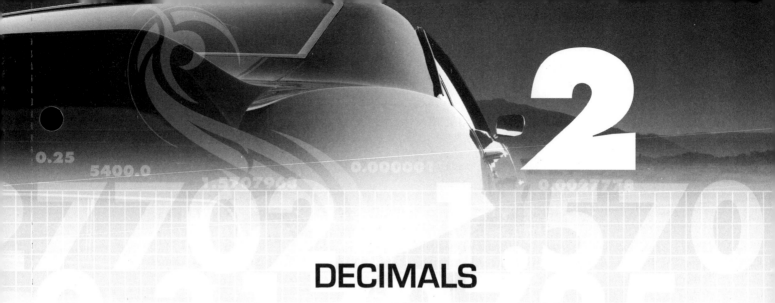

DECIMALS

Objectives: After studying this chapter, you should be able to:

- Add, subtract, multiply, and divide decimal numbers.
- Use estimates and mental math to solve problems with decimals and then compare the results with exact calculations.
- Apply the basic mathematics operations of decimals to solve practical problems.
- Calculate trip distances and fuel economy.
- Determine maintenance costs and payment schedules.
- Calculate piston displacement and total engine displacement.
- Determine component measurements and specification compliance.

Estimating Sums

You will often use a calculator to help you solve a problem. Estimating will help you decide if the calculator has given you the correct solution.

Round each number to the nearest whole number. Add the whole numbers. The answer to an addition problem is called the *sum*.

■ EXAMPLE 2–1

Estimate 42.3 + 25.62.

Solution
42.3 rounds to 42.
25.62 rounds to 26.
42 + 26 = 68
So, 42.3 + 25.62 is about 68.

■ EXAMPLE 2–2

Estimate 37 + 16.81.

Solution
37 is a whole number.
16.81 rounds to 17.
37 + 17 = 54
37 + 16.81 is about 54.

■ EXAMPLE 2–3

Estimate 14.7 + 216.93 + 26.4.

Solution
14.7 rounds to 15.
216.93 rounds to 217.
26.4 rounds to 26.
15 + 217 + 26 = 258
So, 14.7 + 16.93 + 26.4 is about 258.

Adding Decimals

In order to answer a question like the one below, you must be able to add decimals.

At a local discount auto parts store, a customer bought a case of multi-weight motor oil for $13.68, an oil filter for $2.74, and a windshield wiper blade for $5.97. What was the total price of these items before taxes?

Step 1 Write the numbers in a column with the decimal points in a line.

Step 2 Add enough 0s so that each number has the same number of places to the right of the decimal point.

Step 3 Add decimals the same way you add whole numbers. The decimal point in the sum goes directly under the other decimal points.

■ EXAMPLE 2–4

Solve 42.3 + 25.62.

Solution

$$
\begin{array}{r}
4\,2.3 \\
+\;2\,5.6\,2
\end{array}
$$

Step 1 Write the numbers so the decimal points are in a line.

$$
\begin{array}{r}
4\,2.3\,0 \\
+\;2\,5.6\,2
\end{array}
$$

Step 2 Add enough 0s so that each number has the same number of places to the right of the decimal point.

$$
\begin{array}{r}
4\,2.3\,0 \\
+\;2\,5.6\,2 \\
\hline
6\,7.9\,2
\end{array}
$$

Step 3 Add as you would whole numbers. The decimal point goes directly under the other decimal points. Therefore, 42.3 + 25.62 = 67.92.

In Example 2–1, we estimated that 42.3 + 25.62 was about 68. The sum in Example 2–4, 67.92, rounds to 68.

■ EXAMPLE 2–5

Solve 14.7 + 216.93 + 26.4.

Solution

$$
\begin{array}{r}
1\,4.7 \\
2\,1\,6.9\,3 \\
+\;2\,6.4
\end{array}
$$

Step 1 Write so decimal points are in line.

$$
\begin{array}{r}
1\,4.7\,0 \\
2\,1\,6.9\,3 \\
+\;2\,6.4\,0
\end{array}
$$

Step 2 Add 0s.

$$
\begin{array}{r}
1\,4.7\,0 \\
2\,1\,6.9\,3 \\
+\;2\,6.4\,0 \\
\hline
2\,5\,8.0\,3
\end{array}
$$

Step 3 Add. The sum of 14.7 + 216.93 + 26.4 is 258.03.

■ EXAMPLE 2–6

Solve 42 + 143.7 + 6.571.

Solution

$$
\begin{array}{r}
4\,2. \\
1\,4\,3.7 \\
+\;6.5\,7\,1
\end{array}
$$

Step 1

$$
\begin{array}{r}
4\,2.0\,0\,0 \\
1\,4\,3.7\,0\,0 \\
+\;6.5\,7\,1
\end{array}
$$

Step 2

$$
\begin{array}{r}
4\,2.0\,0\,0 \\
1\,4\,3.7\,0\,0 \\
+\;6.5\,7\,1 \\
\hline
1\,9\,2.2\,7\,1
\end{array}
$$

Step 3 This shows that the sum of 42 + 143.7 + 6.571 is 192.271.

■ EXAMPLE 2–7

At the beginning of this section on adding decimals, the following question was asked. Answer the question.

At a local discount auto parts store, a customer bought a case of multi-weight motor oil for $13.68, an oil filter for $2.74, and a windshield wiper blade for $5.97. What was the total price of these items before taxes?

Solution

To find the total price of these items before taxes, we need to add $13.68 + 2.74 + 5.97$.

$$
\begin{array}{r}
1\,3.6\,8 \\
2.7\,4 \\
+\ \ 5.9\,7 \\
\end{array}
$$

Step 1 Write the numbers so the decimal points are in line. In this case, no 0s are needed to make sure that each number has the same number of places to the right of the decimal point. *We proceed to Step 3.*

$$
\begin{array}{r}
1\,3.6\,8 \\
2.7\,4 \\
+\ \ 5.9\,7 \\
\hline
2\,2.3\,9 \\
\end{array}
$$

Step 3 The total price of these items, before taxes, is $22.39.

NAME _____ DATE _____ SCORE _____

Estimate the sums. Do not work problems 1–5.

1. 5.73 + 13.4 = _____

2. 12 + 32.7 + 21.4 = _____

3.   14.61
       2.5
  + 29.1

4. 291.
  + 5.63

5.   22.395
    31.
  + 5.6

Write these problems in columns with the decimal points in the correct places.

6. 4.321 + 16.9

7. 49 + 5.7

8. 192 + 5.36 + 18.5

Add.

9.   5.3
  + 3.4

10.   23.4
   + 16.3

11.   2.13
    3.24
  + 4.62

12.   15.4
     5.9
  + 28.3

13.   3.6
  + 4.7

14.   43.6
   + 5.8

15.   17.6
  + 5.21

16.   134.3
  + 214.61

17. 142.62
    34.1
  + 5.024

18.   4.37
  + 19.8

19.   3.72
  + 25.537

20.   5.4
   16.
    9.37
 + 236.5

21. 24.37 + 5.4 = _____

22. 16.4 + 3.59 = _____

23. 5.61 + 14.37 + 251.4 = _____

24. 5.791 + 62.3 = _____

25. 0.83 + 9.4 + 162.1 = _____

26. 842.3 + 56.9875 + 4.2 = _____

27. The following parts were used to tune up an engine: spark plugs, $12.34; distributor cap, $6.25; and air filter, $3.27.

    (a) Estimate the total cost of the parts.

27a. _____

    (b) What was the actual cost of the parts?

27b. _____

28. An air-conditioning compressor requires 2 horsepower (HP) to operate; the power steering pump, 0.25 HP; the air-injection pump, 0.125 HP; the alternator, 1.125 HP; and the water pump/coolant fan, 4.5 HP. What is the total horsepower required for these components?

28. _____

29. A technician spent 1.5 hours to complete a tune-up, 0.2 hours to clean the battery, 0.3 hours to change the engine oil and filter, 0.4 hours to change and align a headlight, and 1.3 hours to service the air-conditioning system. What was the total time spent to service this automobile?

29. _____

30. The distance from home to school is 3.2 km, from school to the racetrack is 4.8 km, from the racetrack to the fairgrounds is 6.7 km, and the distance from the fairgrounds back home is 11.6 km. What is the total distance traveled?

30. _____

31. A trip from the midwest to California and back is approximately 5674 miles. The fuel tank was filled on different occasions with the following amounts: 7.5 gallons, 18.2 gallons, 23 gallons, 13.8 gallons, 17.1 gallons, 15.6 gallons, 3.4 gallons, 19.9 gallons, 19.3 gallons, 7.6 gallons, 16.1 gallons, and 13 gallons. How much fuel was required for the trip?

31. _____

32. A combustion chamber has a volume of 69.7 cm^3, the dished piston has a volume of 9.8 cm^3, and the head gasket has a volume of 4.9 cm^3. Determine the total combustion volume of this cylinder.

32. _____

33. An engine rebuild requires the following parts at the prices shown:

Roller camshaft	$229.94
Roller valve lifters	165.25
Forged pistons	374.94
Moly piston rings	43.11
Connecting rod bearings	32.36
Crankshaft main bearings	46.19
High-volume oil pump	40.06
Overhaul gasket set	95.63
Engine block plug set	13.99
Bronze valve guide liners	45.00
Valve springs	65.23
Reground crankshaft	119.94
Timing components	34.09
Camshaft bearings	24.92

    (a) What is the total cost for these parts?

33a. _____

    (b) There is an $23.58 shipping charge on this order and a $25 core charge on the crankshaft. What is the total cost of this order?

33b. _____

    (c) After the crankshaft has been returned for the core refund with an accompanying shipping charge of $16.85, what is the final cost of the parts for this engine rebuild?

33c. _____

34. An engine is about to be rebored. The piston diameter measures out at 4.028 inches. This engine will be fitted with hypereutectic pistons, which require 0.0015-inch clearance between the piston and the cylinder wall. What should be the final finished diameter of the cylinder block bore?

34. _____

35. If the same engine is to be turbocharged, it must be fitted with forged pistons, which require a clearance of 0.0045 inch. What must the finished bore diameter be for this application?

35. _____

Estimating Differences

In order to answer a question like the following, you must be able to subtract decimals.

A customer's total bill is $13.32. The customer gives you $24.07. How much change should the customer get?

Round each decimal to the nearest whole number. Subtract the whole numbers. The answer to a subtraction problem is called the *difference*.

■ EXAMPLE 2–8

Estimate $37.2 - 15.54$.

Solution

37.2 rounds to 37.
15.54 rounds to 16.
$37 - 16 = 21$
So, $37.2 - 15.54$ is about 21.

■ EXAMPLE 2–9

Estimate $43 - 18.72$.

Solution

43 is already a whole number.
18.72 rounds to 19.
$43 - 19 = 24$
Therefore, $43 - 18.72$ is about 24.

Subtracting Decimals

Step 1 Write the numbers in a column with the decimal points in a line. Put the first number on top.

Step 2 Add enough 0s so that each number has the same number of places to the right of the decimal point.

Step 3 Subtract decimals the same way you subtract whole numbers.

■ EXAMPLE 2–10

Solve $15.2 - 7.13$.

Solution

$$
\begin{array}{r}
1\,5.2 \\
-\ 7.1\,3 \\
\hline
\end{array}
$$

Step 1 Write the numbers so the decimal points are in a line.

$$
\begin{array}{r}
1\,5.2\,0 \\
-\ 7.1\,3 \\
\hline
\end{array}
$$

Step 2 Add enough 0s so that each number has the same number of places to the right of the decimal point.

$$
\begin{array}{r}
1\,5.2\,0 \\
-\ 7.1\,3 \\
\hline
8.0\,7 \\
\end{array}
$$

Step 3 Subtract as you would whole numbers. The decimal point goes directly under the other two decimal points.
So, $15.2 - 7.13 = 8.07$.

■ EXAMPLE 2–11

Solve $37.2 - 15.54$.

Solution

$$
\begin{array}{r}
3\,7.2 \\
-\ 1\,5.5\,4 \\
\hline
\end{array}
$$

Step 1 Write so that the decimal points are in a line.

$$
\begin{array}{r}
3\,7.2\,0 \\
-\ 1\,5.5\,4 \\
\hline
\end{array}
$$

Step 2 Add 0s.

$$
\begin{array}{r}
3\,7.2\,0 \\
-\ 1\,5.5\,4 \\
\hline
2\,1.6\,6 \\
\end{array}
$$

Step 3 Subtract. Therefore, $37.2 - 15.54 = 21.66$.

In Example 2–8, we estimated that $37.2 - 15.54$ was about 21. The difference in this example is close to 21, so our answer seems reasonable.

■ EXAMPLE 2–12

Solve $169.45 - 27.1$.

Solution

$$
\begin{array}{r}
1\,6\,9.4\,5 \\
-\ 2\,7.1 \\
\hline
\end{array}
$$

Step 1 Write so that the decimal points are in a line.

$$
\begin{array}{r}
1\,6\,9.4\,5 \\
-\ 2\,7.1\,0 \\
\hline
\end{array}
$$

Step 2 Add 0s.

$$
\begin{array}{r}
1\,6\,9.4\,5 \\
-\ 2\,7.1\,0 \\
\hline
1\,4\,2.3\,5
\end{array}
$$

Step 3 Subtract.

■ EXAMPLE 2–13

Solve 53 − 25.41.

Solution

$$
\begin{array}{r}
5\,3. \\
-\ 2\,5.4\,1
\end{array}
$$

Step 1

$$
\begin{array}{r}
5\,3.0\,0 \\
-\ 2\,5.4\,1
\end{array}
$$

Step 2

$$
\begin{array}{r}
5\,3.0\,0 \\
-\ 2\,5.4\,1 \\
\hline
2\,7.5\,9
\end{array}
$$

Step 3

■ EXAMPLE 2–14

At the beginning of this section on subtracting decimals, the following question was asked. Solve the problem.

A customer's total bill is $13.32. The customer gives you $24.07. How much change should the customer get?

Solution

To find the correct amount of change to give the customer, we need to subtract $13.32 from $24.07.

$$
\begin{array}{r}
\$\,2\,4.0\,7 \\
-\ 1\,3.3\,2
\end{array}
$$

Step 1 Write the numbers so that the decimal points are in a line. In this case, no 0s are needed to make sure that each number has the same number of places to the right of the decimal point. *We proceed to Step 3.*

$$
\begin{array}{r}
\$\,2\,4.0\,7 \\
-\ 1\,3.3\,2 \\
\hline
1\,0.7\,5
\end{array}
$$

Step 3 Subtract. The customer should get $10.75 in change.

NAME _____ DATE _____ SCORE _____

Estimate the differences. Do not work problems 1–5.

1. 42.37 − 6.54

2. 4 − 1.38

3.   37.420
− 12.312

4.   143.00
− 18.81

5. 21.871
− 4.000

Rewrite these problems in columns with the decimal points in the correct places.

6. 27.142 − 16.3

7. 81 − 0.431

8. 53.718 − 7

Subtract.

9.   3.9
− 2.7

10.   24.6
− 13.2

11.   4.73
− 1.52

12.   57.38
− 32.25

13. 17.6
− 9.3

14. 7.923
− 3.711

15. 26.38
− 21.97

16. 42.81
− 16.91

17. 7.
− 2.4

18. 27.
− 8.32

19. 235.7
− 16.

20. 137.84
− 83.9

21. 42.3 − 11.2 = _____

22. 58.349 − 23.133 = _____

23. 89.71 − 16 = _____

24. 192.36 − 98 = _____

25. 184 − 52.37 = _____

26. 201 − 135.371 = _____

27. The odometer reading was 68 432.6 before a trip and 69 561.5 after the trip. What was the distance of the trip?

27. _____

28. Five quarts of oil and an oil filter cost $17.35. However, with a coupon, these items may be purchased for $3.43 less. What is the final cost of the oil and filter?

28. _____

29. A cylinder measures 3.1256 inches at the top and 3.1197 inches at the bottom. How much taper does the cylinder have?

29. _____

30. A turbine shaft has 1.7-mm end play. Specifications call for only 0.61 mm. What size selective thrust washer (spacer) must be used to bring the turbine shaft end play within specified limits?

30. _____

31. Differential ring-gear backlash is specified at 0.006 inch and is measured at 0.035 inch. What is the difference between the specified backlash and the measured backlash?

31. _____

32. A valve guide bore measures 0.0453 inch and the valve stem measures 0.0426 inch. What is the valve stem clearance?

32. _____

33. The maximum diameter of a brake drum is listed as 11.090 inches. The brake drum measures 11.063 inches. After machining 0.022 inch from the drum to remove score marks, you must determine if the drum is still usable.

   (a) What does the drum now measure?

33a. _____

   (b) Can it be reused?

33b. _____

34. An engine cylinder bore measures 4.001 inches. The piston, which is matched to that cylinder, measures 3.9955 inches. What is the piston clearance for that cylinder?

34. _____

35. A new piston has a diameter of 3.625 inches. This is 0.0143 larger than a worn piston. What is the size of the worn piston?

35. _____

36. When reboring cylinders you must first determine the exact rebore size. Before reboring, the cylinder measures 3.375 inches in diameter. The new pistons are 0.030 inch oversize and must be installed with 0.002-inch clearance. To what size must the cylinder be bored?

36. _____

Estimating Products

The answer to a multiplication problem is called the *product*.

Round each decimal to the nearest whole number. Multiply the whole numbers.

■ EXAMPLE 2–15

Estimate 13.81×5.2.

Solution

13.81 rounds to 14.
5.2 rounds to 5.
$14 \times 5 = 70$
So, 13.81×5.2 is about 70.

■ EXAMPLE 2–16

Estimate 0.83×12.4.

Solution

0.83 is about 1.
12.4 is about 12.
$1 \times 12 = 12$
So, 0.83×12.4 is about 12.

Multiplying Decimals

The multiplication of decimals is used to solve the following type of problem.

A general rule for engine piston compression ring end gap suggests that the minimum ring end gap should be 0.004 mm for each millimeter of cylinder bore diameter. If the cylinder bore is 95.25 mm, what should the minimum compression ring end gap be?

Step 1 Multiply decimals the same way that you multiply whole numbers.

Step 2 Count the total number of decimal places to the right of the decimal point in the numbers that you multiply.

Step 3 Count the same number of places from the right in your answer and mark the decimal point. Add 0s if the answer does not have enough digits.

■ EXAMPLE 2–17

Solve 14.5×7.9.

Solution

$$
\begin{array}{r}
1\,4.5 \\
\times\ 7.9 \\
\hline
1\,3\,0\,5 \\
1\,0\,1\,5\,0 \\
\hline
1\,1\,4\,5\,5
\end{array}
$$

Step 1 Multiply decimals the same way you multiply whole numbers.

$$
\begin{array}{rl}
1\,4.5 & \leftarrow 1 \text{ decimal place} \\
\times\ 7.9 & \leftarrow 1 \text{ decimal place} \\
\hline
1\,3\,0\,5 & \underline{2} \text{ total decimal places} \\
1\,0\,1\,5\,0 \\
\hline
1\,1\,4\,5\,5
\end{array}
$$

Step 2 Count the total number of decimal places to the right of the decimal point in the numbers you multiplied.

$$
\begin{array}{rl}
1\,4.5 \\
\times\ 7.9 \\
\hline
1\,3\,0\,5 \\
1\,0\,1\,5\,0 \\
\hline
1\,1\,4.5\,5 & \leftarrow 2 \text{ decimal places}
\end{array}
$$

Step 3 Count the same number of places from the right in your answer and mark the decimal point.
So, $14.5 \times 7.9 = 114.55$.

■ EXAMPLE 2–18

Solve 23.7×1.92.

Solution

$$
\begin{array}{r}
2\,3.7 \\
\times\ 1.9\,2 \\
\hline
4\,7\,4 \\
2\,1\,3\,3\,0 \\
2\,3\,7\,0\,0 \\
\hline
4\,5\,5\,0\,4
\end{array}
$$

Step 1 Multiply.

$$
\begin{array}{rl}
2\,3.7 & \leftarrow 1 \text{ decimal place} \\
\times\ 1.9\,2 & \leftarrow 2 \text{ decimal places} \\
\hline
4\,7\,4 & \underline{3} \text{ total decimal places} \\
2\,1\,3\,3\,0 \\
2\,3\,7\,0\,0 \\
\hline
4\,5\,5\,0\,4
\end{array}
$$

Step 2 Count decimal places in numbers multiplied.

```
      2 3.7
   ×  1.9 2
      4 7 4
    2 1 3 3 0
    2 3 7 0 0
    4 5.5 0 4   ← 3 decimal places
```

Step 3 Count decimal places in answer.
So, 23.7 × 1.92 = 45.504.

■ EXAMPLE 2–19

Solve 0.23 × 0.045.

Solution

```
      0.2 3
   ×  0.0 4 5
      1 1 5
      9 2 0
    1 0 3 5
```

Step 1

```
      0.2 3    ← 2 decimal places
   ×  0.0 4 5  ← 3 decimal places
      1 1 5      5 total decimal places
      9 2 0
    1 0 3 5
```

Step 2

```
      0.2 3
   ×  0.0 4 5
      1 1 5
      9 2 0
  0.0 1 0 3 5   ← 5 decimal places
  └──────────── Add 0 to get
                5th decimal place
```

Therefore, 0.23 × 0.045 = 0.01035.

Step 3

■ EXAMPLE 2–20

Solve 0.83 × 12.4.

Solution

```
      0.8 3
   ×  1 2.4
      3 3 2
    1 6 6
    8 3
  1 0 2 9 2
```

Step 1

```
      0.8 3   ← 2 decimal places
   ×  1 2.4   ← 1 decimal place
      3 3 2      3 total decimal places
    1 6 6
    8 3
  1 0 2 9 2
```

Step 2

```
      0.8 3
   ×  1 2.4
      3 3 2
    1 6 6
    8 3
  1 0.2 9 2   ← 3 decimal places
```

Step 3 Compare this result to the estimate in Example 2–16. This answer, 10.292, is close to the estimate of 12, so our answer seems reasonable.

■ EXAMPLE 2–21

At the beginning of this section on multiplying decimals, the following problem was given. Solve the problem.

A general rule for engine piston compression ring end gap suggests that the minimum ring end gap should be 0.004 mm for each millimeter of cylinder bore diameter. If the cylinder bore is 95.25 mm, what should the minimum compressing ring end gap be?

Solution

In order to determine the minimum compression ring end gap for this cylinder, we need to determine the product of 95.25 and 0.004, or 95.25 × 0.004.

```
      9 5.2 5
   ×  0.0 0 4
    3 8 1 0 0
```

Step 1 Multiply.

```
      9 5.2 5   ← 2 decimal places
   ×  0.0 0 4   ← 3 decimal places
    3 8 1 0 0      5 total decimal places
```

Step 2 Count decimal places in the numbers multiplied.

```
      9 5.2 5
   ×  0.0 0 4
   .3 8 1 0 0   ← 5 decimal places
```

Step 3 Count decimal places in answer.
The minimum compression ring end gap should be 0.38100 mm.

NAME _____ DATE _____ SCORE _____

Estimate the products. Do not work problems 1–3.

1. 3.2×5.7 **2.** 12.4×3.2 **3.** 8.21×12.8

Place the decimal point in the correct place.

4.   14.21
    $\times$ 2.7
    3 8 3 6 7

5.   7.24
    $\times$ 0.32
    2 3 1 6 8

6.   0.24
    $\times$ 0.3
    7 2

Multiply.

7.   14.3
    $\times$ 7

8.   9.7
    $\times$ 6

9.   4.3
    $\times$ 1.8

10.   5.4
    $\times$ 7.8

11.   3.57
    $\times$ 2.5

12.   3.68
    $\times$ 8.47

13.   9.07
    $\times$ 2.5

14.   25.08
    $\times$ 4.5

15.   0.058
    $\times$ 2.4

16.   0.032
    $\times$ 4.007

17.   21.47
    $\times$ 0.007

18.   3.075
    $\times$ 9.508

19.   8.9053
    $\times$ 0.037

20.   35.875
    $\times$ 14.88

21.   32.075
    $\times$ 2.445

22.   38.625
    $\times$ 4.884

23. A gear makes 65.2 revolutions each minute. How many revolutions will it make in 17 minutes?

23. _____

24. A wheel has an outside circumference of 2.4 meters. If the wheel rotates 660 times, how far will the wheel have traveled?

24. _____

25. A vehicle gets 34.6 miles per gallon of fuel. How far can the vehicle travel on 17.5 gallons of fuel?

25. _____

26. A round trip to work and back is 6.6 km. How many kilometers are traveled in a month that has 23 working days?

26. _____

27. If a gallon of diesel fuel weighs 8.4 pounds, how much does 14.7 gallons weigh?

27. _____

28. A general rule for engine piston compression ring end gap suggests that the minimum end gap should be 0.004 inch for each inch of cylinder bore diameter. If the cylinder bore is 3.5 inches, what should the minimum compression ring end gap be?

28. _____

29. Determine the displacements of the following engines. You may round the answers to the nearest whole number.

Displacement = area × stroke × number of cylinders

Engine		Area (square inches)	Stroke (inches)
(a)	V-8	12.57	3.48
(b)	V-8	14.93	3.85
(c)	V-6	10.10	2.99
(d)	I-4	11.82	3.19

Displacement

29a. _____

29b. _____

29c. _____

29d. _____

30. Determine the displacements of the following engines. Give answers in both cm^3 and liters (L). To change cm^3 to liters, multiply the number of cm^3 by 0.001.

Engine		Area (square centimeters)	Stroke (centimeters)
(a)	V-8	80.11	9.7
(b)	V-6	66.48	8.4
(c)	I-4	52.28	7.6
(d)	V-10	81.07	9.86

Displacement

30a. _____ cm^3
_____ L

30b. _____ cm^3
_____ L

30c. _____ cm^3
_____ L

30d. _____ cm^3
_____ L

31. If an engine is operating at 2800 rpm in fourth gear (1:1 transmission input/output ratio), what will the engine speed be when the transmission is shifted into overdrive (0.7:1 ratio) while the vehicle speed remains the same?

31. _____

32. Fuel costs $3.949 per gallon and sells for $4.129 a gallon.

(a) What is the profit per gallon?

32a. _____

(b) What is the total profit for 725 gallons?

32b. _____

33. It is estimated that, on average, 7.5 gallons of crude oil are consumed in the production of each automotive tire. If a manufacturer produces 2,952,523 automobiles in a particular year, each equipped with five tires, how many gallons of crude oil is consumed to produce the tires for all these vehicles?

33. _____

34. The new federal Corporate Average Fuel Economy (CAFE) standards mandate higher mpg ratings each year to achieve a goal of 35 mpg by 2020. This new mpg standard is expected to save one billion gallons of fuel each year when the CAFE goal is reached.

Our highway system is maintained through fuel taxes collected from the fuel we purchase. Highway fuel taxes average approximately 45.6 cents per gallon nationwide. How much less tax will be collected if one billion fewer gallons of fuel are sold?

34. _____

Dividing Decimals

In order to answer the following question, you must be able to divide decimals.

During a trip of 506.9 miles, a vehicle uses 18.5 gallons of gasoline. How many miles per gallon does this vehicle get?

There are four steps to follow when dividing decimals. The answer to a division problem is called the *quotient*. The *divisor* is the number that does the dividing.

■ EXAMPLE 2–22

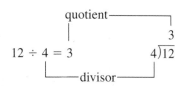

Step 1 Count the number of decimal places to the right of the decimal point in the divisor.

Step 2 Check to see if there are any places to the right of the decimal point in the other number. If not, add 0s at the right of the other number to get the same number of places.

Step 3 Move the decimal point to the right in both numbers. Move the decimal point the same number of places in both numbers. Move both decimal points the number of places that you counted in Step 1.

Step 4 Divide decimals in the same way that you divide whole numbers. You may want to add some 0s to the right of the decimal point in order to get some numbers in the answer that are to the right of the decimal point.

■ EXAMPLE 2–23

Solve $57.2 \div 14.3$.

Solution

$$14.3\overline{)57.2}$$
↑
1 decimal place

Step 1 Count decimal places in divisor.

$$14.3\overline{)57.2}$$
↑
1 decimal place

Step 2 Does the other number have the same number of decimal places? It does, so go to Step 3.

$$14.3\overline{)57.2}$$

Step 3 Move decimal points to the right.

$$\overset{4.}{143\overline{)572.}}$$

Step 4 Divide as you divide whole numbers.
So, $57.2 \div 14.3 = 4$.

■ EXAMPLE 2–24

Solve $96 \div 2.4$.

Solution

$$2.4\overline{)96}$$
↑
1 decimal place

Step 1 Check decimal places in divisor.

$$2.4\overline{)96}$$
↑
0 decimal place

Step 2 Check decimal places in the other number.

$$2.4\overline{)96.0}$$
↑ ↑
1 decimal place

Step 2b Add 0s to the other number.

$$24.\overline{)960.}$$

Step 3 Move decimal points to the right.

$$\overset{40.}{24.\overline{)960.}}$$

Step 4 Divide.
So, $96 \div 2.4 = 40$.

■ EXAMPLE 2–25

Solve $2.556 \div 2.13$.

Solution

$$2.13\overline{)2.556}$$
↑
2 decimal places

Step 1

$$2.13\overline{)2.556}$$

↑
3 decimal places

Step 2

$$213.\overline{)255.6}$$

Step 3

$$\begin{array}{r} 1.2 \\ 213.\overline{)255.6} \end{array}$$

Step 4   Therefore, 2.556 ÷ 2.13 = 1.2.

■ **EXAMPLE 2–26**

The following question was asked at the beginning of this section on dividing decimals. Answer the question.

During a trip of 506.9 miles, a vehicle uses 18.5 gallons of gasoline. How many miles per gallon does this vehicle get?

Solution
To answer this question we need to divide the number of miles traveled by the amount of fuel used.

$$18.5\overline{)506.9}$$

↑
1 decimal place

Step 1   Count decimal places in the divisor.

$$18.5\overline{)506.9}$$

↑
1 decimal place

Step 2   Does the other number have the same number of decimal places? It does, so go to Step 3.

$$185.\overline{)5069.}$$

Step 3   Move the decimal point to the right.

$$\begin{array}{r} 27.4 \\ 185.0\overline{)5069.0} \end{array}$$

↑
1 decimal place

Step 4   Divide. Add 0s to the right of the decimal point to get the right number of decimal places in the answer.

So, 506.9 ÷ 18.5 = 27.4. This car got 27.4 miles per gallon during this trip.

NAME _____ DATE _____ SCORE _____

Divide. Round each answer to the nearest hundredth.

1. $.2\overline{)42.8}$

2. $.5\overline{)1.235}$

3. $.8\overline{)1.84}$

4. $3.6\overline{)32.4}$

5. $1.3\overline{)7.02}$

6. $.05\overline{)1.955}$

7. $1.6\overline{)4.96}$

8. $.53\overline{)4.664}$

9. $.06\overline{)18.0426}$

10. $.05\overline{)2.4}$

11. $.007\overline{)392.63}$

12. $.027\overline{)2.349}$

13. $9.012 \div .4$

14. $205 \div .5$

15. $473.22 \div .9$

16. $15.98 \div 1.7$

17. $71.04 \div 7.4$

18. $4.95 \div 3.5$

19. $24 \div .012$

20. $.140625 \div .009$

21. During a trip of 371 miles, a vehicle uses 13.8 gallons of fuel. How many miles per gallon does this vehicle get?

21. _____

22. A trip of 1858 kilometers is planned during a 4-day time period. How many kilometers must be completed each day to complete the trip within the 4-day limit?

22. _____

23. John bought a used vehicle on eBay. On his drive back home he discovered the engine used 2.5 quarts of oil during the 4565-mile trip. How often was a quart of oil added?

23. _____

24. A vehicle travels 8925 meters. If the outside circumference of the tires is 1.75 meters, how many revolutions does the wheel make while traveling this distance?

24. _____

25. A gallon of diesel fuel weighs 8.4 pounds. If the fuel in a full tank weighs 177 pounds, how many gallons of fuel does it contain?

25. _____

26. To determine compression ratio (CR), the piston displacement (swept volume) plus the combustion volume is divided by the combustion volume. Thus, we have the formula

$$CR = \frac{\text{piston displacement} + \text{combustion volume}}{\text{combustion volume}}$$

Determine the following compression ratios.

Piston Displacement	Combustion Volume
(a) 716.83 cm³	75.45 cm³
(b) 853.60 cm³	77.60 cm³
(c) 659.40 cm³	62.80 cm³

Compression Ratios

26a. _____

26b. _____

26c. _____

27. An engine has a stroke of 3.25 inches, a bore area of 11.8 in.² (square inches) and a combustion volume of 3.953 in.³ (cubic inches).

    (a) What is the piston displacement?

    27a. _____

    (b) What is the compression ratio?

    27b. _____

28. A store can purchase oil for $14,400 per 500 cases (24 quarts per case) or $46,452 per truck load (1975 cases per load).

    (a) What is the cost per quart at 500 cases?

    28a. _____

    (b) What is the cost per quart at a truck load?

    28b. _____

    (c) How much can be saved per quart when buying by the truck load?

    28c. _____

    (d) How much can be saved in total when buying by the truck load?

    28d. _____

29. A car owner's repair bill comes to $2598.48. One-half of the bill is paid in cash. The rest is paid in 12 equal payments. What is the amount of each of the 12 payments?

29. _____

30. A carton of heater hose contains 50 feet of hose and costs $62.50. What is the cost per foot?

30. _____

31. There is a proposal to increase the state gasoline tax by $0.20 per gallon. If Mary's car averages 20 mpg and she drives 15,000 miles per year, how much extra would this new tax cost her for one year of driving?

31. _____

32. If the state in problem 31 has 3,381,238 drivers who share Mary's driving data, how much extra revenue can it expect to collect if this proposed hike in the gas tax is enacted?

32. _____

33. Assuming 15,000 miles driven per year and given the data below, calculate the total yearly cost and the cost per mile driven.

Principle on car loan	$ 3423
Finance charges on car loan	367
Fuel and oil	1283
Insurance	817
Maintenance and repair	637
License, parking, and misc.	546

Total Yearly Cost

33. Total cost _____

    Cost per mile _____

34. Considering the data in problem 33, if the average American earns approximately $15.50 per hour after taxes, how many hours must he or she work to cover his or her car costs? Express your answer in terms of months and days. Assume that the person works an average of 8 hours each day and 22 days a month.

34. _____

FRACTIONS

Objectives: After studying this chapter, you should be able to:

- Add, subtract, multiply, and divide fractions.
- Convert fractions to decimals and decimals to fractions.
- Use the rules for multiplying and dividing negative numbers.
- Apply basic mathematics operations of fractions to solve practical problems.
- Determine the decimal equivalents of fractional tool and drill sizes.
- Determine tread depth and shim thickness.
- Determine fluid capacities.
- Use Ohm's Law to calculate electrical values.
- Convert fractional time periods to decimal equivalents.

The next four sections discuss working with fractions. To do the work in these sections, you need to know some basic facts about fractions.

Mixed Numbers

In order to answer a question like the one below, you must be able to convert mixed numbers to fractions.

A piece of electrical wire is 25⅔ feet long. How many 2¾-foot long pieces can be cut from this one piece of wire?

A mixed number is really the sum of a whole number and a fraction. This means that 5⅔ is the same as 5 + ⅔. To change a mixed number to a fraction, multiply the whole number and the bottom number (denominator) of the fraction, add this product to the top number (numerator), and place the total number over the bottom number.

■ EXAMPLE 3–1

Change 5⅔ to a fraction.

Solution

$$5 \times 3 = 15$$

Step 1 Multiply the whole number 5 and the denominator 3.

$$15 + 2 = 17$$

Step 2 Add that product to the numerator of the fraction.

$$\frac{17}{3}$$

Step 3 Put that answer over the denominator. So, 5⅔ = ¹⁷⁄₃.

■ EXAMPLE 3–2

Change 7⅜ to a fraction.

Solution

$$7 \times 8 = 56$$

Step 1

$$56 + 3 = 59$$

Step 2

$$\frac{59}{8}$$

Step 3 So, $7\frac{3}{8} = \frac{59}{8}$.

■ EXAMPLE 3–3

Change $-2\frac{1}{4}$ to a fraction.

Solution

Remember that $2\frac{1}{4}$ means the same as $2 + \frac{1}{4}$. So, $-2\frac{1}{4}$ is a short way of writing $-(2 + \frac{1}{4})$.

$$2 \times 4 = 8$$

Step 1

$$8 + 1 = 9$$

Step 2

$$\frac{9}{4}$$

Step 3 So, $-2\frac{1}{4} = -\frac{9}{4}$.

■ EXAMPLE 3–4

At the beginning of this section on mixed numbers, the following question was asked. Answer the question.

A piece of electrical wire is $25\frac{2}{3}$ feet long. How many $2\frac{3}{4}$-foot long pieces can be cut from this one piece of wire?

Solution

We have to change two mixed numbers to fractions. We will first change $25\frac{2}{3}$.

$$25 \times 3 = 75$$
$$75 + 2 = 77$$

$$\frac{77}{3}$$

Steps 1, 2, & 3 So, $25\frac{2}{3} = \frac{77}{3}$. Next, we change $2\frac{3}{4}$ to a fraction.

$$2 \times 4 = 8$$
$$8 + 3 = 11$$
$$\frac{11}{4}$$

Steps 1, 2, & 3

So, $2\frac{3}{4} = \frac{11}{4}$. To finish this example, we need to know how to divide fractions. We will finish this example in Example 3–56.

Equivalent Fractions

Two fractions are equivalent, or have the same value, if their cross products are equal.

■ EXAMPLE 3–5

Are $\frac{2}{3}$ and $\frac{6}{9}$ equivalent fractions?

Solution

$$\frac{2}{3} \times\!\!\!\!\diagup \frac{6}{9}$$

Step 1 Find the cross products.

$$2 \times 9 = 18$$
$$3 \times 6 = 18$$

Step 2 Are the cross products equal? Yes, so the fractions are equivalent.

$\frac{2}{3} = \frac{6}{9}$ is another way of saying that the fractions are equivalent.

■ EXAMPLE 3–6

Are $\frac{3}{8}$ and $\frac{12}{32}$ equivalent?

Solution

$$\frac{3}{8} \times\!\!\!\!\diagup \frac{12}{32}$$

Step 1 Find the cross products.

$$3 \times 32 = 96$$
$$8 \times 12 = 96$$

Step 2 Are the cross products equal? Yes, both equal 96, so the fractions are equivalent, and we write $\frac{3}{8} = \frac{12}{32}$.

■ EXAMPLE 3–7

Are $\frac{2}{5}$ and $\frac{10}{24}$ equivalent?

Solution

$$\frac{2}{5} \diagdown \frac{10}{24}$$

Step 1 Find the cross products.

$$2 \times 24 = 48$$
$$5 \times 10 = 50$$

Step 2 Are the cross products equal? No, one is 48 and the other is 50. The fractions are not equivalent.

$\frac{2}{5} \neq \frac{10}{24}$ is one way to say the fractions are not equivalent.

Making Equivalent Fractions

In order to add or subtract fractions that do not have the same denominators (bottom numbers), you have to change the fractions to equivalent fractions that *do* have the same denominator. Therefore, you need to make an equivalent fraction that has a known denominator. All you need to do is to find the missing top number (numerator). The method shown in these examples will work only if the denominator of the fraction with the missing numerator is larger than the other denominator.

■ EXAMPLE 3–8

Find a fraction with a denominator of 12 that is equivalent to $\frac{5}{4}$. This can be written as $\frac{5}{4} = \frac{?}{12}$.

Solution

$$12 \div 4 = 3$$

Step 1 Divide the larger denominator by the smaller one.

$$5 \times 3 = 15$$

Step 2 Multiply the answer in Step 1 by the numerator. This product is the missing numerator.

So, $\frac{5}{4} = \frac{15}{12}$.

■ EXAMPLE 3–9

Solve $\frac{3}{4} = \frac{?}{64}$.

Solution

$$64 \div 4 = 16$$

Step 1

$$16 \times 3 = 48$$

Step 2
So, $\frac{3}{4} = \frac{48}{64}$.

Reducing Fractions

When you are working with fractions, it is often helpful if the fraction is in its lowest terms. A fraction is in lowest terms if the number 1 is the only whole number that can divide *both* the numerator and denominator of the fraction.

To reduce a fraction, find a whole number that will divide *both* the numerator and denominator of the fraction. Then divide both of these numbers by that whole number. Repeat this method until the number 1 is the only whole number you can find that will divide both the numerator and the denominator of the fraction.

■ EXAMPLE 3–10

Reduce $\frac{4}{8}$.

Solution

$$4$$

Step 1 4 divides into both 4 and 8.

$$\frac{4 \div 4}{8 \div 4} = \frac{1}{2}$$

Step 2 Divide both the numerator and denominator by 4.
So, $\frac{4}{8}$ reduces to $\frac{1}{2}$ and $\frac{4}{8} = \frac{1}{2}$.

■ EXAMPLE 3–11

Reduce $\frac{12}{15}$.

Solution

$$3$$

Step 1 3 divides into both 12 and 15.

$$\frac{12 \div 3}{15 \div 3} = \frac{4}{5}$$

Step 2 Divide the numerator and denominator by 3.
So, $\frac{12}{15}$ reduces to $\frac{4}{5}$ and $\frac{12}{15} = \frac{4}{5}$.

■ EXAMPLE 3–12

Reduce $\frac{30}{48}$.

Solution

3

Step 1 3 divides into both 30 and 48.

$$\frac{30 \div 3}{48 \div 3} = \frac{10}{16}$$

Step 2

2

Step 1 again. 2 divides into both 10 and 16.

$$\frac{10 \div 2}{16 \div 2} = \frac{5}{8}$$

Step 2 again.
So, $^{30}\!/_{48} = ^{5}\!/_{8}$.

■ EXAMPLE 3–13

Reduce $^{-16}\!/_{24}$.

Solution

8

Step 1

$$\frac{-16 \div 8}{24 \div 8} = \frac{-2}{3}$$

Step 2
So, $^{-16}\!/_{24} = -^{2}\!/_{3}$.

Comparing Fractions

In order to compare two fractions that do not have the same denominators, you have to change the fractions to equivalent fractions that *do* have the same denominator. Then you compare the numerators of the two fractions with the same denominators. The fraction with the larger numerator is the larger fraction.

If two fractions do not have the same denominator, divide the larger denominator by the smaller denominator. If the remainder is 0, then use the larger denominator as the common denominator. If the remainder is not 0, then multiply the two denominators to get a common denominator.

■ EXAMPLE 3–14

Which fraction is larger: $^{3}\!/_{4}$ or $^{7}\!/_{8}$?

$$8 \div 4 = 2$$

Step 1 Divide the larger denominator by the smaller one. The quotient is 2 with a remainder of 0, so we will use 8 as the common denominator. We write $^{3}\!/_{4}$ as an equivalent fraction with a denominator of 8.

$$3 \times 2 = 6$$

Step 2 Multiply the answer in Step 1 by the numerator of the fraction with the smaller denominator. This product is the numerator of a fraction equivalent to the fraction with the smaller denominator.
So, $^{3}\!/_{4} = ^{6}\!/_{8}$.

Step 3 Compare $^{6}\!/_{8}$ and $^{7}\!/_{8}$. These have the same denominator, 8. Their numerators are 6 and 7. Since 7 is larger than 6, $^{7}\!/_{8}$ is larger than $^{6}\!/_{8} = ^{3}\!/_{4}$.
Thus, $^{7}\!/_{8}$ is larger than $^{3}\!/_{4}$.

■ EXAMPLE 3–15

Which fraction is larger: $^{3}\!/_{4}$ or $^{5}\!/_{7}$?

$$7 \div 4 = 1 \text{ remainder } 3$$

Step 1 Divide the larger denominator by the smaller one. The quotient is 1 with a remainder of 3, so we will use $4 \times 7 = 28$ as the common denominator. We write $^{3}\!/_{4}$ as an equivalent fraction with a denominator of 28.

$$3 \times 7 = 21$$

Step 2a Multiply the numerator of one fraction by the denominator of the other. This product is the numerator of a fraction equivalent to the first fraction.
So, $^{3}\!/_{4} = ^{21}\!/_{28}$.

$$5 \times 4 = 20$$

Step 2b Multiply the numerator of the second fraction by the denominator of the first. This product is the numerator of a fraction equivalent to the second fraction.
So, $^{5}\!/_{7} = ^{20}\!/_{28}$.

Step 3 Compare $^{21}\!/_{28}$ and $^{20}\!/_{28}$. These have the same denominator, 28. Their numerators are 21 and 20. Since 21 is larger than 20, $^{21}\!/_{28} = ^{3}\!/_{4}$ is larger than $^{20}\!/_{28} = ^{5}\!/_{7}$.
Thus, $^{3}\!/_{4}$ is larger than $^{5}\!/_{7}$.

Changing Fractions to Decimals

In order to answer a question like the one below, you must be able to change a fraction to a decimal.

A technician asked you to bring a $^{3}\!/_{8}$-mm shim. The shims are marked in decimals. What decimal size (in millimeters) shim should you get?

There are times when it is helpful to change a fraction to a decimal. Decimals are often easier to use than fractions.

To change a fraction to a decimal, divide the denominator into the numerator.

EXAMPLE 3–16

Change $\frac{3}{4}$ to a decimal.

Solution

$3 \div 4$

$$
\begin{array}{r}
.75 \\
4\overline{)3.00} \\
\underline{2\,8} \\
20 \\
\underline{20}
\end{array}
$$

So, $\frac{3}{4}$ is equivalent to the decimal 0.75.

EXAMPLE 3–17

Change $\frac{5}{8}$ to a decimal.

Solution

$5 \div 8$

$$
\begin{array}{r}
.625 \\
8\overline{)5.000} \\
\underline{4\,8} \\
20 \\
\underline{16} \\
40 \\
\underline{40}
\end{array}
$$

The fraction $\frac{5}{8}$ is the same as the decimal 0.625.

EXAMPLE 3–18

Change $2\frac{3}{4}$ to a decimal.

Solution

First, change $2\frac{3}{4}$ to a fraction.

$2 \times 4 = 8$

$8 + 3 = 11$

$$\frac{11}{4}$$

Steps 1, 2, & 3

Now change $\frac{11}{4}$ to a decimal.

$11 \div 4$

$$
\begin{array}{r}
2.75 \\
4\overline{)11.00} \\
\underline{8} \\
3\,0 \\
\underline{2\,8} \\
20 \\
\underline{20}
\end{array}
$$

This means that $2\frac{3}{4}$ is the decimal 2.75.

EXAMPLE 3–19

Change $-4\frac{3}{5}$ to a decimal.

Solution

First, change $-4\frac{3}{5}$ to a fraction.

$4 \times 5 = 20$

$20 + 3 = 23$

$$\frac{-23}{5}$$

Steps 1, 2, & 3

Now, change $\frac{-23}{5}$ to a decimal.

$23 \div 5$

$$
\begin{array}{r}
4.6 \\
5\overline{)23.0} \\
\underline{20} \\
3\,0 \\
\underline{3\,0}
\end{array}
$$

So, $-4\frac{3}{5}$ is the decimal -4.6.

EXAMPLE 3–20

At the beginning of this section on changing fractions to decimals, the following question was asked. Answer the question.

A technician asked you to bring a $\frac{3}{8}$-mm shim. The shims are marked in decimals. What decimal size (in millimeters) shim should you get?

Solution

We need to change $\frac{3}{8}$ to a decimal, so we need to solve $3 \div 8$.

$$
\begin{array}{r}
.375 \\
8\overline{)3.000} \\
\underline{2\,4} \\
60 \\
\underline{56} \\
40 \\
\underline{40}
\end{array}
$$

You will need to get a 0.375-mm shim.

NAME _____ DATE _____ SCORE _____

Change the following mixed numbers to fractions.

1. $2\frac{1}{3}$ **2.** $4\frac{1}{6}$ **3.** $3\frac{5}{8}$ **4.** $12\frac{3}{4}$

5. $-3\frac{1}{8}$ **6.** $-4\frac{2}{3}$ **7.** $-1\frac{3}{4}$ **8.** $-11\frac{3}{8}$

Check to see if the following fractions are equivalent.

9. $\frac{3}{4}$ and $\frac{6}{8}$ **10.** $\frac{3}{5}$ and $\frac{15}{25}$ **11.** $\frac{18}{12}$ and $\frac{3}{2}$ **12.** $\frac{4}{7}$ and $\frac{14}{22}$

13. $\frac{-3}{5}$ and $\frac{-4}{7}$ **14.** $\frac{2}{3}$ and $\frac{-4}{6}$ **15.** $\frac{-7}{8}$ and $\frac{-13}{16}$ **16.** $\frac{-5}{8}$ and $\frac{11}{18}$

Complete the problems so that the two fractions are equivalent.

17. $\frac{4}{3}$ and $\frac{?}{15}$ **18.** $\frac{7}{8}$ and $\frac{?}{32}$ **19.** $\frac{3}{8}$ and $\frac{?}{64}$ **20.** $\frac{3}{16}$ and $\frac{?}{64}$

21. $\frac{-5}{4}$ and $\frac{?}{12}$ **22.** $\frac{-7}{8}$ and $\frac{?}{24}$ **23.** $\frac{-5}{6}$ and $\frac{?}{72}$ **24.** $\frac{-7}{5}$ and $\frac{?}{40}$

Reduce the following fractions to lowest terms.

25. $\frac{4}{64}$ **26.** $\frac{10}{16}$ **27.** $\frac{-24}{64}$ **28.** $\frac{-52}{64}$

Change each of the following fractions to decimals.

29. $\frac{1}{4}$ **30.** $\frac{1}{2}$ **31.** $\frac{-3}{8}$ **32.** $\frac{5}{16}$

33. $\frac{7}{8}$ **34.** $-3\frac{1}{3}$ **35.** $-2\frac{1}{4}$ **36.** $1\frac{5}{8}$

37. Tire tread depth is measured in 64ths of an inch. An auto technician checked the tread depth of a tire. The depth was ⁸⁄₆₄ in. What is this depth in lowest terms?

37. _____

38. Front suspension alignment shims come in ¹⁄₃₂-, ¹⁄₁₆-, and ⅛-inch thicknesses.

    (a) What is the decimal size of the smallest shim?

    38a. _____

    (b) What is the decimal size of the largest shim?

    38b. _____

39. A technician asks you to bring a ¾-mm shim. The shims are marked in decimals. What (decimal mm) size shim should you get?

39. _____

40. What (decimal mm) size shim should you get if a 1¼-mm shim is wanted?

40. _____

41. The service manager records how long each technician has worked on a job in hours written as decimals. If you spent ¾ hr to change a front engine mount, 1½ hr to replace the engine oil pan, and 2¼ hr to replace the clutch, how would the service manager record the total time?

41. _____

42. The toe specifications for a vehicle are ¼° ± ⅛°. Rewrite these specifications as decimals.

42. _____

43. List the following drill sizes in their decimal equivalents.

    (a) ³⁄₃₂ inch

    43a. _____

    (b) ⅛ inch

    43b. _____

    (c) ³⁄₁₆ inch

    43c. _____

    (d) ²⁹⁄₆₄ inch

    43d. _____

44. Which is larger, a ¹³⁄₃₂-inch S.A.E. socket size or a ⅜-inch S.A.E. socket size?

44. _____

Addition of Fractions

In order to add fractions, they must have the same denominator. If they do not have the same denominator, one or more numbers in the denominator must be changed so that all the numbers in the denominator are the same.

■ EXAMPLE 3–21

Solve $\frac{2}{5} + \frac{4}{5}$.

Solution

These two numbers can be added because they both have the same denominator, 5.

■ EXAMPLE 3–22

Solve $\frac{2}{3} + \frac{5}{8}$.

Solution

These two numbers have different denominators, 3 and 8. These numbers cannot be added until they are changed to equivalent fractions that have the same denominator.

Common Denominators

In order to add fractions, they must have the same numbers in the denominator, or a *common denominator*. There are many ways to find a common denominator. The method below is a quick way to find a common denominator.

Divide the smaller denominator into the larger denominator. If the answer is a whole number, then the larger number is the common denominator. If the answer is not a whole number, then multiply the denominators. This product is a common denominator.

■ EXAMPLE 3–23

What is a common denominator of $\frac{2}{3}$ and $\frac{5}{12}$?

Solution

$$12 \div 3 = 4$$

Step 1 Divide the larger denominator, 12, by the smaller denominator, 3. The answer, 4, is a whole number.

So, 12 is a common denominator of these two fractions.

■ EXAMPLE 3–24

What is a common denominator of $\frac{3}{4}$ and $\frac{5}{9}$?

Solution

$$9 \div 4 = 2\frac{1}{4}$$

Step 1 The answer $2\frac{1}{4}$ is not a whole number.

$$9 \times 4 = 36$$

Step 2 Multiply the denominators. The product 36 is a common denominator.

So, 36 is a common denominator.

Adding Fractions

In order to add fractions, follow these steps. You may not have to use all of them.

Step 1 Change any mixed numbers to fractions.

Step 2 Check the denominators. If they are the same, go to Step 5. If they are different, go to Step 3.

Step 3 Find a common denominator.

Step 4 Change each fraction to an equivalent fraction with the number in the denominator as the common denominator.

Step 5 Add the numerators. The number in the denominator is the common denominator.

Step 6 If possible, reduce the fraction.

■ EXAMPLE 3–25

Add $\frac{3}{8} + 1\frac{5}{16}$.

Solution

$$1\frac{5}{16} = \frac{21}{16}$$

Step 1 Change mixed numbers to fractions.

$$8 \neq 16$$

Step 2 The denominators are different.

$$16$$

Step 3 A common denominator is 16.

$$\frac{3}{8} = \frac{6}{16}$$

Step 4 $\frac{3}{8}$ is equivalent to $\frac{6}{16}$; $\frac{21}{16}$ does not have to be changed.

$$\frac{6}{16} + \frac{21}{16} = \frac{6 + 21}{16} = \frac{27}{16}$$

Step 5 Add the numerators. The denominator is 16.

$$\frac{27}{16} = 1\frac{11}{16}$$

Step 6 $\frac{27}{16}$ is equivalent to $1\frac{11}{16}$.

So, $\frac{3}{8} + 1\frac{5}{16} = \frac{27}{16} = 1\frac{11}{16}$.

■ EXAMPLE 3–26

Solve $\frac{2}{5} + \frac{3}{8}$.

Solution

Step 1 There are no mixed numbers.

$$5 \neq 8$$

Step 2

$$40$$

Step 3

$$\frac{2}{5} = \frac{16}{40} \quad \text{and} \quad \frac{3}{8} = \frac{15}{40}$$

Step 4

$$\frac{16}{40} + \frac{15}{40} = \frac{16 + 15}{40} = \frac{31}{40}$$

Step 5 So, $\frac{2}{5} + \frac{3}{8} = \frac{31}{40}$.

■ EXAMPLE 3–27

Solve $\frac{3}{8} + \frac{1}{8}$.

Solution

Step 1 There are no mixed numbers.

$$8 = 8$$

Step 2 The denominators are equal. *Go to Step 5.*

$$\frac{3}{8} + \frac{1}{8} = \frac{3 + 1}{8} = \frac{4}{8}$$

Step 5

$$\frac{4}{8} = \frac{1}{2}$$

Step 6 So, $\frac{3}{8} + \frac{1}{8} = \frac{1}{2}$.

■ EXAMPLE 3–28

Solve $2\frac{5}{8} + 3\frac{1}{4}$.

Solution

$$2\frac{5}{8} = \frac{21}{8}$$

$$3\frac{1}{4} = \frac{13}{4}$$

Step 1 Change mixed numbers to fractions.

$$8 \neq 4$$

Step 2 The denominators are different.

$$8$$

Step 3 A common denominator is 8.

$$\frac{13}{4} = \frac{26}{8}$$

Step 4 $\frac{13}{4}$ is equivalent to $\frac{26}{8}$. The fraction $\frac{21}{8}$ does not have to be changed.

$$\frac{21}{8} + \frac{26}{8} = \frac{21 + 26}{8} = \frac{47}{8}$$

Step 5 Add the numerators. The denominator is 8.

$$\frac{47}{8} = 5\frac{7}{8}$$

Step 6 $\frac{47}{8}$ is equivalent to $5\frac{7}{8}$.
So, $2\frac{5}{8} + 3\frac{1}{4} = 5\frac{7}{8}$.

■ EXAMPLE 3–29

Solve $\frac{1}{5} + \frac{4}{5}$.

Solution

Step 1 There are no mixed numbers.

$$5 = 5$$

Step 2 The denominators are equal. *Go to Step 5.*

$$\frac{1}{5} + \frac{4}{5} = \frac{1 + 4}{5} = \frac{5}{5}$$

Step 5

$$\frac{5}{5} = 1$$

Step 6 So, $\frac{1}{5} + \frac{4}{5} = 1$.

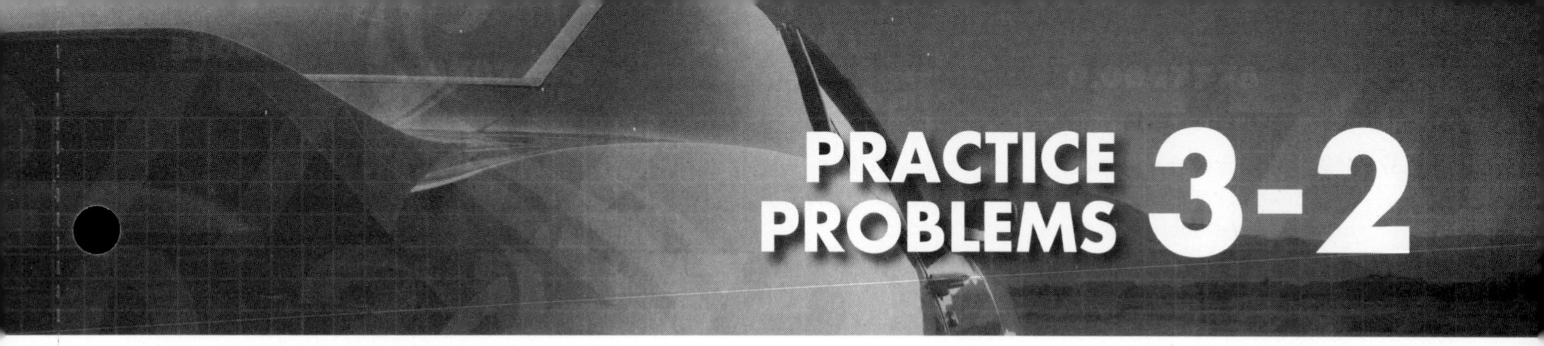

PRACTICE PROBLEMS 3-2

NAME ______________________________ DATE __________ SCORE __________

Find a common denominator for the two fractions in each problem.

1. $\frac{2}{3}$ and $\frac{5}{6}$ **2.** $\frac{1}{4}$ and $1\frac{3}{4}$ **3.** $\frac{5}{8}$ and $\frac{7}{16}$ **4.** $\frac{4}{15}$ and $\frac{2}{3}$

5. $\frac{7}{8}$ and $\frac{4}{3}$ **6.** $\frac{2}{5}$ and $\frac{1}{4}$ **7.** $5\frac{2}{3}$ and $\frac{3}{4}$ **8.** $\frac{4}{5}$ and $\frac{1}{6}$

Change each fraction to an equivalent fraction with a common denominator. *(Note that you found the common denominators in the above Problems 1 through 8.)*

9. $\frac{2}{3}$ and $\frac{5}{6}$ **10.** $\frac{1}{4}$ and $1\frac{3}{4}$ **11.** $\frac{5}{8}$ and $\frac{7}{16}$ **12.** $\frac{4}{15}$ and $\frac{2}{3}$

13. $\frac{7}{8}$ and $\frac{4}{3}$ **14.** $\frac{2}{5}$ and $\frac{1}{4}$ **15.** $5\frac{2}{3}$ and $\frac{3}{4}$ **16.** $\frac{4}{5}$ and $\frac{1}{6}$

Add each of the following. Change any answers that are improper fractions to mixed numbers. *(Problems 17 through 24 are the same as Problems 9 through 16.)*

17. $\frac{2}{3} + \frac{5}{6}$ **18.** $\frac{1}{4} + 1\frac{3}{4}$ **19.** $\frac{5}{8} + \frac{7}{16}$ **20.** $\frac{4}{15} + \frac{2}{3}$

21. $\frac{7}{8} + \frac{4}{3}$ **22.** $\frac{2}{5} + \frac{1}{4}$ **23.** $5\frac{2}{3} + \frac{3}{4}$ **24.** $\frac{4}{5} + \frac{1}{6}$

25. $\frac{3}{8} + \frac{5}{16}$ **26.** $4\frac{7}{16} + \frac{5}{16}$ **27.** $\frac{1}{8} + \frac{1}{64}$ **28.** $2\frac{3}{8} + \frac{7}{8}$

29. $\frac{1}{16} + \frac{5}{8}$ **30.** $4\frac{5}{6} + 2\frac{1}{3}$ **31.** $5\frac{1}{8} + 4\frac{1}{32}$ **32.** $6\frac{1}{2} + 2\frac{3}{8}$

33. $4\frac{2}{3} + \frac{1}{6}$ **34.** $1\frac{5}{6} + 2\frac{1}{2}$ **35.** $3\frac{3}{4} + 5\frac{5}{6}$ **36.** $3\frac{1}{2} + 4\frac{2}{3}$

37. The engine crankcase capacity is 5½ quarts with the oil filter, the automatic transmission capacity is 9¾ quarts, and rear axle capacity is 3¼ quarts. How many quarts of lubricants are in the power train?

37. _____

38. With the steering wheel centered, the toe-in reading is ³⁄₃₂ on one front wheel and ³⁄₁₆ on the other. What is the total toe-in?

38. _____

39. A worker drove 2½ miles to work, ¾ of a mile from work to a store, and 2⁷⁄₁₀ from the store to home. How far did the worker drive that day?

39. _____

40. A driver fills the fuel tank each Saturday. The first week the tank takes 38½ liters; the second, 46³⁄₁₀; the third, 39¾; and the fourth, 42¼. How many liters of fuel were used during that month?

40. _____

41. A technician spends ¼ hour replacing the battery and air filter, ½ hour replacing the spark plugs, 1⅜ hours packing wheel bearings, and 2¾ hours replacing the brake linings. What was the total time spent servicing this car?

41. _____

42. A technician spent 2¾ hours on one car, 3½ on another, 1⅓ on a third, and 2⅗ on a fourth. How many hours were spent on the four cars?

42. _____

43. An auto technician uses tubing in the lengths of 8⁵⁄₁₆ in, 7⅜ in, 2¼ in, 2 in, and 12½ in. What is the total length of the tubing used by the technician?

43. _____

44. A frame is ⁵⁄₁₆ in thick. A crossmember is ⁹⁄₃₂ in thick. What is the total thickness of these two pieces?

44. _____

Subtraction of Fractions

In order to answer a question like the one below, you must be able to subtract fractions.

An electrical wire is 15¼ inches long. A piece 9⅜ inches long is needed to repair a circuit. What length of wire will be left after the needed piece is cut off?

In subtracting fractions, use the following steps. Steps 1 through 4 and Step 6 were discussed in this chapter in the section on adding fractions.

Step 1 Change any mixed numbers to fractions.

Step 2 Check the denominators. If they are the same, go to Step 5. If they are different, go to Step 3.

Step 3 Find a common denominator.

Step 4 Change each fraction to an equivalent fraction with the number in the denominator as the common denominator.

Step 5 Subtract the numerators. The number in the denominator is the common denominator.

Step 6 If possible, reduce the fraction.

■ EXAMPLE 3–30

Solve $1\frac{3}{16} - \frac{5}{8}$.

Solution

$$1\frac{3}{16} = \frac{19}{16}$$

Step 1 Change mixed numbers to fractions.

$$8 \neq 16$$

Step 2 The denominators are different.

$$16$$

Step 3 A common denominator is 16.

$$\frac{5}{8} = \frac{10}{16}$$

Step 4 $\frac{5}{8}$ is equivalent to $\frac{10}{16}$.

$$\frac{19}{16} - \frac{10}{16} = \frac{19 - 10}{16} = \frac{9}{16}$$

Step 5 Subtract the numerators. The denominator is 16.

Step 6 $\frac{9}{16}$ cannot be reduced.
So, $1\frac{3}{16} - \frac{5}{8} = \frac{9}{16}$.

■ EXAMPLE 3–31

Solve $\frac{7}{8} - \frac{3}{5}$.

Solution

Step 1 There are no mixed numbers.

$$8 \neq 5$$

Step 2

$$40$$

Step 3

$$\frac{7}{8} = \frac{35}{40} \quad \text{and} \quad \frac{3}{5} = \frac{24}{40}$$

Step 4

$$\frac{35}{40} - \frac{24}{40} = \frac{35 - 24}{40} = \frac{11}{40}$$

Step 5

Step 6 $\frac{11}{40}$ cannot be reduced.
So, $\frac{7}{8} - \frac{3}{5} = \frac{11}{40}$.

■ EXAMPLE 3–32

Solve $\frac{7}{16} - \frac{3}{16}$.

Solution

Step 1 There are no mixed numbers.

$$16 = 16$$

Step 2 The denominators are equal. *Go to Step 5.*

$$\frac{7}{16} - \frac{3}{16} = \frac{7 - 3}{16} = \frac{4}{16}$$

Step 3

$$\frac{4}{16} = \frac{1}{4}$$

Step 4 So, $\frac{7}{16} - \frac{3}{16} = \frac{1}{4}$.

■ EXAMPLE 3–33

Solve $\frac{3}{4} - \frac{5}{8}$.

Solution

Step 1 There are no mixed numbers.

$$4 \neq 8$$

Step 2

$$8$$

Step 3 A common denominator is 8.

$$\frac{3}{4} = \frac{6}{8}$$

Step 4

$$\frac{6}{8} - \frac{5}{8} = \frac{6-5}{8} = \frac{1}{8}$$

Step 5 So, ¾ − ⅝ = ⅛.

■ EXAMPLE 3–34

Solve ⅔ − ⅖.

Solution

Step 1 There are no mixed numbers.

$$3 \neq 5$$

Step 2

$$15$$

Step 3

$$\frac{2}{3} = \frac{10}{15}$$

$$\frac{2}{5} = \frac{6}{15}$$

Step 4

$$\frac{10}{15} - \frac{6}{15} = \frac{10-6}{15} = \frac{4}{15}$$

Step 5 So, ⅔ − ²⁄₁₅ = ⁴⁄₁₅.

■ EXAMPLE 3–35

At the beginning of this section on subtraction of fractions, you were asked the following question. Answer the question.

An electrical wire is 15¼ inches long. A piece 9⅜ inches long is needed to repair a circuit. What length of wire will be left after the needed piece is cut off?

Solution

We need to subtract: 15¼ − 9⅜.

$$15\frac{1}{4} = \frac{61}{4}$$

$$9\frac{3}{8} = \frac{75}{8}$$

Step 1 Change mixed numbers to fractions.

$$4 \neq 8$$

Step 2 The denominators are different.

$$8$$

Step 3 A common denominator is 8.

$$\frac{61}{4} = \frac{122}{8}$$

Step 4 $\frac{61}{4}$ is equivalent to $\frac{122}{8}$.

$$\frac{122}{8} - \frac{75}{8} = \frac{122-75}{8} = \frac{47}{8}$$

Step 5 Subtract the numerators. The denominator is 8.

$$\frac{47}{8} = 5\frac{7}{8}$$

Step 6 $\frac{47}{8}$ can be written as the mixed number 5⅞. There will be 5⅞ inches of wire left.

NAME _____ DATE _____ SCORE _____

Find a common denominator for the two fractions in each problem.

1. ¾ and ⅝

2. 1⅓ and ⅚

3. ⅞ and ⅝

4. 2³⁄₁₆ and ⅝

5. ⁹⁄₁₀ and ⅔

6. 2⅓ and ⁴⁄₇

7. 3⅗ and ⅚

8. ⅞ and ⅗

Change each fraction to an equivalent fraction with a common denominator. *(You found a common denominator for these numbers in Problems 1 through 8.)*

9. ¾ and ⅝

10. 1⅓ and ⅚

11. ⅞ and ⅝

12. 2³⁄₁₆ and ⅝

13. ⁹⁄₁₀ and ⅔

14. 2⅓ and ⁴⁄₇

15. 3⅘ and ⅚

16. ⅞ and ⅗

Subtract each of the following. Change any answers that are improper fractions to mixed numbers. *(Problems 17 through 24 are the same as Problems 9 through 16.)*

17. ¾ − ⅝

18. 1⅓ − ⅚

19. ⅞ − ⅝

20. 2³⁄₁₆ − ⅝

21. ⁹⁄₁₀ − ⅔

22. 2⅓ − ⁴⁄₇

23. 3⅘ − ⅚

24. ⅞ − ⅗

25. ⁵⁄₁₆ − ⅛

26. 1¼ − ⅜

27. 3¼ − ½

28. 4⅝ − ⁷⁄₁₆

29. ⁵⁄₃₂ − ⅛

30. 2⅝ − 1⁷⁄₁₆

31. 4½ − 1⅝

32. 5⁹⁄₁₆ − 2²³⁄₃₂

33. 4⅔ − 3½

34. 2⅚ − 1⅓

35. 5⅜ − 3⁵⁄₁₆

36. 7¹³⁄₃₂ − 4¾

37. An electrical wire is 17⅝ inches long. A piece 12⅞ inches long is needed to repair a circuit. What length of wire will be left after the needed piece is cut off?

37. _____

38. An electrical wire is 27⅛ inches long. A piece 21⅝ inches long is needed. What length of wire will be left after the needed piece is cut off?

38. _____

39. How much wire is left if a piece 2 feet 4⁵⁄₁₆ inches long is removed from a piece that is 6 feet ⅛ inch long?

39. _____

40. A car is 19′ 7¾″ long. A garage is 24′ 9½″ long inside. How much clearance is left if a workbench 2½′ wide is placed at the end of the garage in front of the car?

40. _____

41. An electrical wire is 18 feet 9¼ inches long. A piece 13 feet 2½ inches is needed to repair a circuit. What length of wire will be left when the job has been completed?

41. _____

42. A 50-foot roll of electrical wire was purchased. From that roll, five pieces were used: one 18′ 9¼″, another 13′ 2½″, two more pieces that were each 3′ 2⅝″, and another at 11′ 7″. What length of wire was left when the job was completed?

42. _____

43. The fuel tank has 16³⁄₁₀ gallons in it. After a trip that consumed 9¾ gallons, how many gallons were left?

43. _____

44. A driver on a trip completes ⅓ of the trip on the first day and ¼ of the total trip on the next day. How much of the trip is left to be completed?

44. _____

45. An engine crankcase holds 4¾ liters. During a trip, the engine consumes ½ liter. How much oil is left after the trip?

45. _____

Multiplication of Fractions

To multiply a fraction and another number, first change the other number to a fraction.

The other number may already be a fraction. But it may also be a whole number, a mixed number, or a decimal. This is a quick review to show how these can be changed to fractions.

Changing Whole Numbers to Fractions

To change a whole number to a fraction so that you can multiply, write the whole number as a fraction with a denominator of 1.

■ EXAMPLE 3–36

Write the whole number 6 as $6/1$. Write 17 as $17/1$.

Changing Mixed Numbers to Fractions

A mixed number is really the sum of a whole number and a fraction. That means that $5\frac{2}{3}$ is the same as $5 + \frac{2}{3}$. To change a mixed number to a fraction, multiply the whole number and the denominator, add this product to the numerator, and place them all over the denominator.

■ EXAMPLE 3–37

Change $5\frac{2}{3}$ to a fraction.

Solution

$$5 \times 3 = 15$$

Step 1 Multiply the whole number by the denominator.

$$15 + 2 = 17$$

Step 2 Add that product to the numerator.

$$\frac{17}{3}$$

Step 3 Put that answer over the denominator.
So, $5\frac{2}{3} = 17/3$.

■ EXAMPLE 3–38

Change $7\frac{3}{8}$ to a fraction.

Solution

$$7 \times 8 = 56$$

Step 1

$$56 + 3 = 59$$

Step 2

$$\frac{59}{8}$$

Step 3 So, $7\frac{3}{8} = 59/8$.

Changing Decimals to Fractions

There are three steps in changing a decimal to a fraction. First, count the number of places to the right of the decimal point. Then remove the decimal point. This is the numerator of the fraction. The denominator is a 1 followed by the same number of 0s as places in the first step.

■ EXAMPLE 3–39

Change 7.34 to a fraction.

Solution

$$7.34$$
$$\uparrow$$
$$2 \text{ places}$$

Step 1 Count the places to the right of the decimal point.

$$734$$

Step 2 Remove the decimal point from the number.

$$\frac{734}{100}$$
$$2 \text{ 0s}$$

Step 3 Write the Step 2 answer over a 1 followed by the same number of 0s as the Step 1 answer.
So, $7.34 = 734/100$.

■ EXAMPLE 3–40

Change 4.875 to a fraction.

Solution

$$4.875$$
$$\uparrow$$
$$3 \text{ places}$$

Step 1

$$4875$$

Step 2

$$\frac{4875}{1000}$$
$$\uparrow$$
$$3\ 0s$$

Step 3   So, $4.875 = {}^{4875}\!/_{1000}$.

■ EXAMPLE 3–41

Change -5.72 to a fraction.

Solution

$$-5.72$$
$$\uparrow$$
$$2\ places$$

Step 1

$$-572$$

Step 2

$$\frac{-572}{100}$$
$$\uparrow$$
$$2\ 0s$$

Step 3   So, $-5.72 = {}^{-572}\!/_{100}$.

Multiplying Fractions

Step 1   Write every number as a fraction.

Step 2   Multiply the numerators together. This is the numerator of the answer.

Step 3   Multiply the denominators together. This is the denominator of the answer.

■ EXAMPLE 3–42

Solve $\frac{3}{4} \times \frac{5}{8}$.

Solution
Step 1   Both numbers are fractions so nothing needs to be done.

$$3 \times 5 = 15$$

Step 2   Multiply the numerators.

$$4 \times 8 = 32$$

Step 3   Multiply the denominators.
So, $\frac{3}{4} \times \frac{5}{8} = {}^{15}\!/_{32}$.

■ EXAMPLE 3–43

Solve $\frac{7}{8} \times 9$.

Solution

$$9 = \frac{9}{1}$$

Step 1   Change 9 to a fraction.

$$7 \times 9 = 63$$

Step 2   Multiply the numerators.

$$8 \times 1 = 8$$

Step 3   Multiply the denominators.
So, $\frac{7}{8} \times 9 = {}^{63}\!/_{8}$. Note that ${}^{63}\!/_{8}$ can be written as $7\frac{7}{8}$.

■ EXAMPLE 3–44

Solve $4\frac{1}{8} \times \frac{3}{4}$.

Solution

$$4\frac{1}{8} = \frac{33}{8}$$

Step 1   Change $4\frac{1}{8}$ to a fraction.

$$33 \times 3 = 99$$

Step 2

$$8 \times 4 = 32$$

Step 3   So, $4\frac{1}{8} \times \frac{3}{4} = {}^{99}\!/_{32}$. Note that ${}^{99}\!/_{32}$ can be written as $3\frac{3}{32}$.

■ EXAMPLE 3–45

Solve $\frac{5}{8} \times 3.25$.

Solution

$$3.25 = \frac{325}{100}$$

Step 1   Change 3.25 to a fraction.

$$5 \times 325 = 1625$$

Step 2

$$8 \times 100 = 800$$

Step 3   So, $\frac{5}{8} \times 3.25 = {}^{1625}\!/_{800}$. This number, ${}^{1625}\!/_{800}$, can be reduced to ${}^{65}\!/_{32}$, which is the same as $2\frac{1}{32}$.

Rules for Multiplying with Negative Numbers

1. A negative number times a positive number gives a negative product.

2. A negative number times a negative number gives a positive product.

■ EXAMPLE 3–46

Solve $^-\!\frac{3}{5} \times 2.45$.

Solution

$$2.45 = \frac{245}{100}$$

Step 1

$$-3 \times 245 = -735$$

Step 2

$$5 \times 100 = 500$$

Step 3 So, $^-\!\frac{3}{5} \times 2.45 = {}^{-735}\!/_{500}$, and this can be reduced to $^{-147}\!/_{100}$.

Remember, it is possible to multiply more than two numbers.

■ EXAMPLE 3–47

Solve $\frac{2}{3} \times \frac{4}{5} \times 1\frac{5}{8}$.

Solution

$$1\frac{5}{8} = \frac{13}{8}$$

Step 1

$$2 \times 4 \times 13 = 104$$

Step 2

$$3 \times 5 \times 8 = 120$$

Step 3 So, $\frac{2}{3} \times \frac{4}{5} \times 1\frac{5}{8} = {}^{104}\!/_{120}$, and this can be reduced to $^{13}\!/_{15}$.

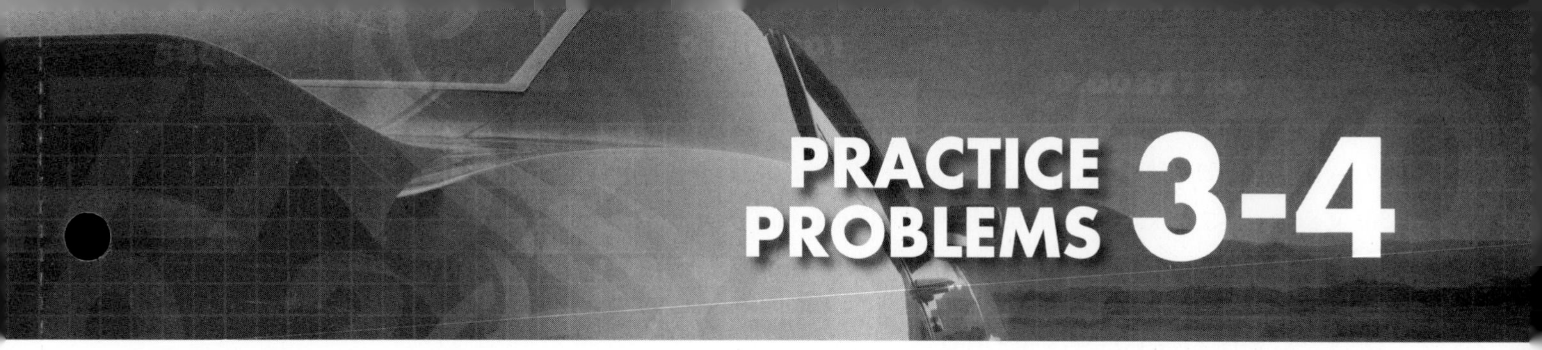

PRACTICE PROBLEMS 3-4

NAME ______________________________ DATE __________ SCORE __________

Change each of these numbers to a fraction.

1. 29 **2.** 15 **3.** -8 **4.** 175

5. $-2\frac{1}{2}$ **6.** $4\frac{3}{4}$ **7.** $13\frac{5}{8}$ **8.** $12\frac{3}{8}$

9. 4.2 **10.** -5.25 **11.** -19.375 **12.** 7.4375

Multiply the following numbers. Change any answers that are improper fractions to mixed numbers.

13. $\frac{2}{3} \times \frac{4}{5}$ **14.** $\frac{3}{4} \times -\frac{5}{8}$ **15.** $\frac{1}{8} \times \frac{3}{4}$ **16.** $\frac{9}{16} \times \frac{1}{2}$

17. $1\frac{1}{2} \times \frac{3}{4}$ **18.** $2\frac{3}{4} \times \frac{5}{8}$ **19.** $1\frac{5}{16} \times \frac{2}{3}$ **20.** $-4\frac{1}{4} \times \frac{3}{8}$

21. $3.2 \times \frac{3}{4}$ **22.** $\frac{3}{8} \times 4.5$ **23.** $-7.25 \times -\frac{7}{3}$ **24.** $2\frac{3}{4} \times 4.65$

25. $-\frac{7}{8} \times 3.8$ **26.** $-4.3 \times -5\frac{2}{3}$ **27.** $3\frac{1}{2} \times 4\frac{2}{5}$ **28.** $-0.64 \times 2\frac{3}{8}$

29. A driver travels 2⁷⁄₁₀ miles each day. How far will this driver travel in 6 days?

30. If a car goes 21¾ km per L of fuel, how far will it go on 17½ liters of fuel?

31. If a car consumes ⅗ liters of oil each day, how much oil will be used for a 2-week trip?

32. A tire rotates 4¾ turns per second. How many times will it turn in 1 minute?

33. A race car averages 3½ miles per minute. How far will this race car travel in 1 hour?

34. How fast (in miles per hour) is the car in problem 33 going?

35. Mary travels 12⅝ miles to work each day.

    (a) How far does she drive to and from work each 5-day week?

    (b) One week she has to work overtime on Saturday. How far did she drive to and from work that week?

    (c) In a normal year she works 49 weeks. If each week is 5 days long, how far can she expect to drive to and from work?

36. A driver begins a 2350-mile trip. The first day she completes ⅓ of the distance, the second day ⅙, the third day ¼, and the fourth day ⅕. How many miles are left to complete?

37. Robert earns $11.52 an hour. One day he was late for work and only worked 7¾ hours. How much did he earn that day?

38. A certain model car requires 23⅜ inches of ½-inch air conditioning hose. How many inches are needed for 5 cars.

29. _____

30. _____

31. _____

32. _____

33. _____

34. _____

35a. _____

35b. _____

35c. _____

36. _____

37. _____

38. _____

Use the drawing of the deck surface of a cylinder block shown in Figure 3–1 for problems 39 and 40.

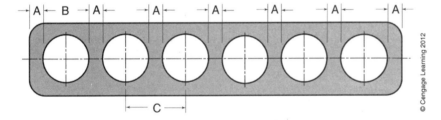

© Cengage Learning 2012

Figure 3–1

39. Each length marked **A** on the cylinder block is ⅝ in. and each cylinder has a diameter of 2¹⁵⁄₁₆ in. What is the total length of the cylinder block?

39. _____

40. If each cylinder has a diameter of 3¹⁷⁄₃₂ in. and distance **A** is ¹³⁄₁₆ in.

    (a) What is the center-to-center length of dimension **C**?

40a. _____

    (b) What is the total length of the cylinder block surface?

40b. _____

Division of Fractions

The answer to a division problem is called the *quotient*. The *divisor* is the number that divides.

$$12 \div 4 \qquad 4\overline{)12}$$
$$\uparrow \qquad \uparrow$$

$$\text{divisor}$$
$$\downarrow$$
$$\frac{5}{8} \div \frac{3}{4}$$

Reciprocals

A *reciprocal* of a given number is a number that when multiplied by the given number equals 1. For example, the reciprocal of ¾ is ⁴⁄₃ and the reciprocal of 2 (²⁄₁) is ½. The division of fractions uses the reciprocal of a number. There are never more than two steps to finding the reciprocal of a number.

Step 1 Change any whole number, mixed number, or decimal to a fraction. If you are not sure that you know how to do this, study the material in the section on multiplying fractions.

Step 2 Find the reciprocal of the number by turning it upside down. In other words, the numerator of the fraction is the denominator of the reciprocal, and the denominator of the fraction is the numerator of the reciprocal.

■ EXAMPLE 3–48

Find the reciprocal of 1⅔.

Solution

$$1\frac{2}{3} = \frac{5}{3}$$

Step 1 Change the mixed number to a fraction.

$$\frac{3}{5}$$

Step 2 Turn the fraction upside down.
So, the reciprocal of 1⅔ is ⅗.

■ EXAMPLE 3–49

Find the reciprocals of 4, 15, ⅜, ¹⁄₇, ⁻⁴⁄₃, and 6.37.

Solution

$$\frac{4}{1}, \frac{15}{1}, \frac{3}{8}, \frac{1}{7}, \frac{-4}{3} \quad \text{and} \quad \frac{637}{100}$$

Step 1 Write each number as a fraction.

$$\frac{1}{4}, \frac{1}{15}, \frac{8}{3}, \frac{7}{1}, \frac{-3}{4} \quad \text{and} \quad \frac{100}{637}$$

Step 2 Turn each number upside down. These are the reciprocals.

Dividing Fractions

To divide a fraction, first find the reciprocal of the divisor. Then multiply the other number by the divisor's reciprocal.

■ EXAMPLE 3–50

Solve ⅝ ÷ ⅔.

Solution
Step 1 Both of the numbers are fractions.

$$\frac{3}{2}$$

Step 2 The divisor is ⅔. The reciprocal is ³⁄₂.

$$\frac{5}{8} \times \frac{3}{2} = \frac{15}{16}$$

Step 3 Multiply the reciprocal of the divisor and the other number.
So, ⅝ ÷ ⅔ = ⅝ × ³⁄₂ = ¹⁵⁄₁₆.

■ EXAMPLE 3–51

Solve ⅞ ÷ ⁹⁄₅.

Solution
Step 1 Both of the numbers are fractions.

$$\frac{5}{9}$$

Step 2 The divisor is ⁹⁄₅. The reciprocal is ⁵⁄₉.

$$\frac{7}{8} \times \frac{5}{9} = \frac{35}{72}$$

Step 3 So, ⅞ ÷ ⁹⁄₅ = ⅞ × ⁵⁄₉ = ³⁵⁄₇₂.

■ EXAMPLE 3–52

Solve 1¾ ÷ 7.

Solution

$$1\frac{3}{4} = \frac{7}{4}$$

$$7 = \frac{7}{1}$$

Step 1 Change the mixed number 1¾ and the whole number 7 to fractions.

$$\frac{1}{7}$$

Step 2 The reciprocal of ⁷⁄₁ is ¹⁄₇.

$$\frac{7}{4} \times \frac{1}{7} = \frac{7}{28} = \frac{1}{4}$$

Step 3 Thus, 1¾ ÷ 7 = ¼.

■ EXAMPLE 3–53
Solve 2⅜ ÷ ¾.

Solution
$$2\frac{3}{8} = \frac{19}{8}$$

Step 1

$$\frac{4}{3}$$

Step 2

$$\frac{19}{8} \times \frac{4}{3} = \frac{76}{24} = \frac{19}{6}$$

Step 3 We see that 2⅜ ÷ ¾ = ¹⁹⁄₆ = 3⅙.

Rules for Dividing with Negative Numbers

1. A negative number divided by a positive number (or vice versa) gives a negative quotient.

2. A negative number divided by a negative number gives a positive quotient.

■ EXAMPLE 3–54
Solve ⅗ ÷ ⁻³⁄₇.

Solution
Step 1 Both of the numbers are fractions.

$$\frac{-7}{3}$$

Step 2 The divisor is ⁻³⁄₇. Its reciprocal is ⁻⁷⁄₃.

$$\frac{3}{5} \times \frac{-7}{3} = \frac{-21}{15} = \frac{-7}{5}$$

Step 3 So, ⅗ ÷ ⁻³⁄₇ = ⁻⁷⁄₅.

■ EXAMPLE 3–55
Solve −3½ ÷ ⅔.

Solution
$$-3\frac{1}{2} = \frac{-7}{2}$$

Step 1

$$\frac{3}{2}$$

Step 2

$$\frac{-7}{2} \times \frac{3}{2} = \frac{-21}{4}$$

Step 3 So, −3½ ÷ ⅔ = ⁻²¹⁄₄ = −5¼.

■ EXAMPLE 3–56
In Example 3–4, we began to solve the following problem.

A piece of electrical wire is 25⅔ feet long. How many 2¾-foot long pieces can be cut from this one piece of wire?

Complete the solution.

Solution
To finish this example, we need to divide fractions. In Example 3–4, we converted the two mixed numbers to fractions. We found that 25⅔ = ⁷⁷⁄₃ and 2¾ = ¹¹⁄₄. Now, we need to solve 25⅔ ÷ 2¾, or ⁷⁷⁄₃ ÷ ¹¹⁄₄.

Step 1 Both of the numbers are now fractions.

Step 2 The divisor is ¹¹⁄₄. Its reciprocal is ⁴⁄₁₁.

Multiplying we get:

$$\frac{77}{3} \times \frac{4}{11} = \frac{308}{33} = \frac{28}{3} = 9\frac{1}{3}$$

Step 3 So, 9 pieces that are each 2¾ feet long can be cut from this length of wire. There will be ⅓ of a piece that is 11 inches long left over.

NAME _____ DATE _____ SCORE _____

In Problems 1 through 4, which number is the divisor?

1. $\frac{3}{4} \div \frac{5}{8}$ **2.** $1\frac{3}{4} \div \frac{2}{3}$ **3.** $4 \div -3\frac{1}{6}$ **4.** $3\frac{1}{7} \div 2\frac{1}{3}$

Write the reciprocals of the divisors for Problems 5 through 12.

5. $\frac{3}{4} \div \frac{5}{8}$ **6.** $1\frac{3}{4} \div \frac{2}{3}$ **7.** $-5\frac{1}{3} \div 2$ **8.** $\frac{3}{5} \div -4$

9. $\frac{3}{8} \div \frac{1}{4}$ **10.** $2\frac{1}{2} \div -3\frac{1}{3}$ **11.** $-3\frac{2}{5} \div -1\frac{1}{2}$ **12.** $4\frac{1}{3} \div 6\frac{1}{3}$

Find the quotients for each of the following problems. Change any answers that are improper fractions to mixed numbers. *(Problems 13 through 20 are the same as Problems 5 through 12.)*

13. $\frac{3}{4} \div \frac{5}{8}$ **14.** $1\frac{3}{4} \div \frac{2}{3}$ **15.** $-5\frac{1}{3} \div 2$ **16.** $\frac{3}{5} \div -4$

17. $\frac{3}{8} \div \frac{1}{4}$ **18.** $2\frac{1}{2} \div -3\frac{1}{3}$ **19.** $-3\frac{2}{5} \div -1\frac{1}{2}$ **20.** $4\frac{1}{3} \div 6\frac{1}{3}$

21. $2\frac{1}{5} \div -3.75$ **22.** $3\frac{3}{8} \div 6.12$ **23.** $5\frac{3}{8} \div 2\frac{1}{2}$ **24.** $3\frac{1}{16} \div 4$

25. $3\frac{1}{3} \div -2\frac{5}{6}$ **26.** $-4\frac{1}{4} \div 2\frac{1}{8}$ **27.** $12\frac{3}{4} \div -17$ **28.** $8\frac{5}{8} \div -0.63$

29. A vehicle is driven 436 miles and uses 17¼ gallons of fuel. How many miles per gallon does this vehicle get?

29. _____

30. A car is driven 589 kilometers and uses 43½ liters of fuel. How many kilometers per liter does this car get?

30. _____

31. Find the current flow in a circuit if the voltage is 13³⁄₁₀ volts and the resistance is 4⅕ ohms. (Current flow = voltage ÷ resistance)

31. _____

32. A piece of electrical wire is 18⅓ feet long. How many 3¾-foot pieces can be cut from this piece?

32. _____

33. A driver has traveled 372 miles in 7⅓ hours. How far did he or she drive each hour? Assume a constant rate of speed.

33. _____

34. A tank of fuel weighs 171 pounds plus the weight of the tank. If a gallon of gasoline weighs 7⅓ pounds, how many gallons of fuel are in the tank?

34. _____

35. A race car travels one lap of the track in ⅚ minute. The track is 2½ miles in length. How long will it take to complete 500 miles?

35. _____

36. Carlos travels 17⅜ miles to work each day.

    (a) How far does he drive to and from work each 5-day week?

    36a. _____

    (b) In a normal year, when he includes vacation and holidays, he works 47⅗ weeks. If each week is 5 days long, how far can Carlos expect to drive to and from work?

    36b. _____

    (c) Carlos's van averages 22¼ miles per gallon (mpg). How many gallons of gasoline can he expect to use driving to and from work?

    36c. _____

    (d) Each gallon of gasoline cost $4.05⁹⁄₁₀. How much can he expect to spend on gas driving to work in one year?

    36d. _____

37. How many pieces of 2³⁄₁₆-inch-long 16-gauge wire can be cut from a piece 27¾ inches long?

37. _____

38. A box can hold 12 wire terminal connectors. If 9¾ boxes of wire terminals cost $56.16, what is the cost for one dozen terminals?

38. _____

39. The CAFE figure for each manufacturer is calculated by using the following formula:

$$CAFE = \frac{\text{Total Production Volume}}{\dfrac{\text{Production Vol. Vehicle A}}{\text{Fuel Economy Vehicle A}} + \dfrac{\text{Production Vol. Vehicle B}}{\text{Fuel Economy Vehicle B}} + \dfrac{\text{Production Vol. Vehicle C}}{\text{Fuel Economy Vehicle C}}}$$

Using the data in the table below, calculate the CAFE number for this one year of production.

Model	Mpg	Production Volume
Vehicle A	22	130,000
Vehicle B	20	120,000
Vehicle C	16	100,000

39. _____

GEOMETRY, ANGLE MEASUREMENT, AND THE METRIC SYSTEM

Objectives: After studying this chapter, you should be able to:

- Use a protractor to measure angles.
- Convert degrees of an angle into minutes and seconds and vice versa.
- Recognize and estimate sizes of angles.
- Distinguish an angle of a given size from a group of angles.
- Add and subtract angular measurements.
- Apply formulas to find the circumference and area of a circle and the volume of a cylinder.
- Use a table of multipliers to convert customary units to metric units and vice versa.
- Use degree relationships and alignment angle values.
- Calculate the diameter, radius, area, and circumference of tires and cylinders.
- Determine cylinder volume and total engine displacement.
- Make conversions between the customary and metric measurement systems.

Angles

An angle is formed by two lines that meet at a point, called the *vertex*. The two lines are the *sides* of the angle. The size of an angle is based on the number of degrees between the sides and not on the lengths of the sides. The two angles in Figure 4–1 are the same size.

Degrees

Angles are measured in degrees. A circle has 360 degrees. The degree sign, °, is used as the symbol for degree. So, a circle has 360°.

Each degree is divided into 60 minutes. The symbol for a minute is the prime sign, ′, So, 30′ = ½°, 60′ = 1°, and 120′ = 2°.

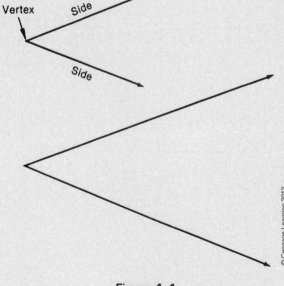

© Cengage Learning 2012

Figure 4–1

Measuring Angles

Angles are measured with a tool called a protractor. Most protractors are in the shape of half a circle. Protractors have two scales. In the protractor shown in Figure 4–2, each scale goes from 0° to 180°.

One scale begins on the left side of the protractor; the other scale begins on the right side. In the middle of the bottom edge of the protractor is a center mark.

In order to measure an angle, place the vertex of the angle at the center mark and the bottom edge of the protractor along one side of the angle. Read the number of degrees where the other side of the angle meets the protractor. Read the scale that has the 0° on the first side of the angle.

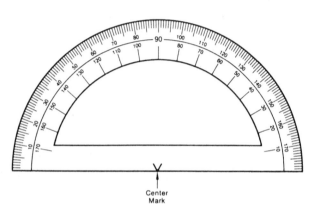

Figure 4–2

■ EXAMPLE 4–1

Measure this angle.

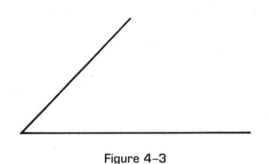

Figure 4–3

Solution

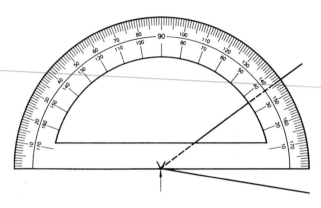

Figure 4–4

Step 1 Place the vertex at the center mark.

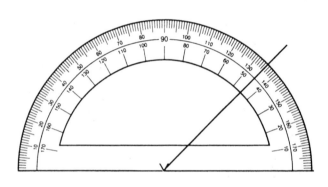

Figure 4–5

Step 2 Place the bottom edge of protractor along one side of the angle. (Make sure that the center mark stays at the vertex.)

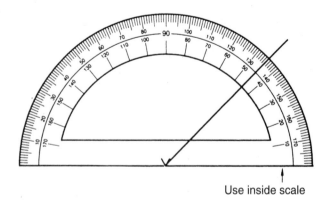

Use inside scale

Figure 4–6

Step 3 Find the scale with 0° on the side of the angle.

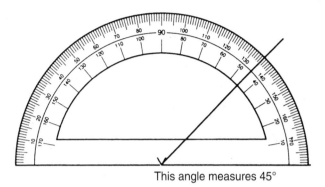

This angle measures 45°

Figure 4–7

Step 4 Read the number on the scale where the other side hits the protractor.

■ EXAMPLE 4–2

Measure this angle.

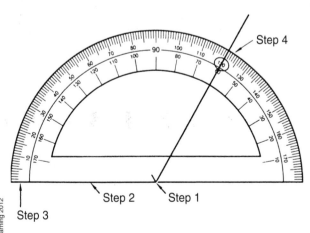

Figure 4–8

Solution

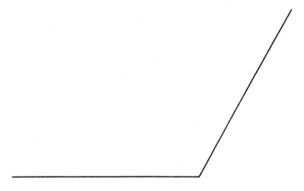

Step 4

Step 2   Step 1

Step 3

Figure 4–9

Step 3 Outside scale has 0° on side of angle.

Step 4 This angle measures 120°.

Fractional Degrees and Minutes

Some companies use degrees and fractional parts of degrees instead of degrees and minutes. To change a fractional degree to minutes, multiply the fraction by 60. (*Remember:* 1° = 60′.)

■ EXAMPLE 4–3

Change ¼° to minutes.

Solution

$$\frac{1}{4} \times \frac{60}{1} = \frac{60}{4}$$

Step 1 Multiply the fractional degree by 60.

$$\frac{60}{4} = 15$$

Step 2 Reduce the answer in Step 1.
So, ¼° = 15′.

■ EXAMPLE 4–4

Change 4⅔° to degrees and minutes.

Solution

$$\frac{2}{3} \times \frac{60}{1} = \frac{120}{3}$$

Step 1

$$\frac{120}{3} = 40$$

Step 2 So, 4⅔° = 4°40′.

To change minutes to fractional degrees, put the number of minutes over 60 and simplify.

■ EXAMPLE 4–5

Change 48′ to degrees.

Solution

$$48' = \frac{48°}{60}$$

Step 1 Write as a fraction with a denominator of 60.

$$\frac{48}{60} = \frac{4}{5}$$

Step 2 Reduce the fraction.
So, $48' = \frac{4}{5}°$.

■ EXAMPLE 4–6

Change $7°36'$ to degrees.

Solution

$$36' = \frac{36°}{60}$$

Step 1

$$\frac{36}{60} = \frac{3}{5}$$

Step 2 Thus, $7°36' = 7\frac{3}{5}°$.

Adding Angular Measurements

The main things to remember when you are adding angular measurements are that you must add like units and that $60' = 1°$.

■ EXAMPLE 4–7

Add $4°20'$ and $12°15'$.

Solution

$$\begin{array}{r} 4°20' \\ + 12°15' \\ \hline 16°35' \end{array}$$

So, $4°20' + 12°15' = 16°35'$.

■ EXAMPLE 4–8

Find the sum of $7°25'$ and $38°45'$.

Solution

$$\begin{array}{r} 7°25' \\ + 38°45' \\ \hline 45°70' \end{array}$$

We see that $7°25' + 38°45' = 46°10'$.

Subtracting Angular Measurements

When you are subtracting, remember to subtract like units. If it is necessary to borrow, remember that $1° = 60'$.

■ EXAMPLE 4–9

Solve $24°35' - 11°15'$.

Solution

$$\begin{array}{r} 24°35' \\ - 11°15' \\ \hline 13°20' \end{array}$$

So, $24°35' - 11°15' = 13°20'$.

■ EXAMPLE 4–10

Find the difference between $35°10'$ and $18°25'$.

Solution

$$\begin{array}{r} 35°10' \\ - 18°25' \\ \hline \end{array}$$

Note that you cannot subtract $25'$ from $10'$, so borrow $1°$ from the $35°$. This changes $35°10'$ to $34°70'$. Now, subtract.

$$\begin{array}{r} 35°10' \\ - 18°25' \end{array} \quad = \quad \begin{array}{r} 34°70' \\ - 18°25' \\ \hline 16°45' \end{array}$$

The answer is $16°45'$.

NAME _____ DATE _____ SCORE _____

Use a protractor to measure these angles.

1.

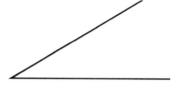

2.

3.

4.

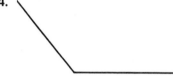

5.

6.

Without using a protractor, estimate the correct size of each angle in Exercises 7 through 12.

7. The size of the angle in Figure 4–10 is

    (a) 30°

    (b) 45°

    (c) 60°

    (d) 120°

8. The size of the angle in Figure 4–11 is

    (a) 30°

    (b) 45°

    (c) 60°

    (d) 120°

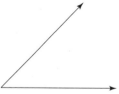

Figure 4–10

7. _____

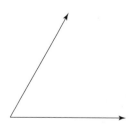

Figure 4–11

8. _____

9. The size of the angle in Figure 4–12 is

    (a) 90°

    (b) 135°

    (c) 150°

    (d) 180°

9. _____

10. The size of the angle in Figure 4–13 is

    (a) 90°

    (b) 135°

    (c) 150°

    (d) 180°

10. _____

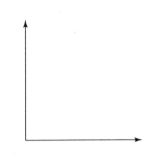

Figure 4–12   **Figure 4–13**

11. What is the size of the angle in Figure 4–14?

12. What is the size of the angle in Figure 4–15?

11. _____

12. _____

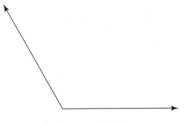 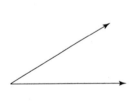

Figure 4–14   **Figure 4–15**

Change the following fractions to degrees and minutes.

13. ¾°   **14.** ⅓°   **15.** 2½°   **16.** 5⅔°

Change the following to fractional degrees.

17. 15′   **18.** 48′   **19.** 7°25′   **20.** 9°24′

Add or subtract as indicated.

21.   25°12′
  + 14°36′
	22. 4°19′
	  + 12°32′
		23. 12°36′
		  − 10°20′
			24. 24°28′
			  − 7° 6′

25.   24°35′
  + 6°45′
	26. 25°48′
	  + 13°32′
		27. 14°15′
		  − 8°30′
			28. 35°20′
			  − 18°48′

29. A vehicle has just made a U turn. It was going east and is now going west. How many degrees has the vehicle turned during the U turn?

29. _____

30. A crankshaft has turned through 5 complete revolutions. How many degrees has the crankshaft turned through its 5 revolutions?

30. _____

31. What does a person mean by saying "I did a 360"?

31. _____

32. The camber angle has been changed from $-.15°$ to $+.45°$. How many degrees or what part of a degree has the camber been changed?

32. _____

33. The caster angle has just been changed from $+\frac{1}{4}°$ to $+1\frac{3}{4}°$. How many degrees has the caster been changed?

33. _____

34. A driver made a right turn, two left turns, a right turn, two more left turns, a U turn, and two right turns. How many degrees did the driver turn if you assume that the right and left turns were $90°$ turns?

34. _____

35. If the driver in Problem 34 was originally driving north, in what direction was the car going after this series of turns was completed?

35. _____

36. (a) A complete engine firing cycle requires the crankshaft to make two complete revolutions. How many degrees does this represent?

36a. _____

(b) The crankshaft–camshaft ratio is 2:1. How many degrees does the camshaft rotate during one firing cycle?

36b. _____

37. Climbing performance in this problem is listed as a percent of grade. A level surface is a 0% grade and a vertical surface is a 100% grade. The climbing performance for an off-road vehicle is listed as follows:

1st gear: 63%
2nd gear: 47%
3rd gear: 34%
4th gear: 20%
5th gear: 12%

What angle in degrees is this vehicle capable of climbing in each gear?

37. 1st gear _____

2nd gear _____

3rd gear _____

4th gear _____

5th gear _____

Circular Measures

A circle has two major parts: the diameter and the radius. A technician may have to use either of these major parts to find two other measurements: the circumference and the area.

Circumference

As shown in Figure 4–16, the circumference, C, is the distance around a circle. The diameter, d or D, passes through the center of the circle and is the longest distance across the circle. In the automotive trade the diameter is also referred to as the bore. A radius, r, is the distance from the center of the circle to the circle itself. The diameter is twice as long as the radius. That is, $d = 2r$ or $r = \frac{1}{2}d$.

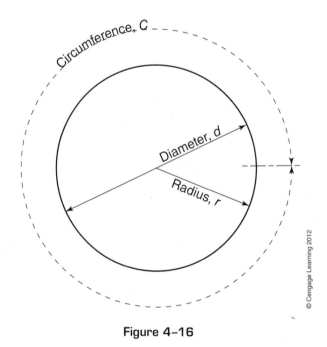

Figure 4–16

© Cengage Learning 2012

The circumference is given by the formula

$$C = \pi d$$
$$\approx 3.1416d$$

The symbol π, pronounced "pie," is a constant. Many calculators have a $\boxed{\pi}$ key. If yours does, then use it; if it does not, then you can use the decimal approximation of 3.1416.

■ EXAMPLE 4–11

A circle has a radius of 12.7 mm. What is its circumference?

Solution

We are given $r = 12.7$, so $d = 2 \times 12.7 = 25.4$. Substituting this in the above formula, we get

$$C = \pi d$$
$$= \pi(25.4)$$
$$\approx 79.796$$

The circumference is about 79.8 mm.

The above answer of 79.796 was obtained using the $\boxed{\pi}$ key. If you use 3.1416 for π, then you will get 79.797.

■ EXAMPLE 4–12

A piston fits into a cylinder with a bore of 3.5 in. What is the bore circumference?

Solution

We are given $d = 3.5$. Substituting this in the above formula, we get

$$C = \pi d$$
$$= \pi(3.5)$$
$$\approx 10.996$$

The circumference is about 11.0 in.

■ EXAMPLE 4–13

What is the circumference of a P205/70R15 tire?

Solution

There are three numbers on this tire. The first number is the section width of the tire in millimeters, measured from sidewall to sidewall. The section width of this tire is 205 mm. The second number is the aspect ratio as a percent. This is the ratio of sidewall height to width. This tire has an aspect ratio of 70 percent. The last number for this tire is 15. It is the diameter of the wheel in inches.

$$\text{Tire diameter} = \frac{\text{section width} \times \text{aspect ratio}}{1270} + \text{rim diameter}$$

For this tire we have

$$\text{Tire diameter} = \frac{205 \times 70}{1270} + 15$$
$$= 26.30$$

So, the diameter of this wheel is 26.3 in.

Now that we know the diameter, we can determine the circumference.

$$C = \pi d$$
$$= \pi(26.3)$$
$$\approx 82.62$$

The circumference is about 82.62 in. To express the circumference in feet, divide 82.62 by 12. Since 82.62 ÷ 12 ≈ 6.89 ft, the circumference of this tire is about 6′10.7″.

Arc Length

An arc is a part of a circle. One arc has been shown in Figure 4–17 by the thick part of the circle. The arc is often named using its endpoints and placing a small arc over the name. So, the arc in Figure 4–17 is $\overset{\frown}{AB}$.

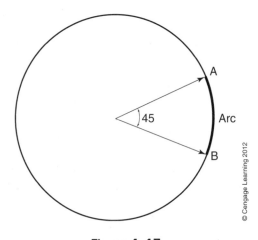

A

45

Arc

B

© Cengage Learning 2012

Figure 4–17

The size of an arc is determined by the number of degrees in its central angle. For example, the arc in Figure 4–17 measures 45°. The arc length is the length of the arc and is the fractional part of the circumference formed by the central angle.

■ EXAMPLE 4–14

Suppose the circle in Figure 4–17 has a radius of 10.5 mm. What is arc length of $\overset{\frown}{AB}$?

Solution

The size of $\overset{\frown}{AB}$ is 45°, so the arc length is $\frac{45°}{360°} = \frac{1}{8}$ the circumference of the circle.

We are given the radius of the circle as $r = 10.5$, so $d = 2 \times 10.5 = 21$. Substituting this in the above formula, we get

$$C = \pi d$$
$$= \pi(21)$$
$$\approx 65.97$$

The circumference of the circle is about 65.97 mm.

The arc length is $\frac{1}{8}$ of this circumference or $\frac{1}{8} \times 65.97 \approx 8.25$ mm.

Area

The area of a circle is used in working with many circular components. The area, A, of a circle is given by either of the following formulas:

$$A = \pi r^2$$
$$= \frac{\pi}{4}d^2$$
$$\approx 0.7854 d^2$$

Most of your math classes probably used the $A = \pi r^2$ formula, but many automotive technicians use $A = 0.7854 d^2$.

■ EXAMPLE 4–15

A circle has a radius of 12.7 mm. What is its area?

Solution

We are given $r = 12.7$, so $d = 2 \times 12.7 = 25.4$. Substituting this in the above formula, $A = 0.7854 d^2$, we get

$$A = 0.7854 d^2$$
$$= 0.7854 \times 25.4 \times 25.4$$
$$\approx 506.71$$

The area is about 506.71 mm^2.

■ EXAMPLE 4–16

A piston fits into a cylinder with a bore of 3.5 in. What is the cross-section area of the piston?

Solution

We are given $d = 3.5$. Substituting this in the above formula, we get

$$A = 0.7854 d^2$$
$$= 0.7854 \times 3.5^2$$
$$\approx 9.62$$

The cross-section area is about 9.62 in^2.

Cylinders

A cylinder is a geometric solid that often has a circular base. Not all cylinders have circular bases; those that are used in the automotive trade do. In this book the term *cylinder* will mean "hollow circular cylinder." A cylinder looks like the objects in Figure 4–18. Notice that a cylinder does not have to be oriented vertically. However, even with the horizontal cylinder at the right of Figure 4–18, the distance between the bases is called the height. The basic shape of a piston is a cylinder.

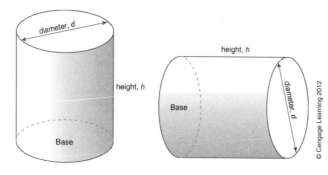

Figure 4-18

Volume

Volume is a measure of how much something holds. The volume of a cylinder is the product of the area of the base and the height of the cylinder. If the base is a circle, the volume is found by the formula

$$V = 0.7854\,d^2h$$

where V is the volume, d is the diameter of the base, and h is the height of the cylinder.

The units for a volume are cubic units such as cubic centimeters (cc or cm^3) or cubic inches (in^3). Some cubic units do not use the word *cubic*. Examples are *gallon*, *liter*, and *quart*.

■ **EXAMPLE 4-17**

The height of a cylinder is 42.8 mm and its base has a radius of 12.7 mm. What is its volume?

Solution

We are given $r = 12.7$, so $d = 2 \times 12.7 = 25.4$. Substituting this in the above formula, $V = 0.7854\,d^2h$, we get

$$\begin{aligned} V &= 0.7854\,d^2h \\ &= 0.7854 \times 25.4^2 \times 42.8 \\ &\approx 21687.13 \end{aligned}$$

The volume is about 21 687.13 mm^3.

■ **EXAMPLE 4-18**

If a cylinder has a bore diameter of 3.5 in. and the piston stroke is 5.2 in., what is the volume displaced as this piston moves through its stroke? The piston stroke is the height of the cylinder.

Solution

We are given $d = 3.5$. Substituting this in the above formula, we get

$$\begin{aligned} V &= 0.7854\,d^2h \\ &= 0.7854 \times 3.5^2 \times 5.2 \\ &\approx 50.03 \end{aligned}$$

The volume is about 50.03 in^3.

NAME _____ DATE _____ SCORE _____

1. If the diameter of a circle is 4.25 cm, what is its circumference?

2. Find the circumference of a circle with a radius of 12.5 cm.

3. A circle has a radius of $4\frac{5}{16}''$. What is its circumference?

4. What is the circumference of a circle with a diameter of 8.25 cm?

5. A circle has a diameter of 6.2 in. An arc on that circle has a 24° central angle.

   (a) What fractional part of the circle is this central angle?

   (b) Determine the length of this arc.

6. A circle has a diameter of 15.7 cm. An arc on that circle has a 72° central angle.

   (a) What fractional part of the circle is this central angle?

   (b) Determine the length of this arc.

7. If the diameter of a circle is 5.36 cm, what is its area?

8. Find the area of a circle with a radius of 17.5 cm.

9. A circle has a radius of $3\frac{11}{32}''$. What is its area?

10. What is the area of a circle with a diameter of 3.75 in?

11. The height of a cylinder is $5\frac{1}{6}''$ and its base has a radius of $3\frac{3}{4}''$. What is its volume?

12. The base of a cylinder is 8.2 cm in diameter and its height is 13.6 cm. What is its volume?

13. What is the circumference of a P175/75R14 tire? Express the answer in

   (a) inches

   (b) feet

14. What is the circumference of a P235/70R15 tire? Express the answer in

   (a) millimeters

   (b) centimeters

   (To change inches to millimeters, multiply by 25.4; to change millimeters to centimeters, multiply by 0.1.)

15. The original tires on a vehicle were P185/75R14 tires. The owners wants to put P235/70R14 tires on the car.

   (a) Which tire has the larger circumference?

   (b) How much larger is this tire, in inches?

1. _____

2. _____

3. _____

4. _____

5a. _____

5b. _____

6a. _____

6b. _____

7. _____

8. _____

9. _____

10. _____

11. _____

12. _____

13a. _____

13b. _____

14a. _____

14b. _____

15a. _____

15b. _____

16. Figure 4–19 shows a timing chain set. The part of the chain that is on each sprocket forms an arc.

    (a) What is the measure of the central angle for the arc of the chain that is on the larger sprocket?

    16a. _____

    (b) What is the arc length of the chain on the larger sprocket?

    16b. _____

    (c) What is the measure of the central angle for the arc of the chain that is on the smaller sprocket?

    16c. _____

    (d) What is the arc length of the chain on the smaller sprocket?

    16d. _____

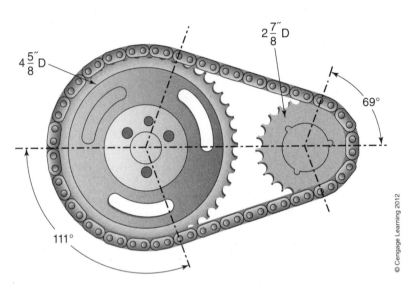

Figure 4–19

© Cengage Learning 2012

17. When a car makes a turn, the wheels on the inside of the curve do not travel as far as those on the outside. Suppose that a car makes a 90° turn and that the rear wheels on the inside of the turn follow a circular path with a radius of 18.5 ft. The outer wheel is 5 ft from the inner wheel.

    (a) What is the arc length of the path taken by the inner wheel?

    17a. _____

    (b) What is the radius of the curve taken by the outer wheel?

    17b. _____

    (c) What is the arc length of the path taken by the outer wheel?

    17c. _____

18. A cylinder has a bore of $3\frac{7}{16}''$. What is its cross-sectional area?

    18. _____

19. (a) A cylinder has a bore of 4.030 in. and the piston stroke is 3.85 in. What is the volume displaced by the piston as it goes through its stroke?

    19a. _____

    (b) If the engine has 8 cylinders, what is the total displacement of the engine?

    19b. _____

The Metric System

The metric system* is becoming more important to workers. Almost every major country uses the metric system. The domestic automobile industry has begun the process of metrication, that is, converting its manufacturing processes and products to metric specifications. This includes vehicle dimensioning and the inclusion of various metric fasteners.

Older technical service publications were printed using the U.S. customary system, while the latest publications are either metric or a combination of both measurement systems. This section gives a brief introduction to the metric system, the different units, and their symbols.

Units of Measure

There are seven base units in the metric system. All other units are formed from these seven. The base units that you will use the most are for length, the meter; for mass or weight, the kilogram; for time, the second; and for electric current, the ampere. The base unit for temperature is the Kelvin, but we will use a variation called the degree Celsius.

Table 4–1
METRIC PREFIXES

Multiples	Prefixes	Symbols
1 000 000	mega	M
1 000	kilo	k
1		
0.01	centi	c
0.001	milli	m
0.000 001	micro	μ

Each unit can be divided into smaller units or made into larger units. To show that a unit has been made smaller or larger, a prefix is placed in front of the base unit. The prefixes are based on powers of 10. The most common prefixes are shown in Table 4–1. For example, 1 kilometer (km) is 1000 m, 1 milligram (mg) is 0.001 g, and 1 megavolt (MV) is 1 000 000 V.

In Table 4–2, some of the metric units for the automotive trades are given.

Table 4–2
METRIC UNITS FOR AUTOMOTIVE TRADES†

Quantity	Unit	Symbol	Use
Length	micrometer millimeter centimeter meter kilometer	μm mm cm m km	paint thickness, surface texture shaft size, motor vehicle dimensions bearing size braking distance, turning circle land distance, maps, odometers
Area	square centimeter square meter	cm^2 m^2	piston head surface, brake and clutch contact area, glass fabrics, roof and floor areas
Temperature	degree Celsius	°C	thermostats, engine operating temperature, oil temperature
Volume/capacity	cubic centimeter cubic meter milliliter liter	cm^3 m^3 mL L	cylinder bore, small engine displacement work or storage space, truck body chemicals, lubricant oils, small liquids lubricant oils, fuel, large engine displacement, gasoline
Mass or weight	gram kilogram metric ton	g kg t	tire balance weights batteries, weights vehicle and load weight
Bending moment, torque, moment of force	newton meter	N·m	torque specifications, fasteners
Pressure/vacuum	kilopascal	kPa	tire pressure, air hose pressure, manifold pressure, cylinder compression
Velocity	kilometers per hour	km/h	vehicle speed
Force, thrust, drag	newton	N	pedal, spring, belt, drawbar
Power	watt	W	engine output, alternator output

*The official name for the metric system is the International System of Units (or, in French, *Le Systèm International d'Unités*). The official abbreviation for the metric system throughout the world is "SI."

†Additional metric definitions for auto trades are found in *Rules for Society of Automotive Engineers (SAE) Use of SI (Metric) Units—TSB 003*, Society of Automotive Engineers, Inc., 400 Commonwealth Drive, Warrendale, PA 15096.

Writing Metrics

You should remember some important rules when using the metric system.

1. The unit symbols that are used are the same in all languages. This means that the symbols used on Japanese cars are the same as those used on German cars.

2. The unit symbols are not abbreviations, and a period is not put at the end of the symbol.

3. Unit symbols are shown in lowercase letters except when the unit is named for a person. Examples of unit symbols that are written in capital letters are joule (J), pascal (Pa), watt (W), and ampere (A). The only exception is the use of L for liter. The symbol L is used for liter to eliminate any confusion between the lowercase unit symbol (l) and the numeral one (1).

4. The symbol is the same for both singular and plural (1 m and 12 m).

5. Numbers with five or more digits are written in groups of three separated by a space instead of a comma. This is done because some countries use the comma as a decimal point. For example, use:

    2 473 or 2473 instead of 2,473

    45 689 instead of 45,689

    47 398 254.263 72 instead of 47,398,254.26372

6. A zero is placed to the left of the decimal point if the number is less than one (0.52 L not .52 L).

7. *Liter* and *meter* are often spelled *litre* and *metre*.

8. The units of area and volume are written by using exponents. For example:

    5 square centimeters are written as 5 cm^2, not as 5 sq cm.

    37 cubic meters are written as 37 m^3, not as 37 cu m.

Changing Between the Metric and Customary System

Many times automotive technicians are called upon to use both measurement systems. Sometimes the measuring tools available may be different than the measuring system in which the specifications are given. In such a case the technician is required to convert from one system to another. We will show two different ways to change between the U.S. customary system and the metric system.

It is best not to change from either the metric or the customary system to the other. However, sometimes a worker has to change from one system to the other. Table 4–3, on page 89, should help you to do this.

■ EXAMPLE 4–19

Express 4.3 kilograms in pounds.

Solution

You are changing from the metric (kilograms) to the customary system.

Look at Table 4–3, under the heading "From Metric to Customary." You are changing a measurement from kilograms to pounds. Look in the column labeled "to change from" until you find the symbol for kilogram—kg. Is the symbol in the next right-hand column the symbol for pound (lb)?

It is, so look in the "multiply by" column opposite these two symbols. The number there is 2.205. So multiply the number of kilograms (4.3) by this number (2.205).

$$4.3 \times 2.205 = 9.4815$$

This means that 4.3 kg is equivalent to 9.481 5 lb.

■ EXAMPLE 4–20

Express 6.5 fluid ounces in liters.

Solution

You are changing from the customary (fluid ounces) to the metric (liters) system.

Look in Table 4–3 under the heading "From Customary to Metric." Your are changing a measurement from fluid ounces to liters. Look under the column labeled "to change from" until you find the symbol for fluid ounces—fl oz. The fl oz symbol is there twice. Now look in the next right-hand column. Do you see the symbol for liters (L)? It is opposite the second fl oz symbol.

If you look in the "multiply by" column opposite these two symbols (fl oz and L), you find 0.029 6. Multiply the number of fluid ounces (6.5) by this number (0.0296).

$$6.5 \times 0.0296 = 0.1924.$$

This means that 6.5 fluid ounces is equivalent to 0.1924 liters.

If you would rather not use Table 4–3 to change between metric units and customary units, the metric–customary equivalents in Table 4–4, on page 90, may be used.

■ EXAMPLE 4–21

Use Table 4–4 to convert 4.3 kg to pounds.

Solution

You are changing from the metric system to the customary system.

Table 4–3
CHANGING UNITS BETWEEN THE METRIC AND CUSTOMARY SYSTEMS

From Metric to Customary				From Customary to Metric			
Quantity	to change from	to	multiply by	Quantity	to change from	to	multiply by
Length	μm	mil	0.039 37	Length	mil	μm	25.4
	mm	in.	0.039 37		in.	mm	25.4
	cm	in.	0.393 7		in.	cm	2.54
	m	ft	3.280 8		ft	m	0.304 8
	km	mile	0.621 37		mile	km	1.609 3
Area	cm^2	sq in.	0.155	Area	sq in.	cm^2	6.451 6
	m^2	sq ft	10.7639		sq ft	m^2	0.092 9
Volume	cm^3	cu in.	0.061	Volume	cu in.	cm^3	16.387
	m^3	cu yd	1.308		cu yd	m^3	0.764 6
	mL	fl oz	0.033 8		fl oz	mL	29.574
	L	fl oz	33.814		fl oz	L	0.029 6
	L	pt	2.113		pt	L	0.473 2
	L	qt	1.056 7		qt	L	0.946 4
	L	gal	0.264 2		gal	L	3.785 4
Mass or weight	g	oz	0.035 3	Mass or weight	oz	g	28.349 5
	kg	lb	2.205		lb	kg	0.453 6
	t	lb	2205		ton	kg	907.2
Bending moment, torque, moment of force	N·m	lbf·in.	8.850 7	Bending moment, torque, moment of force	lbf·in.	N·m	0.113
	N·m	lbf·ft	0.737 6		lbf·ft	N·m	1.355 8
Pressure/vacuum	kPa	psi	0.145	Pressure/vacuum	psi	kPa	6.894 8
Velocity	km/h	mph	0.621 4	Velocity	mph	km/h	1.609 3
Force, thrust, drag	N	lbf	0.224 8	Force, thrust, drag	lbf	N	4.448 2
Power	W	W	1	Power	W	W	1
Temperature	°C	°F	(1.8 × °C) + 32	Temperature	°F	°C	(°F − 32) ÷ 1.8

Look in Table 4–4, on page 90, under the heading "From Metric to Customary." The next to the last entry in the left column has the entry "1 kg ≈ 2.2 lb." Since 1 kg ≈ 2.2 lb, then 4.3 kg would be ≈ 2.2 × 4.3 = 9.46 lb. This means that 4.3 kg is approximately equivalent to 9.46 lb. This is approximately the same result we got in Example 4–19.

■ EXAMPLE 4-22

Use Table 4–4 to convert 6.5 fluid ounces to liters.

Solution

You are changing from the customary system to the metric system.

Look in Table 4–4 under the heading "From Customary to Metric." The eighth entry in this column has the entry "1 fl oz ≈ 29.6 ml." Since 1 fl oz ≈ 29.6 ml, then 6.5 fl oz would be ≈ 6.5 × 29.6 = 192.4 ml. This means that 6.5 fl oz is approximately equivalent to 192.4 ml.

But we want the result in liters, not milliliters. There are 1000 ml in one liter, so 1 ml = 0.001 L. To convert 192.4 ml to liters, multiply 192.4 × 0.001 = 0.1924. This is the desired result. We see that 6.5 fl oz is approximately equivalent to 0.1924 L. This is the same result we got in Example 4–20.

A third method for converting between the two measurement systems uses Figure 4–20 on page 91. This table is used to convert inches to millimeters but may also be used to convert millimeters to inches.

■ EXAMPLE 4-23

Using Figure 4–20, convert 0.247 inches to millimeters.

Solution

You are converting from inches to millimeters. In the left column of Figure 4–20, locate the first two digits to the right of the decimal point for the given inches. In this

Table 4–4
METRIC–CUSTOMARY EQUIVALENTS

From Metric to Customary	From Customary to Metric
1 cm ≈ 0.39 in.	1 in. = 2.54 cm
1 m ≈ 3.28 ft	1 foot ≈ 0.305 m
1 km ≈ 0.62 mi	1 mile ≈ 1.61 km
1 cm^2 ≈ 0.16 sq in. (in.^{2})	1 in.^{2} ≈ 6.5 cm^2
1 m^2 ≈ 10.8 sq ft (ft^2)	1ft^2 ≈ 0.09 m^2
1 cm^3 ≈ 0.06 cu in. (in.^{3})	1 in.^{3} ≈ 16.4 cm^3
1 m^3 ≈ 35.3 cu ft (ft^3)	1ft^3 ≈ 0.03 cm^3
1 L ≈ 33.8 fl oz	1fl oz ≈ 29.6 ml
1 L ≈ 2.1 pt	1 pt ≈ 0.47 L
1 L ≈ 1.06 qt	1 qt ≈ 0.95 L
1 L ≈ 0.26 gal	1 gal ≈ 3.79 L
1 gram ≈ 0.035 oz	1 oz ≈ 2.83 g
1 kg ≈ 2.2 lb	1 lb ≈ 0.45 kg
1 kPa ≈ 0.145 psi	1 psi ≈ 6.895 kPa

© Cengage Learning 2012

case, you would locate 0.24 in the left column of Figure 4–20. This is listed as 0.240 as shown in Figure 4–21, on page 92. Now, across the top, locate the last digit. In this case, you locate the column headed 0.0007. Now, find where the 0.24 row and the 0.007 column intersect. This number, 6.2738, means that 0.247″ ≈ 6.2738 mm.

■ EXAMPLE 4–24

Using Figure 4–20, convert 22.4028 mm to inches.

Solution

You are converting from millimeters to inches. In the body of Figure 4–20, locate the given value of millimeters. In this case, locate 22.4028 in the body of the table, as shown in Figure 4–22, on page 92. This number is at the intersection of the row labeled 0.880 and the column labeled 0.002. Add these together:

$$\begin{array}{r} 0.880'' \\ + \ 0.002'' \\ \hline 0.882'' \end{array}$$

Step 1 The result shows that 22.4028 mm ≈ 0.882″.

■ EXAMPLE 4–25

Using Figure 4–20, convert 3.247″ to millimeters.

Solution

Again, you are converting from inches to millimeters. Think of 3.247″ as 3″ + 0.247″.

From Figure 4–20 we see that 1 inch is exactly 25.4 mm, so 3″ = 3 × 25.4 mm = 76.2 mm. In Example 4–23, we determined that 0.247″ ≈ 6.2738 mm. So,

$$\begin{aligned} 3.247'' = 3'' + 0.247'' &\approx 76.2 \text{ mm} + 6.2738 \text{ mm} \\ &= 82.4738 \text{ mm} \end{aligned}$$

This means that 3.247″ is equivalent to 82.4738 mm.

■ EXAMPLE 4–26

Using Figure 4–20, convert 118.9482 mm to inches.

Solution

You are converting from millimeters to inches. Before Figure 4–20 can be used, we must first determine how many inches are in 118.9482. To do this, divide 118.9482 by 25.4. You do not have to get the exact answer. All you want is the number of inches. So, estimate 118.9482 ÷ 25.4.

118.9482 rounds to 120.
25.4 rounds to 25.
120 ÷ 25 is between 4 and 5.

So, 118.9482 mm is between 4 and 5 inches.

4 inches = 4 × 25.4 mm = 101.6 mm
118.9482 mm − 101.6 mm = 17.3482 mm

In the body of Figure 4–20, locate 17.3482 mm. This number is at the intersection of the row labeled 0.680 and the column labeled 0.003. Add all these together

$$\begin{array}{r} 4'' \\ 0.680'' \\ + \ 0.003'' \\ \hline 4.683'' \end{array}$$

The result shows that 118.9482 mm ≈ 4.683″.

There may be times when you have to combine two or more of the conversion factors in Table 4–3.

■ EXAMPLE 4–27

A hybrid vehicle gets 21 km/L. What is this in miles per gallon (mpg)?

Solution

Two conversions will be needed, kilometers to miles and liters to gallons. From Table 4–3, we see that 1km = 0.621 37 mi and 1 L = 0.2642 gal. We will write these as fractions so that the units cancel.

(Continued on page 92)

WEIGHTS AND MEASURES

Decimals of an Inch to Millimeters
(Based on 1 inch = 25.4 millimeters, exactly)

Millimeters

Inches	0.000	0.001	0.002	0.003	0.004	0.005	0.006	0.007	0.008	0.009
0.500	12.7000	12.7254	12.7508	12.7762	12.8016	12.8270	12.8524	12.8778	12.9032	12.9286
0.510	12.9540	12.9794	13.0048	13.0302	13.0556	13.0810	13.1064	13.1318	13.1572	13.1826
0.520	13.2080	13.2334	13.2588	13.2842	13.3096	13.3350	13.3604	13.3858	13.4112	13.4366
0.530	13.4620	13.4874	13.5128	13.5382	13.5636	13.5890	13.6144	13.6398	13.6652	13.6906
0.540	13.7160	13.7414	13.7668	13.7922	13.8176	13.8430	13.8684	13.8938	13.9192	13.9446
0.550	13.9700	13.9954	14.0208	14.0462	14.0716	14.0970	14.1224	14.1478	14.1732	14.1986
0.560	14.2240	14.2494	14.2748	14.3002	14.3256	14.3510	14.3764	14.4018	14.4272	14.4526
0.570	14.4780	14.5034	14.5288	14.5542	14.5796	14.6050	14.6304	14.6558	14.6812	14.7066
0.580	14.7320	14.7574	14.7828	14.8082	14.8336	14.8590	14.8844	14.9098	14.9352	14.9606
0.590	14.9860	15.0114	15.0368	15.0622	15.0876	15.1130	15.1384	15.1638	15.1892	15.2146
0.600	15.2400	15.2654	15.2908	15.3162	15.3416	15.3670	15.3924	15.4178	15.4432	15.4686
0.610	15.4940	15.5194	15.5448	15.5702	15.5956	15.6210	15.6464	15.6718	15.6972	15.7226
0.620	15.7480	15.7734	15.7988	15.8242	15.8496	15.8750	15.9004	15.9258	15.9512	15.9766
0.630	16.0020	16.0274	16.0528	16.0782	16.1036	16.1290	16.1544	16.1798	16.2052	16.2306
0.640	16.2560	16.2814	16.3068	16.3322	16.3576	16.3830	16.4084	16.4338	16.4592	16.4846
0.650	16.5100	16.5354	16.5608	16.5862	16.6116	16.6370	16.6624	16.6878	16.7132	16.7386
0.660	16.7640	16.7894	16.8148	16.8402	16.8656	16.8910	16.9164	16.9418	16.9672	16.9926
0.670	17.0180	17.0434	17.0688	17.0942	17.1196	17.1450	17.1704	17.1958	17.2212	17.2466
0.680	17.2720	17.2974	17.3228	17.3482	17.3736	17.3990	17.4244	17.4498	17.4752	17.5006
0.690	17.5260	17.5514	17.5768	17.6022	17.6276	17.6530	17.6784	17.7038	17.7292	17.7546
0.700	17.7800	17.8054	17.8308	17.8562	17.8816	17.9070	17.9324	17.9578	17.9832	18.0086
0.710	18.0340	18.0594	18.0848	18.1102	18.1356	18.1610	18.1864	18.2118	18.2372	18.2626
0.720	18.2880	18.3134	18.3388	18.3642	18.3896	18.4150	18.4404	18.4658	18.4912	18.5166
0.730	18.5420	18.5674	18.5928	18.6182	18.6436	18.6690	18.6944	18.7198	18.7452	18.7706
0.740	18.7960	18.8214	18.8468	18.8722	18.8976	18.9230	18.9484	18.9738	18.9992	19.0246
0.750	19.0500	19.0754	19.1008	19.1262	19.1516	19.1770	19.2024	19.2278	19.2532	19.2786
0.760	19.3040	19.3294	19.3548	19.3802	19.4056	19.4310	19.4564	19.4818	19.5072	19.5326
0.770	19.5580	19.5834	19.6088	19.6342	19.6596	19.6850	19.7104	19.7358	19.7612	19.7866
0.780	19.8120	19.8374	19.8628	19.8882	19.9136	19.9390	19.9644	19.9898	20.0152	20.0406
0.790	20.0660	20.0914	20.1168	20.1422	20.1676	20.1930	20.2184	20.2438	20.2692	20.2946
0.800	20.3200	20.3454	20.3708	20.3962	20.4216	20.4470	20.4724	20.4978	20.5232	20.5486
0.810	20.5740	20.5994	20.6248	20.6502	20.6756	20.7010	20.7264	20.7518	20.7772	20.8026
0.820	20.8280	20.8534	20.8788	20.9042	20.9296	20.9550	20.9804	21.0058	21.0312	21.0566
0.830	21.0820	21.1074	21.1328	21.1582	21.1836	21.2090	21.2344	21.2598	21.2852	21.3106
0.840	21.3360	21.3614	21.3868	21.4122	21.4376	21.4630	21.4884	21.5138	21.5392	21.5646
0.850	21.5900	21.6154	21.6408	21.6662	21.6916	21.7170	21.7424	21.7678	21.7932	21.8186
0.860	21.8440	21.8694	21.8948	21.9202	21.9456	21.9710	21.9964	22.0218	22.0472	22.0726
0.870	22.0980	22.1234	22.1488	22.1742	22.1996	22.2250	22.2504	22.2758	22.3012	22.3266
0.880	22.3520	22.3774	22.4028	22.4282	22.4536	22.4790	22.5044	22.5298	22.5552	22.5806
0.890	22.6060	22.6314	22.6568	22.6822	22.7076	22.7330	22.7584	22.7838	22.8092	22.8346
0.900	22.8600	22.8854	22.9108	22.9362	22.9616	22.9870	23.0124	23.0378	23.0632	23.0886
0.910	23.1140	23.1394	23.1648	23.1902	23.2156	23.2410	23.2664	23.2918	23.3172	23.3426
0.920	23.3680	23.3934	23.4188	23.4442	23.4696	23.4950	23.5204	23.5458	23.5712	23.5966
0.930	23.6220	23.6474	23.6728	23.6982	23.7236	23.7490	23.7744	23.7998	23.8252	23.8506
0.940	23.8760	23.9014	23.9268	23.9522	23.9776	24.0030	24.0284	24.0538	24.0792	24.1046
0.950	24.1300	24.1554	24.1808	24.2062	24.2316	24.2570	24.2824	24.3078	24.3332	24.3586
0.960	24.3840	24.4094	24.4348	24.4602	24.4856	24.5110	24.5364	24.5618	24.5872	24.6126
0.970	24.6380	24.6634	24.6888	24.7142	24.7396	24.7650	24.7904	24.8158	24.8412	24.8666
0.980	24.8920	24.9174	24.9428	24.9682	24.9936	25.0190	25.0444	25.0698	25.0952	25.1206
0.990	25.1460	25.1714	25.1968	25.2222	25.2476	25.2730	25.2984	25.3238	25.3492	25.3746
1.000	25.4000	...	...	...	...	...	...	...	...	...

WEIGHTS AND MEASURES

Decimals of an Inch to Millimeters
(Based on 1 inch = 25.4 millimeters, exactly)

Millimeters

Inches	0.000	0.001	0.002	0.003	0.004	0.005	0.006	0.007	0.008	0.009
0.000	.....	0.0254	0.0508	0.0762	0.1016	0.1270	0.1524	0.1778	0.2032	0.2286
0.010	0.2540	0.2794	0.3048	0.3302	0.3556	0.3810	0.4064	0.4318	0.4572	0.4826
0.020	0.5080	0.5334	0.5588	0.5842	0.6096	0.6350	0.6604	0.6858	0.7112	0.7366
0.030	0.7620	0.7874	0.8128	0.8382	0.8636	0.8890	0.9144	0.9398	0.9652	0.9906
0.040	1.0160	1.0414	1.0668	1.0922	1.1176	1.1430	1.1684	1.1938	1.2192	1.2446
0.050	1.2700	1.2954	1.3208	1.3462	1.3716	1.3970	1.4224	1.4478	1.4732	1.4986
0.060	1.5240	1.5494	1.5748	1.6002	1.6256	1.6510	1.6764	1.7018	1.7272	1.7526
0.070	1.7780	1.8034	1.8288	1.8542	1.8796	1.9050	1.9304	1.9558	1.9812	2.0066
0.080	2.0320	2.0574	2.0828	2.1082	2.1336	2.1590	2.1844	2.2098	2.2352	2.2606
0.090	2.2860	2.3114	2.3368	2.3622	2.3876	2.4130	2.4384	2.4638	2.4892	2.5146
0.100	2.5400	2.5654	2.5908	2.6162	2.6416	2.6670	2.6924	2.7178	2.7432	2.7686
0.110	2.7940	2.8194	2.8448	2.8702	2.8956	2.9210	2.9464	2.9718	2.9972	3.0226
0.120	3.0480	3.0734	3.0988	3.1242	3.1496	3.1750	3.2004	3.2258	3.2512	3.2766
0.130	3.3020	3.3274	3.3528	3.3782	3.4036	3.4290	3.4544	3.4798	3.5052	3.5306
0.140	3.5560	3.5814	3.6068	3.6322	3.6576	3.6830	3.7084	3.7338	3.7592	3.7846
0.150	3.8100	3.8354	3.8608	3.8862	3.9116	3.9370	3.9624	3.9878	4.0132	4.0386
0.160	4.0640	4.0894	4.1148	4.1402	4.1656	4.1910	4.2164	4.2418	4.2672	4.2926
0.170	4.3180	4.3434	4.3688	4.3942	4.4196	4.4450	4.4704	4.4958	4.5212	4.5466
0.180	4.5720	4.5974	4.6228	4.6482	4.6736	4.6990	4.7244	4.7498	4.7752	4.8006
0.190	4.8260	4.8514	4.8768	4.9022	4.9276	4.9530	4.9784	5.0038	5.0292	5.0546
0.200	5.0800	5.1054	5.1308	5.1562	5.1816	5.2070	5.2324	5.2578	5.2832	5.3086
0.210	5.3340	5.3594	5.3848	5.4102	5.4356	5.4610	5.4864	5.5118	5.5372	5.5626
0.220	5.5880	5.6134	5.6388	5.6642	5.6896	5.7150	5.7404	5.7658	5.7912	5.8166
0.230	5.8420	5.8674	5.8928	5.9182	5.9436	5.9690	5.9944	6.0198	6.0452	6.0706
0.240	6.0960	6.1214	6.1468	6.1722	6.1976	6.2230	6.2484	6.2738	6.2992	6.3246
0.250	6.3500	6.3754	6.4008	6.4262	6.4516	6.4770	6.5024	6.5278	6.5532	6.5786
0.260	6.6040	6.6294	6.6548	6.6802	6.7056	6.7310	6.7564	6.7818	6.8072	6.8326
0.270	6.8580	6.8834	6.9088	6.9342	6.9596	6.9850	7.0104	7.0358	7.0612	7.0866
0.280	7.1120	7.1374	7.1628	7.1882	7.2136	7.2390	7.2644	7.2898	7.3152	7.3406
0.290	7.3660	7.3914	7.4168	7.4422	7.4676	7.4930	7.5184	7.5438	7.5692	7.5946
0.300	7.6200	7.6454	7.6708	7.6962	7.7216	7.7470	7.7724	7.7978	7.8232	7.8486
0.310	7.8740	7.8994	7.9248	7.9502	7.9756	8.0010	8.0264	8.0518	8.0772	8.1026
0.320	8.1280	8.1534	8.1788	8.2042	8.2296	8.2550	8.2804	8.3058	8.3312	8.3566
0.330	8.3820	8.4074	8.4328	8.4582	8.4836	8.5090	8.5344	8.5598	8.5852	8.6106
0.340	8.6360	8.6614	8.6868	8.7122	8.7376	8.7630	8.7884	8.8138	8.8392	8.8646
0.350	8.8900	8.9154	8.9408	8.9662	8.9916	9.0170	9.0424	9.0678	9.0932	9.1186
0.360	9.1440	9.1694	9.1948	9.2202	9.2456	9.2710	9.2964	9.3218	9.3472	9.3726
0.370	9.3980	9.4234	9.4488	9.4742	9.4996	9.5250	9.5504	9.5758	9.6012	9.6266
0.380	9.6520	9.6774	9.7028	9.7282	9.7536	9.7790	9.8044	9.8298	9.8552	9.8806
0.390	9.9060	9.9314	9.9568	9.9822	10.0076	10.0330	10.0584	10.0838	10.1092	10.1346
0.400	10.1600	10.1854	10.2108	10.2362	10.2616	10.2870	10.3124	10.3378	10.3632	10.3886
0.410	10.4140	10.4394	10.4648	10.4902	10.5156	10.5410	10.5664	10.5918	10.6172	10.6426
0.420	10.6680	10.6934	10.7188	10.7442	10.7696	10.7950	10.8204	10.8458	10.8712	10.8966
0.430	10.9220	10.9474	10.9728	10.9982	11.0236	11.0490	11.0744	11.0998	11.1252	11.1506
0.440	11.1760	11.2014	11.2268	11.2522	11.2776	11.3030	11.3284	11.3538	11.3792	11.4046
0.450	11.4300	11.4554	11.4808	11.5062	11.5316	11.5570	11.5824	11.6078	11.6332	11.6586
0.460	11.6840	11.7094	11.7348	11.7602	11.7856	11.8110	11.8364	11.8618	11.8872	11.9126
0.470	11.9380	11.9634	11.9888	12.0142	12.0396	12.0650	12.0904	12.1158	12.1412	12.1666
0.480	12.1920	12.2174	12.2428	12.2682	12.2936	12.3190	12.3444	12.3698	12.3952	12.4206
0.490	12.4460	12.4714	12.4968	12.5222	12.5476	12.5730	12.5984	12.6238	12.6492	12.6746

Figure 4-20

(From MACHINERY'S HANDBOOK, 23rd edition, by Industrial Press Inc. Used with permission.)

WEIGHTS AND MEASURES

Decimals of an Inch to Millimeters
(Based on 1 inch = 25.4 millimeters, exactly)

Inches	0.000	0.001	0.002	0.003	0.004	0.005	0.006	0.007	0.008	0.009
					Millimeters					
0.000	...	0.0254	0.0508	0.0762	0.1016	0.1270	0.1524	0.1778	0.2032	0.2286
0.010	0.2540	0.2794	0.3048	0.3302	0.3556	0.3810	0.4064	0.4318	0.4572	0.4826
0.020	0.5080	0.5334	0.5588	0.5842	0.6096	0.6350	0.6604	0.6858	0.7112	0.7366
0.030	0.7620	0.7874	0.8128	0.8382	0.8636	0.8890	0.9144	0.9398	0.9652	0.9906
0.040	1.0160	1.0414	1.0668	1.0922	1.1176	1.1430	1.1684	1.1938	1.2192	1.2446
0.050	1.2700	1.2954	1.3208	1.3462	1.3716	1.3970	1.4224	1.4478	1.4732	1.4986
0.060	1.5240	1.5494	1.5748	1.6002	1.6256	1.6510	1.6764	1.7018	1.7272	1.7526
0.070	1.7780	1.8034	1.8288	1.8542	1.8796	1.9050	1.9304	1.9558	1.9812	2.0066
0.080	2.0320	2.0574	2.0828	2.1082	2.1336	2.1590	2.1844	2.2098	2.2352	2.2606
0.090	2.2860	2.3114	2.3368	2.3622	2.3876	2.4130	2.4384	2.4638	2.4892	2.5146
0.100	2.5400	2.5654	2.5908	2.6162	2.6416	2.6670	2.6924	2.7178	2.7432	2.7686
0.110	2.7940	2.8194	2.8448	2.8702	2.8956	2.9210	2.9464	2.9718	2.9972	3.0226
0.120	3.0480	3.0734	3.0988	3.1242	3.1496	3.1750	3.2004	3.2258	3.2512	3.2766
0.130	3.3020	3.3274	3.3528	3.3782	3.4036	3.4290	3.4544	3.4798	3.5052	3.5306
0.140	3.5560	3.5814	3.6068	3.6322	3.6576	3.6830	3.7084	3.7338	3.7592	3.7846
0.150	3.8100	3.8354	3.8608	3.8862	3.9116	3.9370	3.9624	3.9878	4.0132	4.0386
0.160	4.0640	4.0894	4.1148	4.1402	4.1656	4.1910	4.2164	4.2418	4.2672	4.2926
0.170	4.3180	4.3434	4.3688	4.3942	4.4196	4.4450	4.4704	4.4958	4.5212	4.5466
0.180	4.5720	4.5974	4.6228	4.6482	4.6736	4.6990	4.7244	4.7498	4.7752	4.8006
0.190	4.8260	4.8514	4.8768	4.9022	4.9276	4.9530	4.9784	5.0038	5.0292	5.0546
0.200	5.0800	5.1054	5.1308	5.1562	5.1816	5.2070	5.2324	5.2578	5.2832	5.3086
0.210	5.3340	5.3594	5.3848	5.4102	5.4356	5.4610	5.4864	5.5118	5.5372	5.5626
0.220	5.5880	5.6134	5.6388	5.6642	5.6896	5.7150	5.7404	5.7658	5.7912	5.8166
0.230	5.8420	5.8674	5.8928	5.9182	5.9436	5.9690	5.9944	6.0198	6.0452	6.0706
0.240	6.0960	6.1214	6.1468	6.1722	6.1976	6.2230	6.2484	6.2738	6.2992	6.3246
0.250	6.3500	6.3754	6.4008	6.4262	6.4516	6.4770	6.5024	6.5278	6.5532	6.5786
0.260	6.6040	6.6294	6.6548	6.6802	6.7056	6.7310	6.7564	6.7818	6.8072	
	6.8580	6.8834					7.0104	7.0358		

WEIGHTS AND MEASURES

Decimals of an Inch to Millimeters
(Based on 1 inch = 25.4 millimeters, exactly)

Inches	0.000	0.001	0.002	0.003	0.004	0.005	0.006	0.007	0.008	0.009
					Millimeters					
0.500	12.7000	12.7254	12.7508	12.7762	12.8016	12.8270	12.8524	12.8778	12.9032	12.9286
0.510	12.9540	12.9794	13.0048	13.0302	13.0556	13.0810	13.1064	13.1318	13.1572	13.1826
0.520	13.2080	13.2334	13.2588	13.2842	13.3096	13.3350	13.3604	13.3858	13.4112	13.4366
0.530	13.4620	13.4874	13.5128	13.5382	13.5636	13.5890	13.6144	13.6398	13.6652	13.6906
0.540	13.7160	13.7414	13.7668	13.7922	13.8176	13.8430	13.8684	13.8938	13.9192	13.9446
0.550	13.9700	13.9954	14.0208	14.0462	14.0716	14.0970	14.1224	14.1478	14.1732	14.1986
0.560	14.2240	14.2494	14.2748	14.3002	14.3256	14.3510	14.3764	14.4018	14.4272	14.4526
0.570	14.4780	14.5034	14.5288	14.5542	14.5796	14.6050	14.6304	14.6558	14.6812	14.7066
0.580	14.7320	14.7574	14.7828	14.8082	14.8336	14.8590	14.8844	14.9098	14.9352	14.9606
0.590	14.9860	15.0114	15.0368	15.0622	15.0876	15.1130	15.1384	15.1638	15.1892	15.2146
0.600	15.2400	15.2654	15.2908	15.3162	15.3416	15.3670	15.3924	15.4178	15.4432	15.4686
0.610	15.4940	15.5194	15.5448	15.5702	15.5956	15.6210	15.6464	15.6718	15.6972	15.7226
0.620	15.7480	15.7734	15.7988	15.8242	15.8496	15.8750	15.9004	15.9258	15.9512	15.9766
0.630	16.0020	16.0274	16.0528	16.0782	16.1036	16.1290	16.1544	16.1798	16.2052	16.2306
0.640	16.2560	16.2814	16.3068	16.3322	16.3576	16.3830	16.4084	16.4338	16.4592	16.4846
0.650	16.5100	16.5354	16.5608	16.5862	16.6116	16.6370	16.6624	16.6878	16.7132	16.7386
0.660	16.7640	16.7894	16.8148	16.8402	16.8656	16.8910	16.9164	16.9418	16.9672	16.9926
0.670	17.0180	17.0434	17.0688	17.0942	17.1196	17.1450	17.1704	17.1958	17.2212	17.2466
0.680	17.2720	17.2974	17.3228	17.3482	17.3736	17.3990	17.4244	17.4498	17.4752	17.5006
0.690	17.5260	17.5514	17.5768	17.6022	17.6276	17.6530	17.6784	17.7038	17.7292	17.7546
0.700	17.7800	17.8054	17.8308	17.8562	17.8816	17.9070	17.9324	17.9578	17.9832	18.0086
0.710	18.0340	18.0594	18.0848	18.1102	18.1356	18.1610	18.1864	18.2118	18.2372	18.2626
0.720	18.2880	18.3134	18.3388	18.3642	18.3896	18.4150	18.4404	18.4658	18.4912	18.5166
0.730	18.5420	18.5674	18.5928	18.6182	18.6436	18.6690	18.6944	18.7198	18.7452	18.7706
0.740	18.7960	18.8214	18.8468	18.8722	18.8976	18.9230	18.9484	18.9738	18.9992	19.0246
0.750	19.0500	19.0754	19.1008	19.1262	19.1516	19.1770	19.2024	19.2278	19.2532	19.2786
0.760	19.3040	19.3294	19.3548	19.3802		19.4310	19.4564	19.4818	19.5072	
	19.5580	19.5834	19.6088				19.8374	19.86..		

Figure 4-21

(Revised from *MACHINERY'S HANDBOOK*, 23rd edition, by Industrial Press Inc. Used with permission.)

WEIGHTS AND MEASURES

Decimals of an Inch to Millimeters
(Based on 1 inch = 25.4 millimeters, exactly)

Inches	0.000	0.001	0.002	0.003	0.004	0.005	0.006	0.007	0.008	0.009
					Millimeters					
0.000	...	0.0254	0.0508	0.0762	0.1016	0.1270	0.1524	0.1778	0.2032	0.2286
0.010	0.2540	0.2794	0.3048	0.3302	0.3556	0.3810	0.4064	0.4318	0.4572	0.4826
0.020	0.5080	0.5334	0.5588	0.5842	0.6096	0.6350			0.7112	0.7366
0.030	0.7620	0.7874	0.8128	0.8382	0.8636					
	1.0160	1.0414	1.0668	1.0...						
0.330	8.3820				8.2804	8.3058				8.6100
0.340	8.6360	8.6614		8.7122	8.7376	8.7630	8.7884	8.8138	8.8392	8.8646
0.350	8.8900	8.9154	8.9408	8.9662	8.9916	9.0170	9.0424	9.0678	9.0932	9.1186
0.360	9.1440	9.1694	9.1948	9.2202	9.2456	9.2710	9.2964	9.3218	9.3472	9.3726
0.370	9.3980	9.4234	9.4488	9.4742	9.4996	9.5250	9.5504	9.5758	9.6012	9.6266
0.380	9.6520	9.6774	9.7028	9.7282	9.7536	9.7790	9.8044	9.8298	9.8552	9.8806
0.390	9.9060	9.9314	9.9568	9.9822	10.0076	10.0330	10.0584	10.0838	10.1092	10.1346
		10.1854	10.2108	10.2362	10.2616	10.2870	10.3124	10.3378	10.3632	10.3886
		10.4648	10.4902	10.5156	10.5410	10.566..				

WEIGHTS AND MEASURES

Decimals of an Inch to Millimeters
(Based on 1 inch = 25.4 millimeters, exactly)

Inches	0.000	0.001	0.002	0.003	0.004	0.005	0.006	0.007	0.008	0.009
					Millimeters					
0.500	12.7000	12.7254	12.7508	12.7762	12.8016	12.8270	12.8524	12.8778	12.9032	12.9286
0.510	12.9540	12.9794	13.0048	13.0302	13.0556	13.0810	13.1064	13.1318	13.1572	13.18..
0.520	13.2080	13.2334	13.2588	13.2842	13.3096	13.3350	13.3604	13.3858	13.4112	13.
0.530	13.4620	13.487..					13.6144	13.6398	13.6652	1...
0.540	13.7160	13...								
		20.9042								
		21.1328	21.1582	21.1836	21.209..					
	21.3614	21.3868	21.4122	21.4376	21.4630	21.4884	21.513..			21.5646
0.850	21.5900	21.6154	21.6408	21.6662	21.6916	21.7170	21.7424	21.7678	21.7932	21.8186
0.860	21.8440	21.8694	21.8948	21.9202	21.9456	21.9710	21.9964	22.0218	22.0472	22.0726
0.870	22.0980	22.1234	22.1488	22.1742	22.1996	22.2250	22.2504	22.2758	22.3012	22.3266
0.880	22.3520	22.3774	22.4028	22.4282	22.4536	22.4790	22.5044	22.5298	22.5552	22.5806
0.890	22.6060	22.6314	22.6568	22.6822	22.7076	22.7330	22.7584	22.7838	22.8092	22.8346
0.900	22.8600	22.8854	22.9108	22.9362	22.9616	22.9870	23.0124	23.0378	23.0632	23.0886
0.910	23.1140	23.1394	23.1648	23.1902	23.2156			23.2918	23.3172	23.3426
0.920	23.3680	23.3934	23.4188	23.4442					23.5712	23.5966
0.930	23.6220	23.6474	23.6728	23...						23.8506

Figure 4-22

(Revised from *MACHINERY'S HANDBOOK*, 23rd edition, by Industrial Press Inc. Used with permission.)

We will work this two ways. In the first method, we first convert the kilometers to miles.

$$21\frac{\text{km}}{\text{L}} = \frac{21\,\text{km}}{\text{L}} \times \frac{0.621\,37\,\text{mi}}{1\,\text{km}}$$
$$= \frac{13.048\,77\,\text{mi}}{\text{L}}$$

Next, we convert the liters to gallons.

$$= \frac{13.048\,77\,\text{mi}}{\text{L}} \times \frac{1\,\text{L}}{0.2642\,\text{gal}}$$
$$\approx 49.39\,\text{mi/gal}$$

This could also be worked using one step, but you must be careful to make sure the units cancel.

$$21\frac{\text{km}}{\text{L}} = \frac{21\,\text{km}}{\text{L}} \times \frac{0.621\,37\,\text{mi}}{1\,\text{km}} \times \frac{1\,\text{L}}{0.2642\,\text{gal}}$$
$$\approx 49.39\,\text{mi/gal}$$

Using either method, this hybrid gets about 49.4 mpg.

NAME _____ DATE _____ SCORE _____

Round all answers to the nearest thousandth.

1. 4 m = _____ ft
2. 16 L = _____ pt
3. 24 kg = _____ lb
4. 27.5 N·m = _____ lbf·ft
5. 88 km/h = _____ mph
6. 17 in. = _____ cm
7. 4.5 qt = _____ L
8. 21 sq ft = _____ m²
9. 5.8 lb = _____ kg
10. 32 psi = _____ kPa

1. _____
2. _____
3. _____
4. _____
5. _____
6. _____
7. _____
8. _____
9. _____
10. _____

Convert each of the following measurements from either inches to millimeters or from millimeters to inches.

11. Valve seat width, 0.066″
12. Valve stem diameter, 0.314″
13. Crankshaft end play, 0.076 mm–0.230 mm
14. Camshaft end play, 0.05 mm–0.13 mm
15. Valve stem diameter, 9.45 mm
16. Valve lift, 0.405″
17. Piston ring end gap, 0.010″–0.020″
18. Crankshaft journal diameter, 2.50″
19. Connecting rod journal diameter, 53.17 mm
20. Connecting rod side play 0.25 mm–0.48 mm
21. Valve spring free length, 1.99″
22. Camshaft journal diameters

    (a) #1, 2.029″–2.030″

    (b) #2, 2.019″–2.020″

    (c) #3, 2.009″–2.010″

    (d) #1, 1.999″–2.000″

11. _____
12. _____
13. _____
14. _____
15. _____
16. _____
17. _____
18. _____
19. _____
20. _____
21. _____

22a. _____

22b. _____

22c. _____

22d. _____

23. Cylinder bore, 87.5 mm

23. _____

24. Cylinder stroke, 92.0 mm

24. _____

25. Valve spring installed height, 14.53 mm

25. _____

26. Two race vehicles are set up for two different racing venues, circle track and road course. The individual wheel weights are given in both pounds and kilograms. Convert to the opposite measurement.

Circle track	Wheel weight		Road course	Wheel weight	
	lb	kg		lb	kg
Right front	775	_____	Right front	_____	250
Left front	950	_____	Left front	_____	230
Right rear	750	_____	Right rear	_____	195
Left rear	880	_____	Left rear	_____	180

27. A hybrid vehicle gets 27.2 km/L. What is this in mpg?

28. A car gets 26.5 mpg. What is this in km/L?

5

INTEGERS AND SIGNED NUMBERS

Objectives: After studying this chapter, you should be able to:

- Add and subtract positive and negative whole numbers, fractions, and decimals.
- Understand the meaning of the absolute value of a number and demonstrate the use of absolute value when adding signed numbers.
- Apply the rules for adding and subtracting signed numbers to practical problems.
- Determine engine coolant protection.
- Understand alignment adjustment values.

Integers are positive and negative whole numbers, including zero. Whole numbers with a plus sign (+) in front of them are called *positive integers*. Positive integers can also be written with no sign in front of them.

■ EXAMPLE 5–1

+4, +3, +17, 8, 42, 1843, and +971 are all positive integers.

Whole numbers with a negative (−) sign in front of them are called *negative integers*. The number 0 is neither positive nor negative.

■ EXAMPLE 5–2

−5, −17, −41, −182, −325 and −43,212 are all negative integers.

Signed numbers are any positive or negative numbers. Integers are signed numbers. Decimals and fractions can also be signed numbers.

■ EXAMPLE 5–3

+5.21, +42.37, 8.1, +972.0358, 24,352.012, 317, $\frac{2}{3}$, $4\frac{5}{7}$, and $\frac{22}{9}$ are all positive signed numbers.

−6.21, −24.95, −6.21, −3751.207, −439, −$\frac{1}{3}$, −$5\frac{1}{7}$, and −$\frac{17}{11}$ are all negative signed numbers.

Adding Signed Numbers

There are two types of problems that involve adding signed numbers. One type is when both numbers have the same sign. The other type is when the numbers have different signs.

Rule 1 To add signed numbers with the same sign, add the numbers and give the sum the sign of the original numbers.

■ EXAMPLE 5–4

Solve $+8 + +7$.

Solution

$$8 + 7 = 15$$

8 and 7 both have a plus sign, so the sum is a positive number.

Thus, $+8 + +7 = +15$.

■ EXAMPLE 5–5

Solve $-9 + -24$.

Solution

$$9 + 24 = 33$$

9 and 24 both have a negative sign, so the sum is negative.

Thus, $-9 + -24 = -33$.

■ EXAMPLE 5–6

Solve $-7.2 + -4.3 + -8 + -32.21$.

Solution

Because each of these numbers is negative, the sum is negative.

$$-7.2 + -4.3 + -8 + -32.21$$
$$= -11.5 + -8 + -32.21$$
$$\text{because } -7.2 + -4.3 = -11.5$$
$$= -19.5 + -32.21$$
$$\text{because } -11.5 + -8 = -19.5$$
$$= -51.71$$

If the sign of a number is ignored, the number that is left is called the *absolute value* of the number.

■ EXAMPLE 5–7

The absolute value of -8 is 8.
The absolute value of -23.7 is 23.7.
The absolute value of $+14.52$ is 14.52.
The absolute value of $-\frac{15}{7}$ is $\frac{15}{7}$.
The absolute value of 0 is 0.

Rule 2 To add signed numbers with different signs, take the absolute value of both numbers, subtract the absolute value of the smaller number from the absolute value of the larger, and give the answer the sign of the number that had the larger absolute value.

■ EXAMPLE 5–8

Solve $-8 + +5$.

Solution

The absolute value of -8 is 8. The absolute value of $+5$ is 5. Since 8 is larger than 5, the answer is negative. $8 - 5 = 3$.

Thus, $-8 + +5 = -3$.

■ EXAMPLE 5–9

Solve $+14 + -9$.

Solution

The absolute value of $+14$ is 14. The absolute value of -9 is 9. Since 14 is larger than 9, the answer is positive. $14 - 9 = 5$.

So, $+14 + -9 = 5$.

■ EXAMPLE 5–10

Solve $16.5 + -23.7$.

Solution

The absolute value of 16.5 is 16.5. The absolute value of -23.7 is 23.7. Since 23.7 is larger than 16.5, the answer is negative. $23.7 - 16.5 = 7.2$.

So, $16.5 + -23.7 = -7.2$.

■ EXAMPLE 5–11

Solve $-9.1 + -14 + 21.26$.

Solution

$$-9.1 + -14 + 21.26$$
$$= -23.1 + 21.26$$
$$\text{because } -9.1 + -14 = -23.1$$
$$= -1.84$$

This means that $-9.1 + -14 + 21.26 = -1.84$.

■ EXAMPLE 5-12

Solve $-14.9 + 23.125 + -28.4 + -3.36$.

Solution

$$
\begin{aligned}
-14.9 + 23.125 &+ -28.4 + -3.36 \\
&= 8.225 + -28.4 + -3.36 \\
&\quad \text{because } -14.9 + 23.125 = 8.225 \\
&= -20.175 + -3.36 \\
&\quad \text{because } 8.225 + -28.4 = -20.175 \\
&= -23.535
\end{aligned}
$$

So, $-14.9 + 23.125 + -28.4 + -3.36 = -23.535$.

■ EXAMPLE 5-13

Solve $^-\!3/4 + 2/3$.

Solution

A common denominator is 12. Change $^-\!3/4$ to $^-\!9/12$ and change $2/3$ to $8/12$. Add.

$$
\begin{aligned}
\frac{-3}{4} + \frac{2}{3} &= \frac{-9}{12} + \frac{8}{12} \\
&= \frac{-9 + 8}{12} \\
&= \frac{-1}{12}
\end{aligned}
$$

Thus, $^-\!3/4 + 2/3 = ^-\!1/12$.

■ EXAMPLE 5-14

Solve $-5\tfrac{1}{3} + (-4\tfrac{1}{2})$.

Solution

Change $-5\tfrac{1}{3}$ from a mixed number to a fraction. Remember that $-5\tfrac{1}{3} = -(5\tfrac{1}{3})$ and so $-5\tfrac{1}{3} = ^-\!16/3$.

Next, change $-4\tfrac{1}{2}$ to a mixed number. $-4\tfrac{1}{2} = ^-\!9/2$. We are now ready to add

$$
\begin{aligned}
-5\tfrac{1}{3} + (-4\tfrac{1}{2}) &= \frac{-16}{3} + \left(\frac{-9}{2}\right) \\
&= \frac{-32}{6} + \frac{-27}{6} \\
&= \frac{-32 + -27}{6} \\
&= \frac{-59}{6} = -9\frac{5}{6}
\end{aligned}
$$

So, $-5\tfrac{1}{3} + (-4\tfrac{1}{2}) = ^-\!59/6 = -9\tfrac{5}{6}$.

NAME _____ DATE _____ SCORE _____

1. +4 + +5

2. +9 + +13

3. +4 + −3

4. −8 + −15

5. +16 + +23

6. +31 + +25

7. −81 + −35

8. −42 + −79

9. +4.3 + +5.6

10. +24.7 + +16.41

11. −3.6 + −5.8

12. −35.4 + −16.8

13. 8 + −2

14. −3 + 5

15. −9 + 3

16. 2 + −9

17. −12 + 17

18. 16 + −28

19. −23 + 12

20. 14 + −31

21. −6.7 + 9.8

22. 15.4 + −23.5

23. −18.5 + 8.3

24. 24.2 + −42.1

25. −7 + 7

26. −8 + −3 + −9

27. −12 + −4 + −7

28. 15.7 + −15.7

29. −8 + −7 + 9

30. −6 + −19 + 3

31. −12 + 9 + −3

32. −23 + 35 + −42

33. −4.1 + −3.2 + −17.4 + 8.1

34. 3.7 + 10.5 + −78.3 + 2.4

35. 125 + 25 + 10 + −3

36. −756 + 67 + −13 + 4

37. 813.5 + 12.3 + 23.42 + −18.15

38. −291.11 + −41.59 + 74.1 + −8.3

39. 952 + 53 + −33 + 46.95

40. −2 + 1,573 + −16.73

41. 135.1 + 4.79 + −8.5

42. −2.341 + −714 + −51.2

43. 4.35 + −51.2 + −1.314

44. 1.2 + 78.27 + −940

Add the following. Change any improper fractions to mixed numbers.

45. $^{-7}/_8 + ^4/_3$ **46.** $^1/_8 + ^{-3}/_{64}$ **47.** $^{-5}/_4 + ^{-2}/_3$ **48.** $^{-3}/_4 + ^1/_2$

49. $4 + -3^1/_6$ **50.** $-5^1/_3 + 2$ **51.** $^3/_5 + -4$ **52.** $-3^2/_5 + -1^1/_2$

53. When you started your car in the morning, the temperature was $-13°$C. By mid-afternoon, it had risen $27°$C. What was the afternoon temperature?

53. _____

54. On the next morning, the temperature was $-32°$C. It rose $21°$C by mid-afternoon. What was the afternoon temperature?

54. _____

55. When you started your car in the morning, the temperature was $-13°$F. By mid-afternoon, it had risen $35°$F. What was the afternoon temperature?

55. _____

56. The next morning, the temperature was $-27°$F when you started your car. It rose $19°$F by mid-afternoon. What was the afternoon temperature?

56. _____

When performing a front-end alignment, a technician must work with the included angle, the steering axis inclination, and the camber. The included angle is formed by an imaginary line through the ball joints (steering axis inclination) on one side, and a true vertical line originating at the center point of the tire where it contacts the road surface. This angle is viewed from the front of the vehicle.

The included angle contains the angle of the steering axis inclination and may also include the camber angle. The camber angle is measured as a positive (+) angle if it inclines away from the center line of the vehicle and as a negative (−) angle if it inclines toward the center line. (See Figure 5–1.)

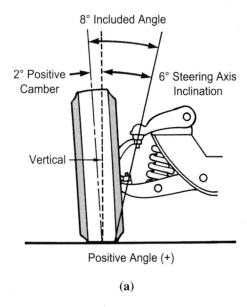

Positive Angle (+)

(a)

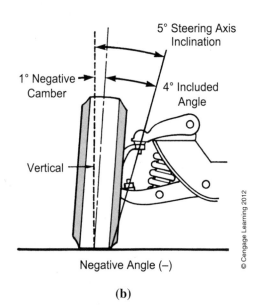

Negative Angle (−)

(b)

© Cengage Learning 2012

Figure 5–1

The camber angle is added to the steering axis inclination angle to determine the included angle. The included angle is the amount of angle between the steering axis and the tire center line. In Figure 5–1(a), the steering axis inclination is $6°$ and the camber angle is positive $2°$. The included angle is $6° + +2° = 8°$. In Figure 5–1(b), the steering axis inclination is $5°$ and the camber angle is negative $1°$ (or $-1°$). The included angle is $5° + -1° = 4°$.

57. What is the included angle if the steering axis angle is $+7°$ and the camber angle is $-2°$?

57. _____

58. What is the included angle if the steering axis angle is $+6°$ and the camber angle is $+1°$?

58. _____

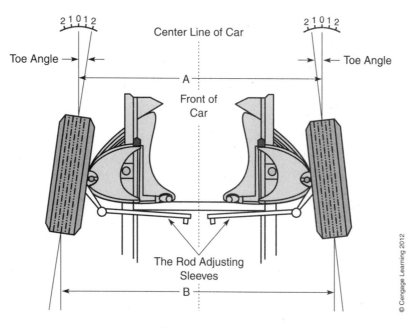

Figure 5–2

Toe is the difference between the distances separating the center lines of the tires at the front and rear of the tires. In Figure 5–2, the (total) toe is the difference in the length of *A* and the length of *B*. If *A* is less than *B*, then the tires are said to toe in. If *B* is less than *A*, then the tires are said to toe out. This means that a negative toe occurs when the front of the tires are closer together than the rear of the tires, and a positive toe (a + sign is normally not used for a positive toe) occurs when the front of the tires is farther apart than the rear of the tires.

With the steering wheel centered, you can also determine the toe on each tire. In the next two exercises, the toe on each tire is determined by computing the distance between the center line of a tire at the front and rear of the tire with the center line of the car.

59. A vehicle has a toe reading on the left front wheel of $-0.23''$ and a toe on the right front wheel of $0.19''$. What is the total toe on the front end?

59. _____

60. On another vehicle, the toe on the left rear wheel is $0.24''$, and the toe on the right rear wheel is $-0.08''$. What is the total toe on the rear end?

60. _____

Subtracting Signed Numbers

In order to answer a question like this one, you must be able to subtract signed numbers.

In performing a front-end alignment, the included angle is 3.2° and the camber angle is −1.6°. What is the steering axis angle?

To subtract two signed numbers, change the sign of the second number and then add.

■ EXAMPLE 5–15

Solve −6 − −9.

Solution

Change −9 to +9 and add.

$$-6 - -9 = -6 + +9 = +3$$

■ EXAMPLE 5–16

Solve −13 − +7.

Solution

Change +7 to −7 and add.

$$-13 - +7 = -13 + -7 = -20$$

■ EXAMPLE 5–17

Solve 14 − 29.

Solution

Change 29 to −29 and add.

$$14 - 29 = 14 + -29 = -15$$

■ EXAMPLE 5–18

Solve 23 − −8.5.

Solution

Change −8.5 to 8.5 and add.

$$23 - -8.5 = 23 + 8.5 = 31.5$$

■ EXAMPLE 5–19

Solve −18.1 − −6.37.

Solution

Change −6.37 to 6.37 and add.

$$-18.1 - -6.37 = -18.1 + 6.37 = -11.73$$

■ EXAMPLE 5–20

Solve 25.3 − 42.156.

Solution

Change 42.156 to −42.156 and add.

$$25.3 - 42.156 = 25.3 + -42.156 = -16.856$$

■ EXAMPLE 5–21

Solve ⁻⅘ − ¾.

Solution

Change ¾ to ⁻¾ and add.

$$\frac{-4}{5} - \frac{3}{4} = \frac{-4}{5} + \frac{-3}{4}$$

$$= \frac{-16}{20} + \frac{-15}{20}$$

$$= \frac{-16 + -15}{20}$$

$$= \frac{-31}{20} \text{ or } -1\frac{11}{20}$$

So, ⁻⅘ − ¾ = −1¹¹⁄₂₀.

■ EXAMPLE 5–22

Solve $-4\frac{2}{3} - (-5\frac{1}{4})$.

Solution

Change $-4\frac{2}{3}$ from a mixed number to a fraction. Remember that $-4\frac{2}{3} = -(4\frac{2}{3})$ and so $-4\frac{2}{3} = \frac{-14}{3}$.

Next, change $-5\frac{3}{4}$ to a mixed number. $-5\frac{3}{4} = \frac{-23}{4}$.

We are now ready to subtract:

$$-4\frac{2}{3} - (-5\frac{3}{4}) = \frac{-14}{3} - (\frac{-23}{4}).$$

Change $\frac{-23}{4}$ to $\frac{23}{4}$ and add.

$$-4\frac{2}{3} - \left(-5\frac{3}{4}\right) = \frac{-14}{3} - \left(\frac{-23}{4}\right)$$

$$= \frac{-14}{3} + \left(\frac{23}{4}\right)$$

$$= \frac{-56}{12} + \frac{69}{12}$$

$$= \frac{-56 + 69}{12}$$

$$= \frac{13}{12} = 1\frac{1}{12}$$

So, $-4\frac{2}{3} - (-5\frac{3}{4}) = \frac{13}{12} = 1\frac{1}{12}$.

■ EXAMPLE 5–23

At the beginning of this section "Subtracting Signed Numbers," the following question was asked. Answer the question.

In performing a front-end alignment, the included angle is 3.2° and the camber angle is −1.6°. What is the steering axis angle?

Solution

To find the steering axis angle, you subtract the camber angle from the included angle. For this problem, we need to solve $3.2° - (-1.6°)$.

Change −1.6° to 1.6° and add.

$$\begin{array}{r} 3.2 \\ +\ 1.6 \\ \hline 4.8 \end{array}$$

The steering axis angle is 4.8°.

■ EXAMPLE 6–10

Simplify $0.8 : \frac{2}{3}$.

Solution

$0.8 = \frac{8}{10} = \frac{4}{5}$.

$$0.8 : \tfrac{2}{3} = \frac{4}{5} : \frac{2}{3} = \frac{4}{5} \div \frac{2}{3} = \frac{4}{5} \times \frac{3}{2} = \frac{12}{10} = \frac{6}{5} \text{ or } 6:5$$

So, $0.8 : \frac{2}{3}$ simplifies to $6:5$.

Ratios that compare a number to 1

Auto mechanic manuals often write a ratio as a number compared to 1. Some examples of this are $4:1$ or $3.2:1$.

To write a ratio as a number compared to 1, divide both numbers by the second number.

■ EXAMPLE 6–11

Write $10:4$ as a ratio compared to 1.

Solution

$$10:4 = \frac{10}{4} = \frac{10 \div 4}{4 \div 4} = \frac{2.5}{1} \text{ or } 2.5:1$$

■ EXAMPLE 6–12

Write $37:9$ as a ratio compared to 1.

Solution

$$37:9 = \frac{37}{9} = \frac{37 \div 9}{9 \div 9} = \frac{4.11}{1} \text{ or } 4.11:1$$

Rates

A rate is a comparison of different kinds of units such as miles and gallons or revolutions and seconds. A ratio is used to compare the numbers of the units.

■ EXAMPLE 6–13

Express the rate of 180 miles in 3 hours as a ratio.

Solution

$$180 : 3 = 60 : 1$$

■ EXAMPLE 6–14

Express the rate of 650 revolutions per minute, or 650 rpm, as a ratio.

Solution

$$650 : 1$$

NAME _____ DATE _____ SCORE _____

Simplify the following ratios.

1. 18:3

2. 24:16

3. $^{25}/_{10}$

4. $^{27}/_{12}$

5. 2/3:5/7

6. $^{3}/_{4}/^{5}/_{9}$

7. 9/4:3/8

8. $^{7}/_{16}/^{14}/_{4}$

9. 2.8:1.4

10. $^{3.5}/_{1.5}$

11. 2.16:1.2

12. $^{3.5}/_{0.75}$

13. ¾:2.75

14. 0.9:1½

15. 0.8/2⅖

16. 1⅓/1.6

Write each of the following as a ratio compared to 1.

17. 9:2

18. $^{7}/_{5}$

19. $^{23}/_{7}$

20. 37:4

Express each of the following rates as a ratio.

21. $1.38 for 16 bolts

22. 725 revolutions per minute

23. 37 miles per gallon

24. 350 kilometers in 4 hours

25. The crankshaft turns 3000 revolutions per minute and the camshaft turns 1500 revolutions per minute. What is the crankshaft-to-camshaft ratio?

26. A drive gear has turned 6000 revolutions while the driven gear has turned 750. What is the gear ratio?

27. A differential ring gear has 46 teeth and the drive pinion has 13. What is the final drive-gear ratio?

28. If the steering wheel is turned 4.5 times and the front wheels move a total of 50 degrees, what is the steering ratio between the steering wheel and the front wheels? (There are 360 degrees in a circle.)

29. An 8-ounce container of fuel injection cleaner is added to an 18-gallon fuel tank. What is the ratio of cleaner to fuel? (There are 32 ounces per quart and 4 quarts in 1 gallon.)

21. _____
22. _____
23. _____
24. _____
25. _____
26. _____
27. _____
28. _____
29. _____

Proportions

The statement that two ratios are equal is called a *proportion*. Two ratios are equal if their cross products are equal.

■ EXAMPLE 6–15

Does $\frac{2}{5} = \frac{8}{20}$?

Solution

$$\frac{2}{5} \times \frac{8}{20}$$

Step 1   Find the cross products.

$$2 \times 20 = 40$$
$$5 \times 8 = 40$$

Step 2   Are the cross products equal? Yes, so the ratios are equal.

■ EXAMPLE 6–16

Does $3:7 = 5:12$?

Solution

Write the ratios as $\frac{3}{7}$ and $\frac{5}{12}$.

$$\frac{3}{7} \times \frac{5}{12}$$

Step 1

$$3 \times 12 = 36$$
$$7 \times 5 = 35$$

Step 2   Are the cross products equal? No, so the ratios are not equal.

Solving Proportions

If three parts of a proportion are known, we can find the fourth and missing term by using the cross products.

■ EXAMPLE 6–17

What is the missing number in the proportion $\frac{5}{7} = \frac{?}{42}$?

Solution

$$5 \times 42 = 210$$

Step 1   Find the cross product that has both numbers.

$$210 \div 7$$

Step 2   Divide this cross product by the third number.

$$210 \div 7 = 30$$

Step 3   The answer to Step 2 is the missing number. So, $\frac{5}{7} = \frac{30}{42}$.

■ EXAMPLE 6–18

Find the missing number in the proportion $\frac{5}{7} = \frac{8}{?}$.

Solution

$$7 \times 8 = 56$$

Step 1   Find the cross product that has both numbers.

$$56 \div 5$$

Step 2   Divide this cross product by the third number.

$$56 \div 5 = \frac{56}{5} = 11\frac{1}{5}$$

Step 3   The answer to Step 2 is the missing number. So, $\frac{5}{7} = 8/11\frac{1}{5}$.

■ EXAMPLE 6–19

Find the missing number in the proportion $2\frac{1}{2}/3\frac{1}{3} = ?/4\frac{1}{3}$.

Solution

$$2\frac{1}{2} \times 4\frac{1}{3} = \frac{5}{2} \times \frac{13}{3} = \frac{65}{6}$$

Step 1

$$\frac{65}{6} \div 3\frac{1}{3} = \frac{65}{6} \div \frac{10}{3}$$

Step 2

$$\frac{65}{6} \div \frac{10}{3} = \frac{65}{6} \times \frac{3}{10} = \frac{195}{60} = \frac{13}{4} = 3\frac{1}{4}$$

Step 3   So, $2\frac{1}{2}/3\frac{1}{3} = 3\frac{1}{4}/4\frac{1}{3}$.

■ EXAMPLE 6–20

Find the missing number in the proportion $\frac{3.5}{2.4} = \frac{0.7}{?}$.

Solution

$$2.4 \times 0.7 = 1.68$$

Step 1

$$1.68 \div 3.5$$

Step 2

$$1.68 \div 3.5 = 0.48$$

Step 3   So, $\frac{3.5}{2.4} = \frac{0.7}{0.48}$.

Direct Proportions

Two quantities are in direct proportion if a change in one ratio causes a similar change in the other ratio. When setting up a direct proportion, the units on the tops and bottoms of both ratios must be the same.

■ EXAMPLE 6–21

If a car travels 240 km in 4 hours, how far can it travel in 6 hours?

Solution

top units
both km

$$\frac{240 \text{ km}}{4 \text{ hours}} = \frac{? \text{ km}}{6 \text{ hours}}$$

bottom units
both hours

$$240 \times 6 = 1440$$

Step 1

$$1440 \div 4$$

Step 2

$$1440 \div 4 = 360$$

Step 3   So, $\frac{240 \text{ km}}{4 \text{ hours}} = \frac{360 \text{ km}}{6 \text{ hours}}$.

The car can travel 360 km in 6 hours.

■ EXAMPLE 6–22

The axle ratio is the direct proportion of the engine speed and the drive-axle speed. If the axle ratio is 6:1 and the engine speed is 2400 revolutions per minute (rpm), what is the drive-shaft speed?

Solution

engine speeds

$$\frac{6}{1} = \frac{2400 \text{ rpm}}{? \text{ rpm}}$$

drive-axle speeds

$$1 \times 2400 = 2400$$

Step 1

$$2400 \div 6$$

Step 2

$$2400 \div 6 = 400$$

Step 3   So, $\frac{6}{1} = \frac{2400 \text{ rpm}}{400 \text{ rpm}}$. The drive-axle speed is 400 rpm.

Inverse or Indirect Proportions

Two quantities are in inverse or indirect proportion if a change in one ratio causes the opposite change in the other ratio.

To work an inverse proportion, set it up the same way that you set up a direct proportion. Then switch the top numbers in the proportion.

■ EXAMPLE 6–23

When two gears are meshed, the speeds that they turn are inversely proportional to the gears' number of teeth. If the drive gear has 60 teeth and a speed of 300 rpm, find the speed of a driven gear that has 20 teeth.

Solution

$$\frac{60 \text{ teeth}}{300 \text{ rpm}} = \frac{20 \text{ teeth}}{? \text{ rpm}}$$

Set up the way that you set up a direct proportion.

$$\frac{20 \text{ teeth}}{300 \text{ rpm}} = \frac{60 \text{ teeth}}{? \text{ rpm}}$$

Switch the top numbers.

$$300 \times 60 = 18,000$$

Step 1

$$18,000 \div 20$$

Step 2

$$18,000 \div 20 = 900$$

Step 3   So, $\frac{20 \text{ teeth}}{300 \text{ rpm}} = \frac{60 \text{ teeth}}{900 \text{ rpm}}$.

This gear with 20 teeth turns at 900 rpm.

■ EXAMPLE 6–24

The speeds of pulleys are inversely proportional to the lengths of their diameters. The diameter of pulley A is 16 cm and the diameter of pulley B is 4 cm. If pulley B turns at 280 rpm, how fast does pulley A turn?

Solution

$$\frac{16 \text{ cm}}{? \text{ rpm}} = \frac{4 \text{ cm}}{280 \text{ rpm}}$$

Set up the way that you set up a direct proportion.

$$\frac{4 \text{ cm}}{? \text{ rpm}} = \frac{16 \text{ cm}}{280 \text{ rpm}}$$

Switch the top numbers.

$$4 \times 280 = 1120$$

Step 1

$$1120 \div 16$$

Step 2

$$1120 \div 16 = 70$$

Step 3   So, $\frac{4 \text{ cm}}{70 \text{ rpm}} = \frac{16 \text{ cm}}{280 \text{ rpm}}$.

Pulley A turns at 70 rpm.

NAME _____ DATE _____ SCORE _____

Use cross products to determine whether the pairs of ratios are equal.

1. $\frac{2}{3}$, $\frac{10}{15}$

2. $\frac{1}{3}$, $\frac{4}{12}$

3. $\frac{145}{25}$, $\frac{29}{5}$

4. $\frac{3}{8}$, $\frac{12}{33}$

Solve each of these problems.

5. $\frac{?}{2} = \frac{6}{4}$

6. $\frac{7}{8} = \frac{?}{32}$

7. $\frac{3}{?} = \frac{9}{24}$

8. $\frac{124}{62} = \frac{158}{?}$

9. $\frac{?}{3.5} = \frac{8}{20}$

10. $\frac{5}{4.5} = \frac{?}{27}$

11. $\frac{2.7}{?} = \frac{13}{169}$

12. $\frac{4.9}{2.8} = \frac{4.2}{?}$

13. $?/2\frac{1}{2} = 4\frac{1}{3}/2\frac{1}{3}$

14. $4\frac{1}{2}/7\frac{1}{8} = ?/8\frac{1}{2}$

15. $5\frac{5}{8}/? = \frac{15}{16}/6\frac{1}{4}$

16. $1\frac{3}{4}/4\frac{7}{8} = 2\frac{1}{5}/?$

17. $\frac{3}{4} = \frac{.16}{.15}$

18. $\frac{7.5}{10.5} = \frac{?}{6.3}$

19. $\frac{20}{?} = \frac{8}{5.6}$

20. $\frac{2.4}{10.8} = \frac{1.6}{?}$

21. $?/3\frac{1}{8} = 7/3\frac{3}{4}$

22. $2\frac{1}{16}/3\frac{1}{8} = ?/4\frac{1}{4}$

23. $6\frac{1}{2}/? = 5\frac{1}{64}/2\frac{1}{8}$

24. $2\frac{7}{64}/4\frac{1}{8} = 3\frac{3}{8}/?$

25. The rear-axle ratio is the ratio of the number of teeth on the ring gear to the number of teeth on the drive pinion gear. If the rear-axle ratio is 3.25 : 1 and the drive pinion gear has 12 teeth, how many teeth does the ring gear have?

25. _____

26. The compression ratio compares the volume of a cylinder with its piston at BDC to the volume when the piston is at TDC. If the compression ratio is 8.7 : 1 and the volume of BDC is 655 cm³, what is the volume at TDC?

26. _____

27. If a fuel pump delivers 0.665 liters of fuel every 10 strokes, how many liters does it deliver in 400 strokes?

27. _____

28. A gear with 15 teeth is meshed with a gear that has 48 teeth. The larger gear turns at 5 rpm. How fast does the smaller gear turn?

28. _____

29. Pulley A has a diameter of 24 cm and pulley B a diameter of 38 cm. If pulley B makes 450 rpm, what is the speed of pulley A?

29. _____

30. An idler gear causes the driven gear to turn in the same direction as the drive gear. Find the speed of the driven gear in Figure 6–1.

30. _____

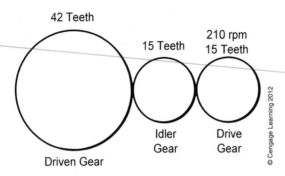

42 Teeth

15 Teeth

210 rpm
15 Teeth

Idler
Gear

Drive
Gear

Driven Gear

© Cengage Learning 2012

Figure 6–1

31. If an engine is operating at 1250 revolutions per minute at 20 miles per hour, how fast will the engine be operating at 55 miles per hour?

31. _____

32. If you can travel 238 miles on 8.7 gallons of fuel, how far can you travel on 12.5 gallons?

32. _____

33. The proper amount of antifreeze for a 50-liter cooling system is 25 liters. How much antifreeze is required for a cooling system with a 14-liter capacity?

33. _____

34. If 18.6 gallons of fuel are used to drive 475 miles, how many gallons are needed to drive 2346 miles?

34. _____

35. A driver added 90 milliliters (mL) of fuel conditioner to 20 liters of fuel. How much is needed to treat 27 L of fuel?

35. _____

36. The flywheel-gear–to–starter-pinion-gear ratio is $18:1$. If the flywheel is turning 350 revolutions per minute, how fast is the starter pinion gear turning?

36. _____

37. The gear ratio between two gears is $4.5:1$. If the large gear has 54 teeth, how many teeth does the small gear have?

37. _____

38. If the engine crankshaft is operating at 3000 rpm and the crankshaft to camshaft ratio is $2:1$, how many revolutions will the camshaft make in 1 minute?

38. _____

39. If ethylene glycol engine coolant and water are to be mixed at a $1:1$ ratio, how many liters of water are required to be mixed with 4 liters of coolant?

39. _____

40. Propylene glycol engine coolant must be mixed with water at a $6:4$ ratio. How many quarts of water are required to mix with 8 quarts of coolant?

40. _____

41. The turbo encapulator and annulus gear have a $6.5:1$ ratio. If the turbo encapulator is operating at 4885 rpm, what is the speed of the annulus gear?

41. _____

Percentages

The word *percent* means the same as per hundred. The symbol for percent is %. So, $35\% = {}^{35}/_{100}$ or 35 per hundred.

Changing Percentages to Decimals

To change a percentage to a decimal, first change it to its meaning of per hundred and then change this fraction to a decimal.

■ EXAMPLE 6–25

Change 15% to a decimal.

Solution

$$15\% = \frac{15}{100}$$

Step 1 Change the percent to per hundred.

$$\frac{15}{100} = 0.15$$

Step 2 Change the fraction to a decimal.

■ EXAMPLE 6–26

Change 14.3% to a decimal.

Solution

$$14.3\% = \frac{14.3}{100}$$

Step 1

$$\frac{14.3}{100} = 0.143$$

Step 2 So, 14.3% = 0.143.

■ EXAMPLE 6–27

Change 4½% to a decimal.

Solution

$$4\tfrac{1}{2}\% = 4.5\% = \frac{4.5}{100}$$

Step 1

$$\frac{4.5}{100} = 0.045$$

Step 2 So, 4½% = 0.045.

■ EXAMPLE 6–28

Change 100% to a decimal.

Solution

$$100\% = \frac{100}{100}$$

Step 1

$$\frac{100}{100} = 1$$

Step 2 So, 100% = 1.

Changing Decimals to Percentages

Changing a decimal to a percentage is just the opposite of changing a percentage to a decimal. Write the decimal as a fraction with a denominator of 100. Then the numerator is the percent.

■ EXAMPLE 6–29

Change 0.15 to a percent.

Solution

$$0.15 = \frac{15}{100}$$

Step 1 Change the decimal to a fraction with a denominator of 100.

$$\frac{15}{100} = 15\%$$

Step 2 Change per hundred to a percent.
So, 0.15 = 15%.

■ EXAMPLE 6–30

Change 1.78 to a percent.

Solution

$$1.78 = \frac{178}{100}$$

Step 1

$$\frac{178}{100} = 178\%$$

Step 2 So, 1.78 = 178%.

■ EXAMPLE 6–31

Change 0.037 to a percent.

Solution

$$0.037 = \frac{37}{1000} = \frac{3.7}{100}$$

Step 1

$$\frac{3.7}{100} = 3.7\%$$

Step 2 So, $0.037 = 3.7\%$.

Changing Percentages to Fractions

If a percentage needs to be written as a fraction, follow the first step in changing a percentage to a decimal and then reduce this fraction.

■ EXAMPLE 6–32

Write 25% as a fraction.

Solution

$$25\% = \frac{25}{100}$$

Step 1 Change the percent to per hundred.

$$\frac{25}{100} = \frac{1}{4}$$

Step 2 Reduce the fraction.

So, $25\% = \frac{1}{4}$.

■ EXAMPLE 6–33

Write 7½% as a fraction.

Solution

$$7\frac{1}{2}\% = 7.5\% = \frac{7.5}{100}$$

Step 1

$$\frac{7.5}{100} = \frac{75}{1000} = \frac{3}{40}$$

Step 2 So, $7\frac{1}{2}\% = \frac{3}{40}$.

Changing Fractions to Percentages

In order to change a fraction to a percentage, first change the fraction to an equivalent fraction with a denominator of 100. The numerator is the percent.

■ EXAMPLE 6–34

Change ⅖ to a percent.

Solution

First, solve $\frac{2}{5} = \frac{?}{100}$.

$$100 \div 5 = 20$$

Step 1 Divide the larger denominator number by the smaller one.

$$2 \times 20 = 40$$

Step 2 Multiply the answer in Step 1 by the numerator. The answer is the missing numerator.

So, $\frac{2}{5} = \frac{40}{100}$.

$$\frac{40}{100} = 40\%$$

Step 3 Change per hundred to percent.

So, $\frac{2}{5} = 40\%$.

■ EXAMPLE 6–35

Change 1³⁄₂₀ to a percentage.

Solution

First, solve $1\frac{3}{20} = \frac{?}{100}$.

$$1\frac{3}{20} = \frac{23}{20} = \frac{?}{100}$$

Change 1³⁄₂₀ to a fraction.

$$100 \div 20 = 5$$

Step 1

$$5 \times 23 = 115$$

Step 2 So, $1\frac{3}{20} = \frac{115}{100}$.

$$\frac{115}{100} = 115\%$$

Step 3 So, $1\frac{3}{20} = 115\%$.

Solving Percentage Problems

All percentage problems can be worked by setting up and solving a proportion. Remember that a proportion states that two ratios are equal. In all percentage problems, one of the ratios is the percent written so that the denominator is 100. The number next to the word "of" is the denominator of the other ratio.

■ EXAMPLE 6–36

What is 20% of 70?

Solution

$$20\% = \frac{20}{100}$$

Step 1   Write the percent with a denominator of 100.

$$\frac{?}{70}$$

Step 2   Write the other ratio so the number next to "of" is the denominator.

$$\frac{20}{100} = \frac{?}{70}$$

Step 3   Set up a proportion using the ratios from Steps 1 and 2. Now solve the proportion.

$$20 \times 70 = 1400$$

Step 4   Find the cross product that has both numbers.

$$1400 \div 100$$

Step 5   Divide this cross product by the third number.

$$1400 \div 100 = 14$$

Step 6   The answer to Step 5 is the answer to the problem.

So, 14 is 20% of 70.

■ EXAMPLE 6–37

What is 125% of 84?

Solution

$$125\% = \frac{125}{100}$$

Step 1

$$\frac{?}{84}$$

Step 2

$$\frac{125}{100} = \frac{?}{84}$$

Step 3

$$84 \times 125 = 10{,}500$$

Step 4

$$10{,}500 \div 100$$

Step 5

$$10{,}500 \div 100 = 105$$

Step 6   We see that 105 is 125% of 84.

■ EXAMPLE 6–38

What percentage of 40 is 12?

Solution

What percentage means $\frac{?}{100}$.

Step 1

$$\frac{12}{40}$$

Step 2

$$\frac{?}{100} = \frac{12}{40}$$

Step 3

$$100 \times 12 = 1200$$

Step 4

$$1200 \div 40$$

Step 5

$$1200 \div 40 = 30$$

Step 6   So, 30% of 40 is 12.

■ **EXAMPLE 6–39**

32 is 16% of what number?

Solution

$$\frac{16}{100}$$

Step 1

$$\frac{32}{?}$$

Step 2

$$\frac{16}{100} = \frac{32}{?}$$

Step 3

$$100 \times 32 = 3200$$

Step 4

$$3200 \div 16$$

Step 5

$$3200 \div 16 = 200$$

Step 6 Thus, 32 is 16% of 200.

PRACTICE PROBLEMS 6-3

NAME ______________________ DATE ________ SCORE ________

Change each of these percents to decimals.

1. 75% **2.** 14% **3.** 2.75% **4.** 4.17$\frac{1}{4}$%

5. 126% **6.** 4.3% **7.** 8$\frac{1}{5}$% **8.** 245%

Change each of these decimals to percents.

9. 0.24 **10.** 0.35 **11.** 0.97 **12.** 0.375

13. 2.59 **14.** 1.70 **15.** 0.045 **16.** 0.0875

Change each of these percents to fractions.

17. 50% **18.** 75% **19.** 80% **20.** 15%

21. 37.5% **22.** 2.5% **23.** 87$\frac{1}{2}$% **24.** 24%

Change each of these fractions to percents.

25. $\frac{3}{4}$ **26.** $\frac{4}{5}$ **27.** $\frac{7}{10}$ **28.** $\frac{13}{20}$

29. 1$\frac{1}{4}$ **30.** $\frac{5}{8}$ **31.** 2$\frac{2}{5}$ **32.** $\frac{3}{16}$

33. What is 25% of 36? **33.** ____________

34. What is 86% of 50? **34.** ____________

35. What is 16% of 80? **35.** ____________

36. What is 70% of 40? **36.** ____________

37. What is 75% of 200? **37.** ____________

38. What is 12% of 54? **38.** ____________

39. What percent of 60 is 12? **39.** ____________

40. What percent of 85 is 68? **40.** ____________

41. 43 is what percent of 50?

41. _____

42. 7 is what percent of 25?

42. _____

43. 9 is what percent of 24?

43. _____

44. What percent of 90 is 5.5?

44. _____

45. 18 is 6% of what number?

45. _____

46. 208 is 32% of what number?

46. _____

47. The directions call for a mixture of 50% antifreeze and 50% water to be used in a cooling system. How much antifreeze must be used in a system that has a capacity of 12 liters?

47. _____

48. A driver was getting 24 miles per gallon until she had the car tuned up. After the tune-up, the fuel mileage increased by 7%. What fuel mileage did she get after the tune-up?

48. _____

49. A driver traveled 325 miles of a 1725 mile trip on the first day. What percentage of the trip was completed?

49. _____

50. An automobile owner is saving money by changing her own oil. The oil required to complete the job costs $18.27 and an oil filter costs $4.67. If the state sales tax is 4.5%, what is the total cost of the items she purchased?

50. _____

51. The materials to complete a tune-up cost as follows: spark plugs, $14; air filter, $5.17; fuel filter, $16.85; and ignition coil, $36.35. The owner has a coupon that allows him a 12% discount on all merchandise. The state sales tax is 4%. How much is his total cost for the tune-up parts if the tax is taken after the discount?

51. _____

52. Many vehicles are used for trailering boats, snowmobiles, campers, or travel trailers. A manufacturer has stated that maximum trailer tongue weight can be no more than 15% of the trailer's weight. If the trailer with load is 4500 lb, what is the maximum allowable tongue weight?

52. _____

53. In a survey, 72 drivers who were involved in accidents were not wearing a seatbelt. These 72 drivers represent 40% of the total number of drivers in the survey.

(a) How many drivers were in the survey?

53a. _____

(b) How many of the drivers in the survey were wearing seatbelts?

53b. _____

54. Only 152,832 of one model of automobiles, or 10.3%, were manufactured without air conditioners.

(a) How many total automobiles of this model were manufactured?

54a. _____

(b) How many of these automobiles were manufactured with air conditioners?

54b. _____

55. An automatic transmission contains approximately 9 quarts of ATF. If only 45% of the ATF drains when you change the transmission oil filter, approximately how many quarts of ATF will you need for a filter change?

56. Engineering studies have shown that little intake flow improvement occurs after valve lift equals 25% of the valve diameter. Determine maximum theoretical valve lift of the following valve diameters based on this rule of thumb.

(a) 1.740 inch

56a. _____

(b) 1.940 inch

56b. _____

(c) 2.020 inch

56c. _____

(d) 2.125 inch

56d. _____

57. A driver is planning a 5250-mile trip. Each day the driver completes the percentage of the total trip as shown in the table below. Determine the miles traveled each day.

	Day	Percentage
(a)	1	7
(b)	2	11
(c)	3	3.5
(d)	4	18
(e)	5	14
(f)	6	12
(g)	7	9.3
(h)	8	2.1
(i)	9	8.1
(j)	10	9
(k)	11	6

57a. _____

57b. _____

57c. _____

57d. _____

57e. _____

57f. _____

57g. _____

57h. _____

55i. _____

55j. _____

55k. _____

58. An auto repair shop charges a customer $213.71 for a turn-signal switch. This price includes a mark-up of 87% from what the shop was charged by the distributor. How much did the repair shop pay for the switch?

58. _____

59. An auto repair shop charges a customer $415.65 for some parts and materials. This price includes a mark-up of 72%. How much did these parts and materials cost the repair shop?

59. _____

60. A customer is charged $352.66 for some parts and materials. This price includes a mark-up of 83%. How much did these parts and materials cost the repair shop?

60. _____

61. While backing out of a narrow space, John hit a hidden post, badly damaging the rear bumper on his car. A new bumper from the dealer cost $238.00. Online, John found a local salvage yard where the cost of a salvaged replacement bumper would be $59.99 if he would remove it himself. If John wanted warranty coverage for the salvaged part, it would cost an additional $5.49. The core charge for the transaction was $3.00 and the sales tax was 6.5%. (A person must pay a core charge if the old or damaged part is not turned in when a new part is purchased.)

(a) If John purchased the salvaged bumper with warranty coverage and brought his damaged bumper to the salvage yard, what would be his final cost for the replacement bumper?

61a. _____

(b) How much money would he save by using a salvaged bumper rather than buying a new one from the dealership?

61b. _____

62. The engine in Jamal's car needed to be overhauled. The dealership quoted him a price of $3658.26. Jamal located a low-mileage engine at a salvage yard for $972.00. The garage installing the new engine told him that the time required for the replacement was 11.8 hours, which was charged at a rate of $79.99 per hour. The garage recommended that the oil filter, the fuel filter, and the antifreeze also be changed during the engine replacement. This would add another $115.63 to the total cost. The state tax rate for parts and materials was 10.3%. The core charge for the engine was $149.99. Jamal decided to keep the old engine to overhaul in an automotive engines class and keep it as a spare.

(a) What was the cost for all parts and materials?

62a. _____

(b) What was the cost for labor?

62b. _____

(c) What was the cost for state tax?

62c. _____

(d) What was the final cost for this engine replacement?

62d. _____

63. A steel fuel tank weighs 37 lb. Engineers have produced a new blow-molded tank that weighs 23% less. How much does the new tank weigh?

63. _____

64. Felica has taken a cross-country trip of 3654 miles. Her car averaged 32.7 miles per gallon. If 10.5% of the fuel used was expended to overcome the frictional rolling resistance of the tires, how many gallons of fuel does this represent?

64. _____

65. One year the passenger car sales in the United States totaled 7,667,066. Of that number, 255,032 were hybrid vehicles. What percent of total sales were hybrid vehicles?

65. _____

66. One year there were 200,622 hybrid cars sold; the next year there were 217,614 sold. What was the percent increase from one year to the next?

66. _____

67. It is estimated that there are 127,721,000 passenger vehicles and 73,775,000 light trucks in use in the United States. Express this relationship in terms of percentage of each type compared to the total number of these vehicles.

67. _____

68. If a vehicle weighs 3000 lb and the driver weighs 150 lb, how much of the car's energy goes to moving the car itself? Express your answer as a percentage of the total energy expended.

68. _____

69. The average fuel tax in the United States is $0.45, compared to $1.03 in Canada, $2.07 in Japan, and $4.24 in Great Britain. Expressed as a percentage compared to the United States, how much more do the people in these other countries pay?

69. Canadians _____

Japanese _____

British _____

BUSINESS AND STATISTICS

Objectives: After studying this chapter, you should be able to:

- Calculate repair costs, discounts, interest, and payment amounts.
- Calculate business profit/loss and employee commissions.
- Calculate tax payments.
- Determine the three averages (mean, median, and mode) of a set of numbers.

Discounts

One use of percents is with discounts. A *discount* is a reduction in the list price and is used to encourage sales. So, every discount exercise deals with money. With most discount questions, the answer that is wanted is the new, or discounted, price. The discount is the amount that is subtracted from the original price to get the discounted price.

Most discount problems need two steps.

- Multiply the old price by the percent discount. This gives the amount of the discount.
- Subtract the discount from the old price. This gives the discounted price.

■ EXAMPLE 7–1

What is the discounted price off an item with a list price of $76.98 if the discount is 22%?

Solution

First, find the discount: 22% of $76.98.

$$22\% = \frac{22}{100}$$

Step 1

$$\frac{?}{\$76.98}$$

Step 2

$$\frac{22}{100} = \frac{?}{\$76.98}$$

Step 3

$$\$76.98 \times 22 = \$1693.56$$

Step 4

$$\$1693.56 \div 100 = \$16.94$$

Step 5   The discount is $16.94.

$$\begin{array}{r} \$76.98 \\ -16.94 \\ \hline \$60.04 \end{array}$$

Step 6   To find the discounted price, subtract the discount from the list price.
  The discounted price is $60.04.

There is a shorter method for finding the discount. Steps 1 through 3 are combined into one step and Steps 4 and 5 are in a second step. In this shortcut, convert the percent discount to a decimal and multiply by the list price.

■ EXAMPLE 7–2

What is the discounted price of an item with a list price of $76.98 if the discount is 22%?

Solution

First, find the discount: 22% of $76.98.

$$22\% = \frac{22}{100} = 0.22$$

Step 1   Write the percent as a decimal.

$$\$76.98 \times 0.22 = \$16.9356$$

Step 2   Multiply the decimal percent by the list price. The discount is $16.94.

$$
\begin{array}{r}
\$76.98 \\
-16.94 \\
\hline
\$60.04
\end{array}
$$

Step 3 To find the discounted price, subtract the discount from the list price.

The discounted price is $60.04. This is the same answer we got in Example 7–1.

At times more than one discount is given. When this happens, use the first discount on the list price to find the first discounted price. Next, use the second discount on the first discounted price to find the final discounted price.

■ EXAMPLE 7–3

A tire normally sells to a dealer for $56.90. The tire company is having a special and will give the dealer a 15% discount, with another 4% discount if 12 or more tires are purchased at the same time. A tire dealer decides to stock 16 of these tires. How much is the dealer charged?

Solution

The normal (list) price of these 16 tires is 16 × $56.90 = $910.40. The first discount is 15%, so

$$15\% = \frac{15}{100} = 0.15$$

Step 1   Write the percent as a decimal.

$$\$910.40 \times 0.15 = \$136.56$$

Step 2   Multiply the decimal percent by the list price. The discount is $136.56.

$$
\begin{array}{r}
\$910.40 \\
-136.56 \\
\hline
\$773.84
\end{array}
$$

Step 3 To find the first discounted price, subtract the discount from the list price.

The first discounted price is $773.84. The second discount is 4%, so

$$4\% = \frac{4}{100} = 0.04$$

Step 4   Write the percent as a decimal.

$$\$773.84 \times 0.04 = \$30.95$$

Step 5   Multiply the decimal percent by the first discounted price. The second discount is $30.95.

$$
\begin{array}{r}
\$773.84 \\
-30.95 \\
\hline
\$742.89
\end{array}
$$

Step 6   To find the second, final discounted price, subtract the second discount from the first discounted price.

The second, or final, discounted price is $742.89.

Suppliers may send bills with a notation such as 1.5% 10/Net 30. This means that there is a 1.5% discount if the bill is paid within 10 days. The total bill must be paid within 30 days.

NAME _____ DATE _____ SCORE _____

1. A technician purchases 12 spark plugs at $2.42 each. A discount of 30% is given. What does the technician pay for the spark plugs?

1. _____

2. Motor oil costs $3.68 a quart. If a case is bought, a discount of 20% is given. What is the discounted cost of a case of 24 quarts of oil?

2. _____

3. One shock absorber costs $42.18. If 4 are bought, a discount of 28% is given. What is the discounted cost of 4 shock absorbers?

3. _____

4. A parts department needs to resupply some brake rotors. One rotor costs $152.25. If 12 are bought, a discount of 22% is given. What is the discounted cost of 12 brake rotors?

4. _____

5. A garage needs to restock its supply of batteries. One battery costs $65.36.

   (a) If 24 batteries are bought, a discount of 17% is given. What is the discounted cost of 24 batteries?

5a. _____

   (b) If the batteries are paid for within 10 days, an additional discount of 5% is given. What is the price for the 24 batteries if the supplier is paid within 10 days?

5b. _____

6. An air filter O-ring kit is priced at $15.99. If you buy it online, you get a discount of 12%. What is the discounted price?

6. _____

7. A clutch kit for a BMW is priced at $354.00. If you buy it online, you get a discount of 12%. What is the discounted price?

7. _____

8. Lug nuts for a Toyota are priced at $7.52 for a package of 4. If 36 lug nuts are bought, a discount of 15% is given. What is the discounted cost of 36 lug nuts?

8. _____

9. An AC cabin filter for a Toyota is priced at $32.40. If you buy it online, you pay $25.34. What is the percent discount?

9. _____

10. An PCV valve for a Toyota is priced at $6.59. If you buy it online, you pay $3.23. What is the percent discount?

10. _____

11. A timing chain for a Toyota is locally priced at $74.94. If you buy it online, you pay $46.83. Shipping costs are $5.50.

   (a) How much is the discount for one timing chain?

11a. _____

   (b) What is the percent discount, without shipping?

11b. _____

   (c) What is the price after shipping has been added?

11c. _____

   (d) If shipping is included, what is the discount?

11d. _____

   (e) If shipping is included, what is the percent discount?

11e. _____

12. A headlight housing is priced at $139.95. If you buy it online, you get a 17.5% discount. Shipping costs are $8.95.

    (a) What is the discounted price for one headlight housing?

    12a. _____

    (b) What is the price after shipping has been added?

    12b. _____

    (c) If shipping is included, what is the discount?

    12c. _____

    (d) If shipping is included, what is the percent discount?

    12d. _____

13. A set of six iridium spark plugs is priced at $131.70. If you buy the set online, you pay $90.00—you save $41.70. The online store has the following four shipping options:

Ground	3-Day	2-Day	Next-Day
$6.73	$17.14	$24.52	$49.10

    (a) What is the percent discount before shipping?

    13a. _____

    (b) What is the percent discount with ground shipping?

    13b. _____

    (c) What is the percent discount with 3-day shipping?

    13c. _____

    (d) What is the percent discount with 2-day shipping?

    13d. _____

    (e) What is the percent discount with next-day shipping?

    13e. _____

14. A water pump for a BMW Z4 roadster is priced at $80.75. If you buy the pump online, you pay $62.42. The online store has the following three shipping options:

Ground	2-Day	Overnight
Free	$12.95	$18.98

    (a) What is the percent discount with ground shipping?

    14a. _____

    (b) What is the percent discount with 2-day shipping?

    14b. _____

    (c) What is the percent discount with next-day (overnight) shipping?

    14c. _____

15. A supplier sends a $2538.72 bill to a repair shop. The bill has the notation 2% 10/Net 30. What will the repair shop have to pay if the entire bill is paid within 10 days?

    15. _____

16. A shop gets a bill for $927.46 for parts it has bought. The bill has the notation 1.5% 10/Net 30. What will the shop have to pay if the entire bill is paid within 10 days?

    16. _____

Profit, Loss, and Commissions

Any business has to be concerned with profit and loss. A business, such as a repair shop, also has to deal with commissions.

Profit and Loss

Net income is the amount left after expenses have been subtracted from the revenue. If the net income is positive, it is called a *profit*; if the net income is negative, it is referred to as a *loss*. Profit is the amount that is left over after all expenses are paid. If the expenses are more than the income, the difference is a loss.

■ EXAMPLE 7–4

Last week, Junior's Auto Repair had a total revenue of $15,842. Expenses were $732.50 for advertising, $4642.73 for parts and supplies, $6415.93 in wages, $416.25 for insurance, and $687.59 in taxes. How much was the profit or loss?

Solution

To determine the net income or loss, we subtract the total expenses from the total revenue.

Total expenses:

$	732.50	Advertising
	4642.73	Parts and Supplies
	6415.93	Wages
	416.25	Insurance
	687.59	Taxes
$12,895.00		

The total expenses were $12,895.00. Because this is less than the revenue, the company made a profit.

Profit:

$15,842	Profit
12,895	Expenses
$ 2947	

The profit was $2947.

The *profit margin* tells how much profit a company makes for every $1 it generates in revenue. Profit margin is calculated as net income after taxes (profit) divided by revenues. Profit margin is displayed as a percentage. The formula cannot be used if the company lost money, because there is no profit.

$$\text{Profit Margin} = \frac{\text{Net Income after Taxes}}{\text{Revenue}} \times 100$$

$$= \frac{\text{Profit}}{\text{Revenue}} \times 100$$

■ EXAMPLE 7–5

Last week, Junior's Auto Repair had a profit of $2947 on total revenue of $15,842. What was the profit margin?

Solution

$$\text{Profit Margin} = \frac{\text{Net Income after Taxes}}{\text{Revenue}} \times 100$$

$$= \frac{2,947}{15,842} \times 100$$

$$\approx 18.6$$

Junior's Auto Repair had a profit margin of 18.6%.

A profit margin of 18.6% means that each dollar of sales that Junior's Auto Repair generates contributes about 18.6 cents to its net income (or bottom line).

Commission

A *commission* for someone in sales is a percentage of the selling price. A commission is paid to the salesperson. A commission for a technician is a percentage of the labor charges and is paid to the technician.

■ EXAMPLE 7–6

Mr. Zuidaberg was charged $487.50 in labor for the work done on his Corvette. If the technician was paid a 42% commission, how much did the technician make on this job?

Solution

We want 42% of $487.50. Since $42\% = \frac{42}{100} = 0.42$, the commission was $0.42 \times 487.50 = 204.75$.

The technician's commission was $204.75.

NAME _____ DATE _____ SCORE _____

1. In January, Junior's Auto Repair had a net income of $1575 and a total revenue of $18,250. What was the profit margin?

   1. _____

2. In February, Junior's Auto Repair had a net income of $2287 and a total revenue of $17,395. What was the profit margin?

   2. _____

3. In March, Junior's Auto Repair had a net income of $2137 and a total revenue of $21,732. What was the profit margin?

   3. _____

4. In April, Junior's Auto Repair had a net income of $1942 and a total revenue of $20,978. What was the profit margin?

   4. _____

5. The total receipts at Junior's Auto Repair for May were $7500. Of this, 42% must go for wages, 12% for rent, 6% for insurance, 5% for heating and air conditioning, 7% for taxes, and 18% for other overhead expenses.

   (a) Did the shop make a profit?

   5a. _____

   (b) What was the profit (or loss), expressed in dollars?

   5b. _____

   (c) If the company made a profit, what was the profit margin?

   5c. _____

6. The total receipts at Junior's Auto Repair for June were $8275. Of this, 42% must go for wages, 15% for parts and supplies, 12% for rent, 6% for insurance, 5% for air conditioning, 7% for taxes, and 18% for other overhead expenses.

   (a) Did the shop make a profit?

   6a. _____

   (b) What was the profit (or loss), expressed in dollars?

   6b. _____

   (c) If Junior's made a profit, what was the profit margin?

   6c. _____

7. The total receipts at Junior's Auto Repair for July were $6250. Of this, 42% must go for wages, $875 for parts and supplies, $900 for rent, $450 for insurance, $425 for air conditioning, 7% for taxes, and 18% for other overhead expenses.

   (a) Did the shop make a profit?

   7a. _____

   (b) What was the profit (or loss), expressed in dollars?

   7b. _____

   (c) If the business made a profit, what was the profit margin?

   7c. _____

8. The total receipts at Junior's Auto Repair for August were $9150. Of this, 42% must go for wages, 12% for parts and supplies, $900 for rent, $450 for insurance, $425 for air conditioning, 7% for taxes, and 18% for other overhead expenses.

   (a) Did the shop make a profit?

   8a. _____

   (b) What was the profit (or loss), expressed in dollars?

   8b. _____

   (c) If the repair shop made a profit, what was the profit margin?

   8c. _____

9. The September–December financial statement for Junior's Auto Repair is shown below. Complete the table.

Junior's Auto Repair
Finances for September–December

	September	October	November	December
Net Sales	$21,378	$23,964	$32,451	$17,642
Variable Expenses: Salaries/Wages (42%)				
Parts (15%)				
Legal/Accounting ($250 + 1.5%)				
Advertising (1%)				
Office Supplies (0.5%)				
Utilities (6%)				
Payroll Taxes (19% of salaries)				
Total Variable Expenses				
Fixed Expenses: Rent	1775	1775	1775	1775
Insurance	215	215	215	215
Licenses/Permits	0	0	0	250
Loan Payments	2135	2135	2135	2135
Total Fixed Expenses				
Total Expenses				
Net Profit (loss) before Taxes				
Taxes (24% of net profit)				
Net Profit (loss) after Taxes				
Profit Margin				

Interest and Payments

People borrow money to buy automobiles, homes, and businesses. Any time money is borrowed or loaned, interest has to be calculated—and this involves percentages. Taxes also involve percentages.

Interest

The amount that is borrowed is called the principal, P, or base. The amount charged for the use of the principal is called the interest rate, r. The interest rate is usually expressed as a percent for a time period, such as $1\frac{1}{2}\%$/month, 7.2%/year, or 8.3% annually. The amount of interest paid or interest due is I. The term is the length of the loan and the amount paid, A, is the principal plus the interest due, so $A = P + I$.

■ EXAMPLE 7–7

Junior's Auto Repair needs to replenish the shop supplies and must borrow $8500 for one year at 7.2%.

(a) How much interest with have to be repaid at the end of the year?

(b) How much will be needed to pay off the entire loan?

Solution

We are given the principal $P = 8500$ and the interest $r = 7.2\%$.

(a) To determine the interest after one year, find 7.2% of $8500.

$$7.2\% = \frac{7.2}{100} = 0.072$$

Step 1 Change 7.2% to a decimal. So, $r = 0.072$.

$$I = P \times r$$
$$= (8500)(0.072) = 612$$

Step 2 Multiply the principal by the interest as a decimal.

The shop will have to pay $612 interest at the end of the year.

(b) To pay off the entire loan, the shop will have to pay $P + I = 8500 + 612 = \$9112$.

Compounding is the process of calculating interest and adding it to existing principal for the next time interval, such as daily, monthly, or yearly (or annually). Compounding is most often used with saving money. This involves a slightly more complicated formula:

$$A = P\left(1 + \frac{r}{n}\right)^{ny}$$

where A is the accumulated amount, P is the principal, r is the annual interest rate, n is the number of payments in one year, and y is the number of years.

■ EXAMPLE 7–8

A technician puts $500 in a savings account for two years. The account pays 5.6% compounded monthly. How much money will be in the account at the end of the two years?

Solution

Here $P = 500$, $r = 5.6\% = 0.056$, $n = 12$, and $y = 2$.

$$A = P\left(1 + \frac{r}{n}\right)^{ny}$$
$$= 500\left(1 + \frac{0.056}{12}\right)^{12\times2}$$
$$= 500\left(1 + \frac{0.056}{12}\right)^{24}$$
$$\approx 559.11$$

At the end of 2 years the technician will have $559.11.

Payments

When you borrow money, you want to know how much you will have to pay each time. To make the formula clearer, we will let $R = \frac{r}{n}$, where r is the annual interest rate and n is the number of payments each year. To find the amount of each payment, P, you need to know the amount of money borrowed, B, and the length of the loan in years, y,

$$P = \frac{BR(1 + R)^{ny}}{(1 + R)^{ny} - 1}$$

■ EXAMPLE 7–9

Junior's Auto Repair needs to replenish the shop supplies and must borrow $16,250 for three years at 7.5%. How much is the monthly payment on this loan?

Solution

We are given the principal $B = 16{,}250$, $r = 7.5\% = \frac{7.5}{100} = 0.075$, $y = 3$, and $n = 12$. From this we determine that

$$R = \frac{r}{n} = \frac{0.075}{12} = 0.00625.$$

$$P = \frac{BR(1 + R)^{ny}}{(1 + R)^{ny-1}}$$
$$= \frac{(16{,}250)(0.00625)(1 + 0.00625)^{36}}{(1 + 0.00625)^{36} - 1}$$
$$\approx 505.48$$

Normally, the payment is rounded up to the nearest cent or dollar. This will make the last payment slightly less than the other payments. In this example, the monthly payment will be either $505.48 or $506.

■ EXAMPLE 7–10

How much of the first monthly payment on the loan in Example 7–9 was in interest and how much went to reduce the principal?

Solution

Each month,

$$R = \frac{r}{n} = \frac{7.5\%}{12} = \frac{0.075}{12} = 0.00625$$

of the remaining principal was in interest. The first month the interest was $16{,}250 \times 0.00625 = 101.56$. The rest of the payment was used to reduce the principal.

Payment	$ 505.48
Interest	−101.56
	403.92

The loan was reduced by $403.92.

Lending companies often provide a chart or computer program that shows how much interest was paid and the unpaid balance left after each payment. Part of one such chart is shown below.

A loan of $16,250 with payments made monthly at 7.5% annually. Each of the first 35 payments is $505.48. The last payment is $505.32.			
Payment	**Principal**	**Interest**	**Unpaid Balance**
1	$403.92	$101.56	$15,846.08
2	406.44	99.04	15,439.64
3	408.98	96.50	15,030.66
4	411.54	93.94	14,619.12
5	414.11	91.37	14,205.01
6	416.70	88.78	13,788.31
⋮	⋮	⋮	⋮
34	496.12	9.36	1,001.40
35	499.22	6.26	502.18
36	502.18	3.14	0.00

© Cengage Learning 2012

■ EXAMPLE 7–11

If the monthly payments in Example 7–9 were $505.48, how much interest had been paid at the end of 3 years for the loan?

Solution

The monthly payments were $505.48, and they were made for 36 months. So, the total amount paid was

$$\$505.48 \times 36 = \$18{,}197.28$$

The $18,197.28 that was paid is for the amount borrowed and the interest. Since the amount borrowed was $16,250, the interest was

$ 18,197.28	
−16,250.00	
1947.28	

Junior's Auto Repair paid $1947.28 in interest for this loan.

NAME _____ DATE _____ SCORE _____

1. How much interest will $900 earn in one year at 4.75%?

   1. _____

2. How much interest will $2100 earn in one year at 5.2%?

   2. _____

3. An auto repair shop needs to buy some new equipment, so it borrows $12,400 for 1 year at 7.75%.

   (a) How much interest will have to be repaid at the end of the year?

   3a. _____

   (b) How much will be needed to pay off the entire loan?

   3b. _____

4. An auto repair shop needs to buy some equipment for the office, so it borrows $8400 for 1 year at 7.4%.

   (a) How much interest with have to be repaid at the end of the year?

   4a. _____

   (b) How much will be needed to pay off the entire loan?

   4b. _____

5. A technician deposits $250 in a savings account for 5 years. During that time, the only money added to the account is the monthly interest. The account earns 1.2% compounded monthly. How much money will be in the account at the end of 5 years?

   5. _____

6. Another technician is able to put $300 in a savings account for 5 years. This account earns 1.25% compounded monthly. How much money will be in the account at the end of 5 years if the only money added to the account is the monthly interest?

   6. _____

7. An auto repair shop needs to remodel the shop, so it borrows $45,750 for 1 year at 10.125%. How much is the monthly payment for this loan?

   7. _____

8. An auto technician wants to buy a new house. The price of the house is $175,750. The technician plans to make a down payment of 10% and then borrow the rest of the money for the house for 30 years at 5.5% compounded monthly.

   (a) How much is the down payment?

   8a. _____

   (b) How much does the technician need to borrow?

   8b. _____

   (c) How much is the monthly payment?

   8c. _____

   (d) How much money will the technician have paid at the end of 30 years?

   8d. _____

9. The technician in the previous problem decides to make the payments every two weeks rather than monthly.

   (a) How many payments will be made each year?

   9a. _____

   (b) How much is each payment?

   9b. _____

   (c) How much money will the technician have paid at the end of 30 years?

   9c. _____

   (d) How much interest will the technician have saved at the end of 30 years compared to the loan in Exercise 8?

   9d. _____

10. The technician in Exercise 8 decides to get a 15-year loan rather than one for 30 years. Payments will be made monthly.

   (a) How many payments will be made each year?

   (b) How much is each monthly payment?

   (c) How much money will the technician have paid at the end of 15 years?

   (d) How much interest will the technician save compared to the loan in Exercise 8?

10a. _____

10b. _____

10c. _____

10d. _____

11. A garage owner borrows $8000 for some new equipment. The loan will be paid back over an 18-month period at 9.9% annual interest.

   (a) How much is the monthly payment, rounded up to the nearest dollar?

11a. _____

   (b) Complete the following payment schedule. The grey cell at the bottom should not be filled in.

A loan of $8000 with payments made monthly at 6% compounded annually. Each of the first 17 paymets are $_____ . The last payment is $_____ .			
Payment	**Principal**	**Interest**	**Unpaid Balance**
1			
2			
3			
4			
5			
6			
7			
8			
9			
10			
11			
12			
13			
14			
15			
16			
17			
18			
Total			

12. A mobile tool truck stops by Junior's Garage each week to see if any of the technicians need new tools. This salesperson not only sells tools but also provides financing for the purchases. During one visit a technician purchased a timing light for $213.76, a DVOM for $436.25, an A/C diagnostic gauge set for $271.15, and a set of micrometers for $406.87. State sales tax is 7.25%.

    (a) What is the cost of the tools before sales tax?

    12a. _____

    (b) What is the total cost of the technician's purchase?

    12b. _____

    (c) If the mobile tool shop allows 1 year to pay for the tools and charges 5.5% for the service, what will be the amount of the monthly payments?

    12c. _____

    (d) Halfway through the above payments the technician decides to purchase a new tool box for $2316.85 from the same mobile tool shop. What is the total cost of the tool box with the state tax added?

    12d. _____

    (e) If the cost of the tool box is added to the unpaid balance of the first purchase and financed for 2 years at 5.5% interest, what will be the amount of the new monthly payments for the 2-year period?

    12e. _____

13. The owner of Junior's Garage decides that the business needs a new building. The garage will have two auto bays and a small office. It will cost $450,000 to construct the building and another $75,000 for the new equipment. The owner can make a 10% down payment.

    (a) How much is the down payment?

    13a. _____

    (b) How much will need to be borrowed?

    13b. _____

    (c) If the loan is for 30 years at 7.1%, how much is the monthly payment?

    13c. _____

    (d) How much money will the owner have paid at the end of 30 years?

    13d. _____

14. The owner in the previous exercise thinks that a 15-year loan at 7.25% might be a better choice.

    (a) How much is the monthly payment?

    14a. _____

    (b) How much money will the owner have paid at the end of 15 years?

    14b. _____

    (c) Compare the answers in 13(d) and 14(b). Which is smaller and by how much?

    14c. _____

Taxes

Both employers and employees must pay taxes. The employee has three federal taxes: withholding (or income), social security, and Medicare. The employer also pays three federal taxes: social security, Medicare, and Federal Unemployment Tax (FUTA). Both the employee and the employer may also have to pay some state taxes.

Employee Taxes

The social security and Medicare taxes are often referred to as FICA (Federal Insurance Contributions Act). These taxes are not the same every year. Table 7–1 and Table 7–2 show some of the rates for 2010.

The withholding rates in Table 7–2 are given for a person who is paid biweekly. Different rates exist for people who are paid on a weekly, semimonthly, or monthly basis.

Table 7–1
FICA Tax Rates—2010

Tax Type	Tax Rate	Maximum Wage Base	Maximum Tax
Social Security	6.20%	$106,800	$6622
Medicare	1.45%	No limit	No limit

Table 7–2
BIWEEKLY Payroll Withholding—2010

(a) SINGLE person (including head of household)—
If the amount of wages (after subtracting withholding allowances) is: The amount of income tax to withhold is:
Not over $233 . . . . $0

Over—	But not over—		of excess over—
$233	—$401	. . .10%	—$233
$401	—$1,387	. . .$16.80 plus 15%	—$401
$1,387	—$2,604	. . .$164.70 plus 25%	—$1,387
$2,604	—$3,248	. . .$468.95 plus 27%	—$2,604
$3,248	—$3,373	. . .$642.83 plus 30%	—$3,248
$3,373	—$6,688	. . .$680.33 plus 28%	—$3,373
$6,688	—$14,450	. . .$1,608.53 plus 33%	—$6,688
$14,450	. . . .	.$4,169.99 plus 35%	—$14,450

(b) MARRIED person—
If the amount of wages (after subtracting withholding allowances) is: The amount of income tax to withhold is:
Not over $529 . . . . $0

Over—	But not over—		of excess over—
$529	—$942	. . .10%	—$529
$942	—$2,913	. . .$41.30 plus 15%	—$942
$2,913	—$3,617	. . .$336.95 plus 25%	—$2,913
$3,617	—$4,771	. . .$512.96 plus 27%	—$3,617
$4,771	—$5,579	. . .$824.53 plus 25%	—$4,771
$5,579	—$8,346	. . .$1,026.53 plus 28%	—$5,579
$8,346	—$14,669	. . .$1,801.29 plus 33%	—$8,346
$14,669	. . . .	.$3,887.88 plus 35%	—$14,669

■ EXAMPLE 7–12

In the last two weeks, Juan, who is married, earned $1638 in commissions. How much were his federal taxes?

Solution

Begin with the FICA taxes listed in Table 7–1. In each case, multiply the tax rate, expressed as a decimal, by his commissions.

Tax Type	Tax Rate	Tax Paid
Social Security	6.20%	$ 101.56
Medicare	1.45%	23.75

For his withholding tax, look in Table 7–2(b) for the line that has his salary. He made $1638, so look in the line that looks like this:

$942   −$2,913 . . $41.30 plus 15%   −$942

According to this, his will pay $41.30 plus 15% of whatever he earned that was over $942. We first determine the amount that he made over $942.

$1,638
−942
‾‾‾‾‾
696

Juan's withholding tax will be $55 + 15% of $696, or

$$41.30 + 0.15 \times 696 = 145.70$$

The total of his federal taxes were 101.56 + 23.75 + 145.70 = $271.01

Employer Taxes

An employer has to pay the same FICA taxes as an employee. In addition, employers have to also pay FUTA. A state unemployment tax may also have to be paid. Notice that the employer pays these unemployment taxes and the employee does not. In 2010, the FUTA tax was 6.2% of the first $7000 of the employee's wages for the year.

■ EXAMPLE 7–13

In the last two weeks, Juan, who is married, earned $1638 in commissions working at Junior's Auto Repair. How much federal taxes did Junior's have to pay on Juan's wages if Juan's total wages for the year before this pay period were
(a) $4500, (b) $12,250, and (c) $6250?

Solution

(a) The FICA taxes are on the first two lines of the following table. They are the same as the ones Juan had to pay. Juan's wages for this year after this pay period are $4500 + $1638 = $6138. This is less than $7000, so Junior's will have to pay the full amount, as shown on the third line of the table.

Tax Type	Tax Rate	Tax Paid
Social Security	6.20%	$101.56
Medicare	1.45%	23.75
FUTA	6.2%	101.56

Junior's Auto Repair had to pay federal taxes of 101.56 + 23.75 + 101.56 = $226.87 on Juan's wages.

(b) The FICA taxes are shown in the following table. They are the same as the ones Juan had to pay. Juan's wages for this year before this pay period were $12,250. This is more than $7000, so Junior's will not have to pay any FUTA.

Tax Type	Tax Rate	Tax Paid
Social Security	6.20%	$101.56
Medicare	1.45%	23.75
FUTA	0	0.00

Junior's Auto Repair had to pay federal taxes of 101.56 + 23.75 = $125.31 on Juan's wages.

(c) Junior's had to pay the FICA taxes shown on the first two lines of the following table. They are the same as the ones Juan had to pay. Juan's wages for this year before this pay period were $6250. Junior's will only have to pay FUTA on $7000 − $6250 = $750. For this pay period, Junior's will have to pay FUTA of 0.062 × 750 = 46.50.

Tax Type	Tax Rate	Tax Paid
Social Security	6.20%	$101.56
Medicare	1.45%	23.75
FUTA	6.2%	46.50

For this pay period, Junior's Auto Repair had to pay federal taxes of 101.56 + 23.75 + 46.50 = $171.81 on Juan's wages.

Property Taxes

People who own property usually have to pay a property tax. Most often this is a tax on the value of a person's or company's real estate. In some areas, property such as automobiles and boats are also subject to a property tax. It is possible to pay both a city and a county property tax.

The property tax is based on the assessed value of the property. This value is usually determined by a government official called an assessor. Tax rates are expressed as percents or in one of the following ways:

- Mills per dollar value or millage (1 mill is $\frac{1}{10}¢ = 0.1¢$ or $\$\frac{1}{1000} = \$0.001$.)
- Dollars per hundred dollars of assessed value
- Dollars per thousand dollars of assessed value
- Dollars on a percentage of assessed value

■ EXAMPLE 7–14

Junior's Auto Repair has an assessed value of $450,000. The tax rate in this community is $1.87 per $100 dollars of assessed value. How much property tax will the owner of Junior's Auto Repair have to pay?

Solution

Because the tax is based on $100 of assessed value, divide the assessed value by $100.

$$450,000 ÷ \$100 = 4500$$

Multiply this amount by the tax rate.

$$4500 × \$1.87 = \$8,415$$

Junior's Auto Repair will have to pay $8415 in property tax.

■ EXAMPLE 7–15

A repair shop has an assessed value of $520,000. The property tax rate in this community is a millage of 29.45 for 40% of the assessed value. How much property tax will the owner of this shop have to pay?

Solution

The tax is based on 40% of dollars of assessed value. This is called the taxable value of the property. This shop has a taxable value of

$$520,000 × 0.40 = \$208,000$$

Because the taxable value is in dollars, the millage must be changed to dollars. If a mill is 0.1¢ = $0.001, then 29.45 mills = 29.45 × $0.001 = $0.02945. Multiply the taxable value by 0.02945 to get the tax.

$$208,000 × 0.02945 = \$6125.60$$

This repair shop will have to pay $6125.60 in property tax.

NAME _____ DATE _____ SCORE _____

1. In the last two weeks, a technician, who is married, earned $2624 in commissions. How much was withheld from the technician's paycheck for

    (a) Social security?   1a. _____

    (b) Medicare?   1b. _____

    (c) Federal income tax?   1c. _____

2. In the last two weeks, a technician, who is single, earned $1964 in commissions. How much was withheld from the technician's paycheck for

    (a) Social security?   2a. _____

    (b) Medicare?   2b. _____

    (c) Federal income tax?   2c. _____

3. In the last two weeks, a technician, who is single, earned $1164 in commissions. How much was withheld from the technician's paycheck for

    (a) Social security?   3a. _____

    (b) Medicare?   3b. _____

    (c) Federal income tax?   3c. _____

4. In the last two weeks, a technician, who is married, earned $2188 in commissions. How much was withheld from the technician's paycheck for

    (a) Social security?   4a. _____

    (b) Medicare?   4b. _____

    (c) Federal income tax?   4c. _____

5. In the last two weeks, a technician, who is married, earned $1579 in commissions. Before this pay period, the employee had earned $3267 this year. How much did the technician's company have to pay in

    (a) Social security?   5a. _____

    (b) Medicare?   5b. _____

    (c) FUTA?   5c. _____

6. In the last two weeks, a technician, who is single, earned $2953 in commissions. Before this pay period, the employee had earned $4755 this year. How much did the technician's company have to pay in

    (a) Social security?   6a. _____

    (b) Medicare?   6b. _____

    (c) FUTA?   6c. _____

7. An automotive repair shop is valued at $475,000. If the tax rate in the community is 25.4 mills based on the assessed valuation, how much is the owner taxed?

7. _____

8. An automotive repair shop is valued at $597,500. If the tax rate in the community is 37.2 mills based on 50% of the assessed valuation, how much is the owner taxed?

8. _____

9. An automotive repair shop is valued at $375,500. The tax rate in this community is $1.87 per $100 of assessed value. How much is the owner taxed?

9. _____

10. An automotive repair shop is valued at $437,000. The tax rate in this community is $9.27 per $1000 of assessed value. How much is the owner taxed?

10. _____

Averages

People often talk about averages, such as the average yearly earnings of an auto technician or something like "the average job today was an oil change." Both of these are examples of people using the word *average*, which can have different meanings. In fact, there are three types of averages: the mean, median, and mode. Together they are called the *measures of central tendency*. In this section, we look at the different types of averages.

Mean

The *mean* or *arithmetic mean* is the average that is probably most used and is the one people usually intend when they say "average."

The arithmetic mean, or mean, is found by adding all the measurements and dividing this by the total number of measurements.

$$\text{Mean} = \frac{\text{Sum of Measurements}}{\text{Number of Measurements}}$$

■ EXAMPLE 7–16

At an automobile engine plant, a quality control technician removes crankshafts from the assembly line at regular intervals. The technician measures a critical dimension on each of these crankshafts. Even though the dimension is supposed to be 182.000 mm, some variation occurs during production. Here are the measurements for one morning's sample:

$$182.120 \quad 182.005 \quad 182.025$$
$$181.987 \quad 181.898 \quad 182.034$$

Find the mean of these measurements.

Solution

$$\text{Mean} = \frac{\text{Sum of Measurements}}{\text{Number of Measurements}}$$

$$= \frac{182.120 + 182.005 + 182.025 + 181.987 + 181.898 + 182.034}{6}$$

$$= \frac{1092.069}{6} = 182.0115 \approx 182.012$$

The mean dimension of these crankshafts is about 182.012 mm.

The mean measurement is written with the same number of decimal places as each of the measurements.

Sometimes the data are reported in terms of a frequency table. Here, to get the sum of the measurements, multiply each measurement by the number of times it occurs.

■ EXAMPLE 7–17

A repair shop pays its employees a hourly wage. The amount the employee is paid depends on experience and training. Junior's Auto Repair has 27 employees. The hourly wages and the number of technicians making that wage are shown in the following table:

Hr. wage ($)	12.44	13.78	15.22	16.54
No.	2	3	4	5
Hr. wage ($)	18.14	19.56	21.37	32.51
No.	7	2	3	1

Find the mean of these hourly wages.

Solution

$$\text{Mean} = \frac{\text{Sum of Measurements}}{\text{Number of Measurements}}$$

$$= \frac{\begin{array}{c}12.44 \cdot 2 + 13.78 \cdot 3 + 15.22 \cdot 4 + 16.54 \cdot 5 \\ + 18.14 \cdot 7 + 19.56 \cdot 2 + 21.37 \cdot 3 + 23.51 \cdot 1\end{array}}{2 + 3 + 4 + 5 + 7 + 2 + 3 + 1}$$

$$= \frac{463.52}{27} \approx 17.1674 \approx 17.17$$

The mean hourly wage of a technician at Junior's is $17.17.

Median

The second average is the median. The *median* is the middle number of a group that is arranged in order from smallest to largest (or from largest to smallest). One-half of the values are larger than the median and one-half are smaller. If there are an even number of items, the median is the mean of the two middle numbers.

■ EXAMPLE 7–18

Given the numbers 9, 8, 3, 2, and 4, determine the median.

Solution

First arrange the numbers in increasing order: 2, 3, 4, 8, 9. There are five numbers, so the middle number is the third number. The third number is 4, so the median is 4.

■ EXAMPLE 7–19

Find the median of 8, 11, 12, 15, 18, 20, 20, and 27.

Solution

These numbers are already in numerical order. There are eight numbers. The median is the mean of the fourth and fifth numbers. The fourth number is 15 and the fifth is 18. Their mean is

$$\frac{15 + 8}{2} = 16.5$$

Thus, the median is 16.5.

■ EXAMPLE 7–20

In Example 7–16, we determined the mean of a critical dimension that was measured in millimeters on six crankshafts. What is the median of these measurements?

Solution

The measurements were 182.120, 182.005, 182.025, 181.987, 181.898, and 182.034. These are not in numerical order, so listing them from lowest to highest, we get 181.898, 181.987, 182.005, 182.025, 182.034, and 182.120.

There are six numbers, so the median is the mean of the third and fourth numbers, 182.005 and 182.025. The mean of these two numbers is

$$\frac{182.005 + 182.025}{2} = 182.015.$$

The median of these six measurements is 182.015 mm.

Both the mean and the median are widely used when referring to an average. The median is a good choice when there are some extreme values. The mode is not used as often.

■ EXAMPLE 7–21

A small company has six technicians. The president of the company has a salary of $112,500. The salaries of the technicians are $37,230, $37,950, $39,125, $42,375, $45,300, and $48,715. Determine the mean and the median of these seven salaries.

Solution

Mean: To find the mean, we add the salaries and divide by seven.

$$\text{Mean} = \frac{\text{Sum of Measurements}}{\text{Number of Measurements}}$$

$$= \frac{\begin{matrix}112,500 + 37,230 + 37,950 \\ + 39,125 + 42,375 + 45,300 + 48,715\end{matrix}}{7}$$

$$= \frac{363,195}{7} = 51,885$$

Median: When the salaries are arranged in increasing order, they are $37,230, $37,950, $39,125, $42,375, $45,300, $48,715, and $112,500. The fourth number is $42,375, so this is the median.

In this example, we have found that the mean salary is $51,885 and the median is $42,375.

Notice in the last example that the mean is higher than six of the seven salaries. The one extreme value for the salary of the company president helps show why the median is often used as the average when there are extreme values.

Mode

The third and final measure of central tendency is the mode. The *mode* is the value that has the greatest frequency. A set of numbers can have more than one mode. If there are two modes, the data are said to be *bimodal*. Not every set of numbers has a mode.

■ EXAMPLE 7–22

Find the mode of 8, 11, 12, 15, 18, 20, 20, 27.

Solution

These are the same numbers we used in Example 7–19. The mode is 20, because that value occurs twice and all the other values occur once.

■ EXAMPLE 7–23

What is the mode for these data: 10, 12, 12, 17, 18, 19, 19, and 20?

Solution

There are two modes, 12 and 19, because each of these values occur twice and all the other values occur once. This is a bimodal set of numbers.

Find the mean, median, and mode for each set of measurements in Exercises 1 through 4

1. 4.2, 2.5, 6.4, 3.6, 7.4, 5.3, 6.9, 2.1, 8.3, 2.7

2. 50.1, 52.4, 52.6, 54.6, 54.8, 54.3, 52.4, 56.7, 58.3, 58.2

3.

80.0	77.0	82.0	73.0	92.0	89.0	100.0	96.0	96.0	94.0
74.0	94.0	94.0	96.0	83.0	84.0	96.0	87.0	84.0	96.0

4.

100	98	96	94	93	90	89	85
82	78	76	66	64	64	78	89
93	96	98	96	93	64	96	83

5. The manager of Junior's Auto Repair wants a better idea of the costs for engine tune-up parts. A sample of 40 customer invoices was taken and the costs of parts, rounded to the nearest dollar, are listed below. Find the mean, median, and mode for these costs.

89	87	102	72	63	112	137	107	68	92
153	94	107	115	98	141	89	107	88	105
173	167	159	97	112	91	111	141	95	112
157	138	133	78	97	115	183	159	97	106

6. The manager of Junior's Auto Repair wanted to determine the amount of time the technicians spent on a 15,000-mile maintenance for a Ford Fusion. Technicians were asked to record the actual amount of time it took to perform this maintenance. The results for the first 20 times are shown in the following table with times rounded to the nearest 0.1 hour (6 minutes). Find the mean, median, and mode for these times.

1.2	1.1	1.3	1.5	1.1	0.9	0.8	1.1	0.9	1.0
1.2	1.3	0.9	1.0	1.2	1.1	1.4	1.6	1.1	0.9

1. mean _____
median _____
mode _____

2. mean _____
median _____
mode _____

3. mean _____
median _____
mode _____

4. mean _____
median _____
mode _____

5. mean _____
median _____
mode _____

6. mean _____
median _____
mode _____

COMPLETING REPAIR ORDERS

Objectives: After studying this chapter, you should be able to:

- Apply addition and multiplication of decimals.
- Apply percent in a business setting.
- Complete a repair order.
- Determine repair labor/time costs.
- Determine business overhead.
- Determine business profit/loss amounts.

When a customer arrives at a repair facility, a repair order is usually filled out giving information about the customer, the car, and the services required. The repair order and the automobile are assigned to a service technician who follows the service instructions.

As parts are received, they are listed on the repair order by the parts department. The service manager or technician lists labor times, and the business office usually finishes the repair order by computing these charges and totaling the repair bill. In some small facilities the service manager, or even the technician, may complete the bill.

In our example, Eric Anderson was planning a long trip in his 2006 Ford Fusion. He had recently noted an advertised lubrication special for $29.95 that included changing the engine oil and oil filter and inspecting and correcting all fluid levels and tire pressures, so he brought his vehicle into the garage. The service writer at the repair facility also suggested several other service procedures and wrote the repair order in Figure 8–1. There were five service procedures to be performed on this Ford Fusion.

Look in Appendix A and locate the Ford Time and Parts Guide. You will notice that the Time and Parts Guide is applicable to several Ford models: the Fusion, the Milan, the Zephyr, and the MKX. In our first exercise we will use the data related to the Ford Fusion. We will use this material to determine the cost of the labor, parts, and materials that are used and the sales tax, and finally total the entire bill.

The objective of this exercise is to fill in and complete the repair order in Figure 8–1. As an example, in Figure 8–1, the first service procedure listed is "Lube Special." Because this procedure is not a standard procedure, but rather an advertised special for $29.95, we enter "SP" in the column titled "Operation Number" immediately to the left of the service instruction "Lube Special" written by the service writer. Because we already know the cost of this procedure, we enter $29.95 to the right in the column titled "Amount." Again, because this is an advertised special there is no time listed in the "Time" column (Figure 8–2) and there is no added cost for the oil used or for the oil filter; these are all included in the "Lubrication Special" price of $29.95.

To complete the rest of the repair order we use the Ford Time and Parts Guide in Appendix A. The next service procedure listed on the repair order is "R & R Air Filter."

Quan	Part No.	Name of Part	Sales Amount	
		Total Parts		

REPAIR ORDER

Name   Eric Anderson

Address   870 Meadowbrook Dr

Phone No.   555-0190   **Date**

Speedometer Reading   26,032   **VIN No.**   53271

Year, Make, and Model	License No. and State	Engine
06 Fusion	My Car	2.3 L

Operation Number	SERVICE INSTRUCTIONS	TIME	AMOUNT
	Lube Special		
	R & R Air Filter		
	Balance 4 Wheels		
	Rotate 4 Wheels		
	R & R Front Brakes		

	Accessories	Amount		
	@		Total Labor	
	@		Total Parts	
Qts. Oil	@		Materials	
Lbs. Grease	@		Accessories	
	Total Materials		Tires, Tubes	
	Total Accessories		Outside Work	
			Total	
			Tax	
			TOTAL AMOUNT	

I hereby authorize the above repair work to be done along with the necessary material, and hereby grant you and/or your employees permission to operate the car, truck or vehicle herein described on streets, highways or elsewhere for the purpose of testing and/or inspection. An express mechanic's lien is hereby acknowledged on above car, truck or vehicle to secure the amount repairs thereto.

NOT RESPONSIBLE FOR LOSS OR DAMAGE TO CARS OR ARTICLES LEFT IN CARS IN CASE OF FIRE, THEFT OR ANY OTHER CAUSE BEYOND OUR CONTROL.

WORK AUTHORIZED BY   *Eric Anderson*   DATE

© Cengage Learning 2012

Figure 8–1

"R & R" stands for remove and replace. If you look in the Alphabetical Index of the Ford Time and Parts Guide you will note that "Air Filter" is to be found on page 1-4. Turning to page 1-4 you will see that the page is divided into two section headings: "Labor 1 Maintenance & Lubrication 1 Labor" and "Parts 1 Maintenance & Lubrication 1 Parts." The number, in this case 1, represents a major category or group of procedures; 1 for "Maintenance & Lubrication," 2 for "Emissions," 3 for "Electrical," and so on. Each of these groups is published into two divisions, "Labor" and "Parts."

Having found page 1-4, Group 1, "Maintenance & Lubrication," begin with the first division of that group, "Labor," and look at the Operation Index located at the extreme upper left corner of the page. "Air Filter, R & R" is listed first and has an operation number 9. Scanning below in the "Labor" section to number 9 you will see "Air Filter, R & R." Below this heading, in the first column, the vehicle models and the applicable engine sizes are listed, if needed. The second column lists the model years that are covered and the third and fourth columns list the amount of time authorized for each particular service procedure. This time is often referred to as the flat rate time and is entered as tenths of an hour. Notice that two times are listed for most of the procedures: a Factory Time and a Motor Time. Disregard the Factory Time; we will use the Motor Time for all repair order exercises and problems.

In our example, "Air Filter, R & R," note that all models and model years use the same flat rate time. The Motor Time amount listed for this procedure is 0.3 hours. Enter this amount in the "Time" column.

This process is repeated for all applicable items listed under "Service Instructions" on the repair order.

All garages do not use the same hourly labor rate; they vary from about $40.00 to over $100.00. In each repair order exercise you are given the applicable hourly rate to use in your calculations. For this first repair order you will use an hourly labor rate of $85.00 per hour. The last column in the service section of the repair order to be completed is the one titled "Amount." The amount to be entered for the first service procedure is determined by multiplying the time amount by the labor amount. When we multiply a time of 0.3 hours for replacing the air filter by a labor rate of $85.00 we see that the cost of labor for "R & R Air Filter" is $25.50. This amount is entered to the extreme right of the "R & R Air Filter" line in the "Amount" column. This process is repeated for each of the service items listed in the repair order (Figure 8–2). When all service procedure items have been finalized, all the individual amounts are added together to determine the total labor amount, and this is entered to the right of

the "Total Labor" heading in the lower right section of the repair order.

If the service procedure requires parts replacement, such as an air filter during the "R & R Air Filter," you must look in the Group 1 Parts section, where the air filter is listed as:

5-Air Filter
Fusion, Milan
2.3L. 06–09   6E5Z9601EA   25.02

The large number you see is the Ford part number for this filter followed by the price.

On the upper left of the repair order is the parts accounting section. The "R & R Air Filter" procedure requires one air filter, so in the "Quan" (quantity) column we enter 1. For the part number enter the last four digits of the part number, followed by the name of the part in the "Name of Part" column. The price of the part is entered in the "Sales Amount" column (Figure 8–3). This part accounting process is repeated for all parts used and totaled in the "Total Parts" space.

In the event that other parts or supplies are used during the service procedures that are not listed in the Ford Time and Parts Guide, they are listed in the dedicated space at the lower left section of the repair order. During the wheel balance procedure, for example, seven wheel weights were used. Multiply the price of the individual wheel weights times the number of weights used and enter the total on the same line in the "Sales Amount" column (Figure 8–3). Add all costs in this section and enter the total amount in the "Total Materials" space.

When the individual totals have been determined, you must total the entire invoice. Record the total parts amount and the total materials amount in their corresponding space in the right lower portion of the repair order (Figure 8–4). Then total the entire invoice amount and record this amount in the space titled "Total." To complete the billing you must add the tax to the Total. Tax is only charged against the parts and materials used in the service operations, labor is tax exempt. The tax rate for this invoice will be 6%.

When all relevant information has been entered in the lower right accounting space on the repair order it is time to calculate the tax amount. Add the total parts cost, $125.99, to the total materials cost of $8.75, for a total parts and materials amount of $134.74. To determine the tax multiply this amount by the tax rate: $134.74 $\times$ 6% = $8.08. Enter the amount of the tax, $8.08, in the space titled "Tax." Now add the total to the tax to determine the total amount of this repair order: $394.19 $\times$ $8.08 = $402.27 (Figure 8–4).

Quan	Part No.	Name of Part	Sales Amount
		Total Parts	

REPAIR ORDER

Name _Eric Anderson_

Address _870 Meadowbrook Dr_

Phone No. _555-0190_   **Date** _____

Speedometer Reading _26,032_   **VIN No.** _53271_

Year, Make, and Model	License No. and State	Engine
06 Fusion	My Car	2.3 L

Operation Number	SERVICE INSTRUCTIONS	TIME	AMOUNT	
SP	Lube Special	—	29	95
1-9	R & R Air Filter	0.3	25	50
1-3	Balance 4 Wheels	0.9	76	50
1-4	Rotate 4 Wheels	0.4	34	00
18-1	R & R Front Brakes	1.1	93	50

Accessories	Amount		
		Total Labor	259	45
@		Total Parts	
@		Materials	
Qts. Oil @		Accessories	
Lbs. Grease @		Tires, Tubes	
Total Materials		Outside Work	
Total Accessories		Total	
		Tax	
		TOTAL AMOUNT	

I hereby authorize the above repair work to be done along with the necessary material, and hereby grant you and/or your employees permission to operate the car, truck or vehicle herein described on streets, highways or elsewhere for the purpose of testing and/or inspection. An express mechanic's lien is hereby acknowledged on above car, truck or vehicle to secure the amount repairs thereto.

NOT RESPONSIBLE FOR LOSS OR DAMAGE TO CARS OR ARTICLES LEFT IN CARS IN CASE OF FIRE, THEFT OR ANY OTHER CAUSE BEYOND OUR CONTROL.

WORK AUTHORIZED BY _Eric Anderson_   DATE

© Cengage Learning 2012

Figure 8–2

Quan	Part No.	Name of Part	Sales Amount	
1	01EA	Air Filter	25	02
1	001A	Front Pads	100	97
		Total Parts	125	99

REPAIR ORDER

Name Eric Anderson

Address 870 Meadowbrook Dr

Phone No. 555-0190   **Date** _____

Speedometer Reading 26,032   **VIN No.** 53271

Year, Make, and Model	License No. and State	Engine
06 Fusion	My Car	2.3 L

Operation Number	SERVICE INSTRUCTIONS	TIME	AMOUNT	
SP	Lube Special	—	29	95
1-9	R & R Air Filter	0.3	25	50
1-3	Balance 4 Wheels	0.9	76	50
1-4	Rotate 4 Wheels	0.4	34	00
18-1	R & R Front Brakes	1.1	93	50

Accessories	Amount				
		Total Labor	259	45
		Total Parts		
		Materials		
		Accessories		
		Tires, Tubes		
		Outside Work		

Quan		@	Sales Amount	
		@		
7	Wheel Wts	@ 1.25	8	75
	Qts. Oil	@		
	Lbs. Grease	@		
	Total Materials		8	75

Total Accessories	

Total	
Tax	
TOTAL AMOUNT	

I hereby authorize the above repair work to be done along with the necessary material, and hereby grant you and/or your employees permission to operate the car, truck or vehicle herein described on streets, highways or elsewhere for the purpose of testing and/or inspection. An express mechanic's lien is hereby acknowledged on above car, truck or vehicle to secure the amount repairs thereto.

NOT RESPONSIBLE FOR LOSS OR DAMAGE TO CARS OR ARTICLES LEFT IN CARS IN CASE OF FIRE, THEFT OR ANY OTHER CAUSE BEYOND OUR CONTROL.

WORK AUTHORIZED BY _____Eric Anderson_____ DATE

Figure 8–3

Quan	Part No.	Name of Part	Sales Amount	
1	01EA	Air Filter	25	02
1	001A	Front Pads	100	97
		Total Parts	125	99

REPAIR ORDER

Name Eric Anderson

Address 870 Meadowbrook Dr

Phone No. 555-0190   **Date** _____

Speedometer Reading 26,032   **VIN No.** 53271

Year, Make, and Model	License No. and State	Engine
06 Fusion	My Car	2.3 L

Operation Number	SERVICE INSTRUCTIONS	TIME	AMOUNT	
SP	Lube Special	—	29	95
1-9	R & R Air Filter	0.3	25	50
1-3	Balance 4 Wheels	0.9	76	50
1-4	Rotate 4 Wheels	0.4	34	00
18-1	R & R Front Brakes	1.1	93	50

					Accessories	Amount
		@				
7	Wheel Wts	@ 1.25	8	75		
	Qts. Oil	@				
	Lbs. Grease	@				
		Total Materials			Total Accessories	

Total Labor	259	45
Total Parts	125	99
Materials	8	75
Accessories	0	00
Tires, Tubes	0	00
Outside Work	0	00
Total	394	19
Tax	8	08
TOTAL AMOUNT	402	27

I hereby authorize the above repair work to be done along with the necessary material, and hereby grant you and/or your employees permission to operate the car, truck or vehicle herein described on streets, highways or elsewhere for the purpose of testing and/or inspection. An express mechanic's lien is hereby acknowledged on above car, truck or vehicle to secure the amount repairs thereto.

NOT RESPONSIBLE FOR LOSS OR DAMAGE TO CARS OR ARTICLES LEFT IN CARS IN CASE OF FIRE, THEFT OR ANY OTHER CAUSE BEYOND OUR CONTROL.

WORK AUTHORIZED BY _____*Eric Anderson*_____ DATE

Figure 8–4

NAME _____ DATE _____ SCORE _____

1. Express each of the following times in tenths of an hour.

   (a) 30 minutes

   (b) 12 minutes

   (c) 48 minutes

   (d) 6 minutes

   (e) 1 hour

   (f) 1 hour 36 minutes

   (g) 3 hours 24 minutes

   1a. _____

   1b. _____

   1c. _____

   1d. _____

   1e. _____

   1f. _____

   1g. _____

2. Express each of the following in minutes and hours.

   (a) 0.3 hour

   (b) 0.7 hour

   (c) 0.9 hour

   (d) 1.2 hours

   (e) 5.6 hours

   2a. _____

   2b. _____

   2c. _____

   2d. _____

   2e. _____

3. If you were working on a car and spent 54 minutes repairing a component, how would you write this in tenths of an hour?

   3. _____

For Exercises 4 through 15, assume that this automobile has come in for some fairly extensive work. The labor rate is $79.50/hour, the sales tax rate is 6.5%, and there is no sales tax on labor charges. Complete and total the repair order in Figure 8–5.

   2007 Milan
   2.3 L engine
   Automatic trans.

Quan	Part No.	Name of Part	Sales Amount	
1		PCV Valve		
1		Lock Switch		
1		Starter		
1		Fuel Rail		
1		Thermostat Hou.		
1		Heater Core		
1		Eng. Mount		
		Total Parts		

REPAIR ORDER

Name Juan Melarro

Address 1326 Manitou Ave.

Phone No. 555-1414   **Date** _____

Speedometer Reading 35,827   **VIN No.** 48236

Year, Make, and Model	License No. and State	Engine
07 Milan	Auto Trans	2.3 L

Operation Number	SERVICE INSTRUCTIONS	TIME	AMOUNT	
2-1	R & R PCV Valve			
3-75	R & R Front Door Lock Switch			
18-17	Adjust Parking Brake			
5-2	R & R Engine Starter			
6-10	R & R Fuel Rail			
8-8	R & R Cooling System Thermostat			
9-8	R & R Heater Core			
10-7	R & R Front Engine Mount			
19-3	4 Wheel Alignment			

				Accessories		Amount	Total Labor		
		@					Total Parts		
1	Qt. Antifreeze	@ 5.67	5 67				Materials		
		@					Accessories		
	Lbs. Grease	@					Tires, Tubes		
		Total Materials	5 67	Total Accessories			Outside Work		
							Total		
							Tax		
							TOTAL AMOUNT		

I hereby authorize the above repair work to be done along with the necessary material, and hereby grant you and/or your employees permission to operate the car, truck or vehicle herein described on streets, highways or elsewhere for the purpose of testing and/or inspection. An express mechanic's lien is hereby acknowledged on above car, truck or vehicle to secure the amount repairs thereto.

NOT RESPONSIBLE FOR LOSS OR DAMAGE TO CARS OR ARTICLES LEFT IN CARS IN CASE OF FIRE, THEFT OR ANY OTHER CAUSE BEYOND OUR CONTROL.

WORK AUTHORIZED BY _Juan melarro_ DATE

Figure 8–5

4. Look up all the procedures listed on the repair order and fill in the column headed "TIME." Multiply the time for each procedure by the hourly labor rate. Fill in the column headed "AMOUNT." Total these amounts and enter your answer on the line labeled "Total Labor."

4. _____

5. Look up the prices and part numbers of the required parts listed and enter them in the columns headed "Sales Amount" and "Part No." Total the sales amount and enter your answer on the line labeled "Total Parts."

5. _____

6. Determine the sales amount for the antifreeze. Enter your answer on the lines labeled "Materials."

6. _____

7. Assume 6.5% sales tax on all parts and materials used for these procedures. Determine the amount of tax and enter your answer on the line labeled "Tax."

7. _____

8. Total the entire bill that the customer is to pay.

8. _____

9. If the technician is paid a 45% commission on all labor charges, how much money did the technician make on this job?

9. _____

10. Considering the number of hours the technician spent on the car, and assuming that the technician's speed was equal to the suggested flat rate time from the *Parts and Time Guide*, how much money did the technician average per hour?

10. _____

11. The amount charged to the customer for parts and materials includes a mark-up of 87%.

   (a) What was the total the customer was charged for parts and materials?

11a. _____

   (b) How much did these parts and materials cost the repair shop?

11b. _____

12. What is the total of the repair shop's expenses for labor, parts, and materials?

12. _____

13. Garage revenue is the actual dollar amount retained by the garage from repairing Mr. Melarro's car. To determine garage revenue, take the total amount changed to Mr. Melarro and subtract the tax, labor paid to the technician, and the cost to the garage of the parts and materials used during the repair. What is the garage revenue from Mr. Melarro's repair?

13. _____

14. Considering Mr. Melarro's bill, if we assume that 78% of the garage revenue (the answer to Exercise 13) is used to pay overhead costs, what amount of this revenue is used to pay garage overhead?

14. _____

15. The garage's profit is the difference between the revenue prior to deducting overhead costs and the overhead. What was the profit of the shop from the service of Mr. Melarro's car?

15. _____

THE AUTOMOBILE ENGINE

Objectives: After studying this chapter, you should be able to:

- Calculate the volume of a cylinder.
- Make engine measurements and read related specifications.
- Calculate engine compression ratio.
- Determine engine valve timing.
- Estimate engine repair costs.
- Become familiar with engine torque and horsepower values.

Proper engine servicing and/or overhaul is not possible without the use of mathematics. When determining service requirements, it is necessary to measure and calculate how much wear has occurred before a decision can be made about the serviceability of a particular component.

The problems in this chapter help to illustrate this point by asking you to make specific service recommendations that are based on engine specifications and your calculations.

The proper service procedure requires ring ridge removal before you remove the pistons if the ring ridge is more than 0.004 inch (in.) [0.1 millimeter (mm)].

After you clean the carbon from the top of the cylinder, you take two measurements. Take the first one at the top in the unworn portion of the cylinder. This is shown in Figure 9–1 at the letter A. Take the other measurement slightly lower at the area of the greatest wear. This is shown in Figure 9–1 at B. The amount of ring ridge is determined by subtracting A from B.

■ EXAMPLE 9–1

If 1 cylinder of a V-6 engine has a displacement of 42 cubic inches (in.^{3}), what is the total cubic-inch displacement of the engine?

Solution

A V-6 engine has 6 cylinders. Each cylinder of this engine displaces 42 in.^{3}. This engine displaces a total of 6 × 42 in.^{3}.

$$6 \times 42 = 252$$

Therefore, this engine displaces a total of 252 in.^{3}.

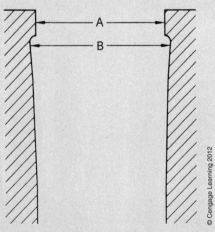

Figure 9–1

157

■ EXAMPLE 9–2

Take two measurements at right angles to each other at the top of a cylinder. If the measurement at A is 3.501 in. and the one at B is 3.508 in., what is the ring ridge on this cylinder? Must it be removed before the pistons are removed?

Solution

Measurement A is 3.501 in. Measurement B is 3.508 in. Subtract A from B.

$$\begin{array}{r} 3.508 \\ -\ 3.501 \\ \hline 0.007 \end{array}$$

The ring ridge is 0.007 in.

0.007 in. is more than 0.004 in.

Therefore, this ring ridge must be removed before extracting the piston.

■ EXAMPLE 9–3

Take two cylinder measurements at right angles to each other to determine the cylinder out-of-round. (See Figure 9–2.) On a Buick, one measurement is 88.990 mm and the other is 89.040 mm. How much is the cylinder out of round?

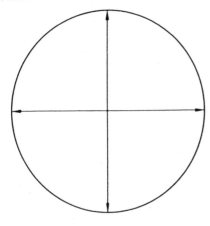

Figure 9–2

Solution

The larger measurement is 89.040 mm. The other measurement is 88.990 mm.

Subtract:

$$\begin{array}{r} 89.040 \\ -\ 88.990 \\ \hline 0.050 \end{array}$$

Therefore, the cylinder is out of round by 0.050 mm.

Now look in Appendix A for the specifications for a Buick. Find the section with the heading "Cylinder Bore." Three cylinder bore measurements are listed. (See Figure 9–3.) The second one is called Out of Round. The table says it is 0.02 max. At the bottom of the specifications page, it says that all dimensions are in millimeters. Therefore, the cylinder should be out of round by no more than 0.02 mm. We have determined that the out-of-round is 0.05 mm. This is more than 0.02 mm; therefore, the out-of-round is not within specifications.

Cylinder Bore:

Diameter	88.992-89.070
Out of Round	.02 Max
Taper-Thrust Side	.02 Max

Figure 9–3

■ EXAMPLE 9–4

The crankshaft and camshaft operate at a 2:1 speed ratio. That is, each time the crankshaft turns two times, the camshaft turns once. If the engine crankshaft is operating at 2600 revolutions per minute (rpm), how fast is the camshaft turning? (See Figure 9–4.)

How many times will it turn in 2 minutes?

Solution

When the crankshaft turns two times, the camshaft turns one time. This is a ratio of 2:1. In 1 minute, the crankshaft turns 2600 times. Now set up the proportion.

$$\frac{2 \text{ crankshaft turns}}{1 \text{ camshaft turn}} = \frac{2600 \text{ crankshaft turns}}{? \text{ camshaft turns}}$$

$$2600 \times 1 = 2600$$

Step 1

$$2600 \div 2$$

Step 2

$$2600 \div 2 = 1300.$$

Step 3   Therefore, the camshaft turns 1300 times in 1 minute. In 2 minutes, the camshaft turns $1300 \times 2 = 2600$ times.

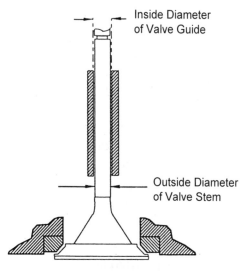

Figure 9–5

Solution

The inside diameter of the guide is 0.3458″. The diameter of the valve stem is 0.3413″.

Subtract

$$0.3458$$
$$-\,0.3413$$
$$\overline{0.0045}$$

The valve stem-to-valve guide clearance is 0.0045″.

Look in Appendix A. Find the specifications for a Ford. Locate the section with the title "Valve Stem to Guide Clearance." There are three clearance measurements given: Intake, Exhaust, and Service Clearance Limit. Which one should you use?

This is an exhaust valve. But because it is being serviced, you should use the service clearance limit of 0.0055″ maximum. We measured the clearance at 0.0045″, which is less than 0.0055″. This exhaust valve stem-to-valve guide clearance is within the service clearance limits.

■ EXAMPLE 9–7

An 8-cylinder engine has a total displacement of 454 in.³. What is the displacement of each piston or cylinder?

Solution

Notice that this is just the reverse of Example 9–1. In Example 9–1 you were given the displacement for one cylinder and asked to find the total displacement of the engine.

The total displacement for these 8 cylinders is 454 in.³. Each cylinder will displace $454 \div 8$ in.³.

Camshaft
Timing Gear
24 Teeth   24 : 12
2 : 1
Crankshaft
Timing Gear
12 Teeth

Figure 9–4

■ EXAMPLE 9–5

Each time the camshaft rotates one complete turn, each valve is pushed open one time. If the camshaft is rotating at 500 rpm, how many times is each valve opened during 1 minute of operation?

Solution

When the camshaft rotates one complete turn, each valve opens one time. In 1 minute, the camshaft turns 500 times. In 1 minute, the valve opens $500 \times 1 = 500$ times.

The clearance between the valve stem and the valve guide is important for proper valve lubrication and reasonable oil consumption. One method used to determine the valve stem-to-guide clearance is to measure both the inside diameter (i.d.) of the valve guide and the outside diameter (o.d.) of the valve stem. (See Figure 9–5.) Then subtract the diameter of the valve stem from the diameter of the valve guide.

■ EXAMPLE 9–6

If the inside diameter of the guide you have measured is 0.3458″ and the outside diameter of the valve stem is 0.3413″, what is the valve stem-to-valve guide clearance?

Is this exhaust valve stem-to-valve guide clearance within service specifications for Ford? (See Appendix A.)

$$
\begin{array}{r}
56.75 \\
8\overline{)454.00} \\
\underline{40} \\
54 \\
\underline{48} \\
60 \\
\underline{56} \\
40 \\
\underline{40} \\
0
\end{array}
$$

So, each cylinder displaces 56.75 in.^{3}.

Cylinder compression readings are often used to determine the mechanical condition of an engine. A good engine has fairly even compression from cylinder to cylinder. If the highest and lowest readings vary more than 25%, the engine should be repaired. This is determined by multiplying the highest reading by 0.75. This answer is compared to the lowest compression reading taken. If the answer is greater than the lowest reading, engine repair is recommended; the engine is considered within limits if the answer is less than the lowest reading.

■ EXAMPLE 9–8

Compression readings for a 4-cylinder engine are shown in Figure 9–6. Determine if the engine compression range is within recommendations.

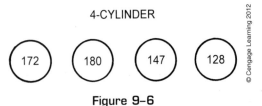

4-CYLINDER

172   180   147   128

© Cengage Learning 2012

Figure 9–6

Solution

The highest engine compression reading in Figure 9–6 is 180. The lowest is 128. Multiply the highest reading, 180, by 0.75.

$$
\begin{array}{r}
180 \\
\times\ 0.75 \\
\hline
135
\end{array}
$$

Because the lowest reading, 128, is below 135, engine repair is recommended.

Engine displacement is a term used to describe the total volume displaced as the pistons move from the top of their stroke, called top dead center (TDC), to the bottom of their stroke, called bottom dead center (BDC). This volume is usually expressed as cubic inches of displacement (CID), or liters if the measurement is in metric terms.

The formula used is:

$$
\text{Displacement} = 0.7854 \times \text{Bore}^2 \times \text{Stroke} \times \text{Number of Cylinders}
$$

■ EXAMPLE 9–9

A popular Ford V-8 engine has a bore of 4.00 inches (101.6 mm) and a stroke of 3.00 inches (76.2 mm). Determine the cubic inch displacement of this engine.

Solution

$$
\begin{aligned}
\text{CID} &= 0.7854 \times 4 \times 4 \times 3 \times 8 \\
&= 301.5936 \text{ in.}^3
\end{aligned}
$$

Ford calls this engine a 302.

Because many metric engine measurements are given in millimeters, entering these figures into the displacement formula results in an answer in cubic millimeters. To convert this into the more practical measurement of cubic centimeters, the answer must be divided by 1000. To obtain a displacement in liters, it must be divided by 1,000,000.

■ EXAMPLE 9–10

What is the displacement in metric units of the Ford V-8 in Example 9–9?

Solution

$$
\text{Displacement} = 0.7854 \times \text{Bore}^2 \times \text{Stroke} \times \text{Number of Cylinders}
$$

$$
\text{Displacement (cc)} = \frac{0.7854 \times 101.6 \times 101.6 \times 76.2 \times 8}{1000}
$$
$$
= 4942.23
$$

The displacement is about 4942 cm^3.

$$
\text{Displacement (Liters)} = \frac{0.7854 \times 101.6 \times 101.6 \times 76.2 \times 8}{1\,000\,000}
$$
$$
= 4.94 \text{ Liters}
$$

Ford rounds this up to 5 liters.

During construction of a high-output engine, it is necessary to accurately determine the compression ratio (CR).

This process can get quite involved and requires several different methods of measurement. Essentially, you compare the volume of the cylinder and combustion area with the piston at BDC to the combustion volume with the piston at TDC.

The bore and stroke dimensions are fairly easily obtained, either through written specifications or by actual measurement. You also need to know the combustion chamber volume as measured by a burette. This information is then put into the following formula:

$$CR = \frac{\text{Cylinder Volume} + \text{Combustion Volume}}{\text{Combustion Volume}}$$

You also need the following two formulas, which you first saw in Chapter 4.

$$\text{Area} = 0.7854d^2$$

$$\text{Volume} = A \times h$$

Most burettes are graduated in milliliters (ml), which are the same as cubic centimeters (cc). All compression-ratio calculation formulas use ccs as the unit of measure. This measurement will need to be multiplied by 0.061 to convert it to cubic inches (assuming all other measurements are in nonmetric units). For example, a 70 cm^3 chamber measurement is multiplied by 0.061 to obtain 4.27 cubic inches. If all dimensions are in metric units, no conversion is needed. Remember, a ratio has no dimensional title; it is only a comparison of two things—volumes in this case.

■ EXAMPLE 9–11

Determine the compression ratio of an engine using the following data. Assume a flat-topped piston.

Bore = 10.24 cm
Stroke = 8.84 cm
Deck height = 0.056 cm
Head gasket thickness = 0.104 cm
Combustion chamber volume (CCV) = 68 cm^3

Solution

To determine the compression ratio of this engine you must first determine:

Cylinder area
Cylinder volume
TDC combustion volume

Cylinder volume is determined by first calculating the circular area of the cylinder:

$$A = 0.7854d^2$$

where d is the diameter. Here the cylinder bore diameter = 10.24 cm, so $d = 10.24$.

$$A = 0.7854 \times \text{bore diameter}^2$$
$$= 0.7854 \times 10.24^2$$
$$= 0.7854 \times 10.24 \times 10.24$$
$$\approx 82.36$$

The circular area of the cylinder is about 82.36 cm^2 (square centimeters).

To determine cylinder volume you must multiply this area by the height (or length) of the piston stroke.

$$V_{\text{cylinder}} = A \times h$$
$$= 82.36 \text{ cm}^2 \text{ (piston area)}$$
$$\times 8.84 \text{ cm (piston stroke height)}$$
$$V_{\text{cylinder}} \approx 728.06 \text{ cm}^3 \text{ (cubic centimeters)}$$

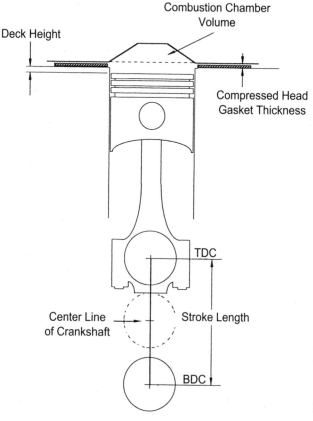

Figure 9–7

The combustion volume includes three separate volumes (Figure 9–7):

Deck height volume
Head gasket volume
Cylinder head combustion chamber volume

Deck height volume (DHV) = $A \times$ Deck height
$$DHV = 82.36 \text{ cm}^2 \times 0.056 \text{ cm}$$
$$= 4.61 \text{ cm}^3$$

Head gasket volume (HGV) = $A \times$ Head gasket thickness
$$HGV = 82.36 \text{ cm}^2 \times 0.104$$
$$= 8.57 \text{ cm}^3$$

Finally, total combustion volume (CV) = DHV + HGV + CCV (combustion chamber volume).

$$CV = DHV + HGV + CCV$$
$$= 4.61 \text{ cm}^3 + 8.57 \text{ cm}^3 + 68 \text{ cm}^3$$
$$= 81.18 \text{ cm}^3$$

Now that you have all the information you need, you can plug the values into the compression ratio (CR) formula.

$$(CR) = \frac{\text{Cylinder Volume} + \text{Combustion Volume}}{\text{Combustion Volume}}$$

$$= \frac{728.06 + 81.18}{81.18}$$

$$= 9.97$$

The compression ratio is about 9.97:1.

In Example 9–11, the data for calculating the compression ratio was in metric form. This will not always happen. The next example uses the U.S. Customary System.

■ EXAMPLE 9–12

Determine the compression ratio of an engine using the following data. Assume a flat-topped piston.

    Bore = 4.030 inches
    Stroke = 3.48 inches
    Deck height = 0.022 inches
    Head gasket thickness = 0.041 inches
    Combustion chamber volume (CCV) = 68 cm^3

Solution

To determine the compression ratio of this engine you must first determine:

    Cylinder volume
    TDC combustion volume

Cylinder volume is determined by first calculating the circular area of the cylinder and then multiplying this number by the height of the piston stroke.

    Cylinder bore diameter = 4.030 in.

$$A = 0.7854 \, d^2$$

where d is the diameter. Here the cylinder bore diameter = 4.030 cm, so d = 4.030.

$$A = 0.7854 \times \text{bore diameter}^2$$
$$= 0.7854 \times 4.030^2$$
$$= 0.7854 \times 4.030 \times 4.030$$
$$\approx 12.76$$

The circular area of the cylinder is about 12.76 in.^{2} (square inches).

To determine cylinder volume, multiply this area by the height (or length) of the piston stroke.

$$V_{\text{cylinder}} = A \times h$$
$$= 12.76 \text{ in.}^2 \times 3.48 \text{ in.}$$
$$V_{\text{cylinder}} \approx 44.40 \text{ in.}^3 \text{ (cubic inches)}$$

The combustion volume includes three separate volumes (Figure 9–7):

    Deck height volume
    Head gasket volume
    Cylinder head combustion chamber volume

Deck height volume (DHV) = $A \times$ Deck height
$$DHV = 12.76 \text{ in.}^2 \times 0.022 \text{ in.}$$
$$= 0.28 \text{ in.}^3$$

Head gasket volume (HGV) = A
$$\times \text{ Head gasket thickness}$$
$$HGV = 12.76 \text{ in.}^2 \times 0.041 \text{ in.}$$
$$= 0.52 \text{ in.}^3$$

Before you can perform the CR calculation you must convert 68 cm^3 (a metric value) to a U.S. Customary System value. Cubic centimeters are converted into cubic inches by multiplying the cubic centimeter value by 0.061, the conversion factor.

$$\text{Cubic inches} = 68 \text{ cm}^3 \times 0.061$$
$$\text{Combustion chamber size} = 4.15 \text{ in.}^3$$

Now you are ready to determine total combustion volume.

Total combustion volume (CV) = DHV + HGV + CCV (combustion chamber volume).

$$CV = DHV + HGV + CCV$$
$$= 0.28 \text{ in.}^3 + 0.52 \text{ in.}^3 + 4.15 \text{ in.}^3$$
$$= 4.95 \text{ in.}^3$$

Now that you have all the information you need, you can plug the values into the compression ratio (CR) formula.

$$(CR) = \frac{\text{Cylinder Volume} + \text{Combustion Volume}}{\text{Combustion Volume}}$$

$$= \frac{44.40 + 4.95}{4.95}$$

$$= 9.97$$

The compression ratio is about 9.97:1.

Engine output can be described in many different ways. One way is to consider output in terms of torque. Torque is defined as a twisting force. The amount of twisting force can be broken down into two additional components, force and distance (see Chapter 6). It is this twisting force, or torque, that drives the automobile.

To further consider engine output we need to add the dimension of time, or speed, described as engine revolutions per minute (rpm). When torque and speed are combined, we commonly use the term *horsepower*. Horsepower is represented in a formula as:

$$\text{Horsepower} = \frac{\text{RPM} \times \text{Torque}}{5252}$$

■ EXAMPLE 9–13

If engine output is measured as 358 foot-pounds of torque at 5500 rpm, what is the horsepower of this engine?

Solution

$$\text{Horsepower} = \frac{5500 \times 358}{5252}$$

$$\approx 374.9$$

This engine has about 375 horsepower.

NAME _____ DATE _____ SCORE _____

1. An engine block weighs 79 kg, cylinder head 17 kg, crankshaft 11 kg, pistons and connecting rods 8 kg, and intake and exhaust manifolds 9 kg. What is the total weight of these parts?

1. _____

2. The valve system of a 4-cylinder engine has 1 camshaft, 8 valve lifters, 8 pushrods, 8 rocker arms, 1 rocker shaft, 8 valve springs, 16 valve stem locks, 8 valve spring retainers, 4 intake valves, and 4 exhaust valves. How many parts are in this valve system?

2. _____

3. A technician installed a new camshaft, cam timing components, and cam followers in 3.6 hr, tuned the engine in 0.8 hr, removed and replaced (R&R) and checked the fuel injectors in 1.3 hr, and installed a new O_2 sensor in 0.5 hr.

    (a) How much time was spent on the car?

    3a. _____

    (b) If the labor rate is $79.50 per hour, what was the labor charge?

    3b. _____

    (c) If the technician is paid 45% of all labor charges, what was the technician paid for this work?

    3c. _____

4. The camshaft in Problem 3 costs $159, cam timing components cost $35, the eight cam followers were $7.32 each, the tune-up kit was $43.95, one fuel injector was $63.84, and an O_2 sensor was $86.72.

    (a) What was the total cost of the parts to the customer?

    4a. _____

    (b) If the parts department has an 87% markup on these parts, how much profit did they make on this job?

    4b. _____

    (c) Considering the 55% profit rate from labor charges that the shop keeps (Problem 3(c)) and the parts department markup, how much money did the shop make on this job?

    4c. _____

5. A valve grinding operation requires the use of a 0.343 in. pilot. A technician located 4 pilots, and they are marked $^{21}/_{64}$, $^5/_{16}$, $^{11}/_{32}$, and $^3/_8$ in. Which pilot should be used?

5. _____

6. One cylinder of a V-8 engine has a displacement of 38.375 in^3. What is the total displacement of the engine?

6. _____

7. An engine has a $3^7/_{16}$ in. bore and a $3\frac{1}{2}$ in. stroke.

    (a) What is the volume of each cylinder?

    7a. _____

    (b) If this is a V-8 engine, what is the total displacement?

    7b. _____

8. One cylinder of a V-6 engine has a displacement of 633.34 cm³.

   (a) What is the total displacement of this engine in cubic centimeters?   8a. _____

   (b) What is the displacement in liters?   8b. _____

9. Each cylinder of a 4-cylinder engine displaces 0.48 L.

   (a) What is the total displacement of this engine?   9a. _____

   (b) What is the displacement in cubic centimeters?   9b. _____
     (1 liter = 1000 cubic centimeters.)

10–15. For Problems 10 through 15, first determine the amount of ring ridge for each cylinder in Figure 9–8. Then decide if the ring ridge needs to be removed before the piston is removed. Remove the ring ridge if it is more than 0.004 in. (0.1 mm).

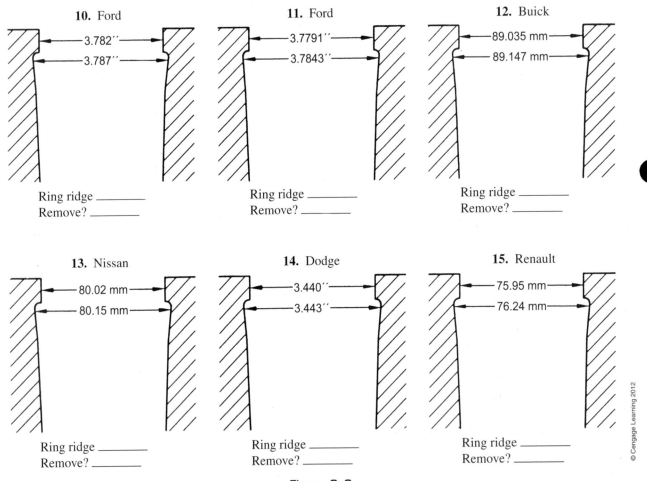

10. Ford

3.782″

3.787″

Ring ridge _____
Remove? _____

11. Ford

3.7791″

3.7843″

Ring ridge _____
Remove? _____

12. Buick

89.035 mm

89.147 mm

Ring ridge _____
Remove? _____

13. Nissan

80.02 mm

80.15 mm

Ring ridge _____
Remove? _____

14. Dodge

3.440″

3.443″

Ring ridge _____
Remove? _____

15. Renault

75.95 mm

76.24 mm

Ring ridge _____
Remove? _____

Figure 9–8

16–18. In Problems 16 through 18, determine the out-of-round for each cylinder in Figure 9–9. Then look in Appendix A for the service specifications for each engine. Finally, state if the out-of-round is within service specification.

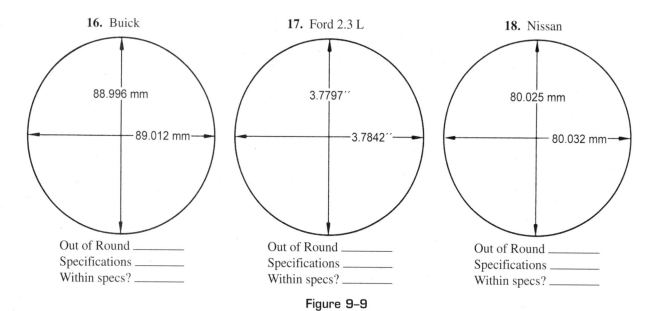

16. Buick

88.996 mm

←———89.012 mm———→

Out of Round _____
Specifications _____
Within specs? _____

17. Ford 2.3 L

3.7797″

←———3.7842″———→

Out of Round _____
Specifications _____
Within specs? _____

18. Nissan

80.025 mm

←———80.032 mm———→

Out of Round _____
Specifications _____
Within specs? _____

Figure 9–9

19. If the crankshaft in a V-8 engine is rotating at 4000 rpm, how many times is each valve opened during any single minute of operation? (See Figure 9–10.)

19. _____

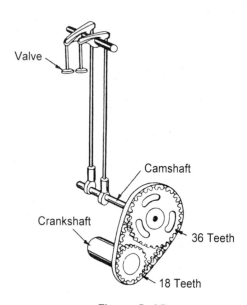

Valve

Camshaft

Crankshaft

36 Teeth

18 Teeth

Figure 9–10

20. If the crankshaft in a 4-cylinder engine is rotating at 4000 rpm, how many times is each valve opened during any single minute of operation?

20. _____

21–23. For Problems 21 through 23, determine the valve stem-to-guide clearance, find the specifications in Appendix A for each, and decide if the service clearance is within the specifications.

	21. **BUICK EXHAUST VALVE**	22. **FORD**	23. **NISSAN INTAKE VALVE**
Inside diameter of valve guide	7.948 mm	0.3447 in.	7.013 mm
Outside diameter of valve stem	7.912 mm	0.3423 in.	6.982 mm
Valve stem-to-guide clearance			
Specifications			
Within specifications?			

The amount of wear that has taken place on a camshaft may be determined by taking two measurements on a cam lobe. One measurement, dimension A, is across the cam lobe and the other, dimension B, is across the base circle (see Figure 9–11). The lobe lift, C, is determined by subtracting the B measurement from the A measurement.

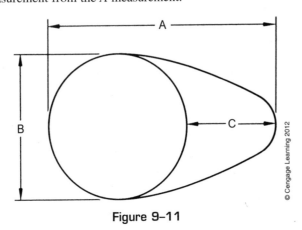

© Cengage Learning 2012

Figure 9–11

24. On a Buick, cam lobe dimension A is 35.90 mm and dimension B is 29.25 mm.

   (a) What is the lobe lift?

   (b) What is the Buick specification?

   (c) Is the lobe lift within specifications?

24a. _____

24b. _____

24c. _____

25–27. For Problems 25 through 27, determine the specifications and decide if the camshaft is within specifications.

	25. **FORD 2.3 L**	26. **DODGE 2.6 L**	27. **NISSAN INTAKE CAM LOBE**
Dimension A	1.515 in.	41.65 mm	1.7515 in.
Dimension B	1.275 in.		
Lobe lift			
Specifications			
Within specifications?			

28. During one crankshaft revolution, the piston travels from its highest position to its lowest position and back to its highest position. The distance from the high position to the low position is called the *stroke*.

(a) If the stroke is 3 inches, how many inches does the piston travel in 1 minute if the engine is operating at 2000 rpm?

28a. _____

(b) How many feet does the piston travel in 1 minute?

28b. _____

(c) How many degrees does the crankshaft turn if the piston goes through two strokes?

28c. _____

29. (a) Consider a crankshaft that is rotating at 2000 rpm. If the engine operates at that speed for 24 hours, how many revolutions does the crankshaft make during that 24-hour period?

29a. _____

(b) If the person driving the car in Problem 29(a) stopped to sleep for 9 hours, spent 30 minutes each for breakfast, lunch, and dinner, and 15 minutes each for two gas stops, how much total time was spent driving?

29b. _____

(c) How many revolutions did the crankshaft make during the trip in Problem 29(b) if we assume it averaged 2500 rpm?

29c. _____

(d) If the driver averaged 55 mph during the driving time, how far did the driver get during that 24-hour period?

29d. _____

30. You can determine the amount of taper within a cylinder by first measuring the diameter of the cylinder at both the top and bottom as shown in Figure 9–12. Then subtract the bottom measurement from the top measurement.

(a) If the top of a cylinder measures 3.7831 in. and the bottom measures 3.7795 in., what is the taper of the cylinder?

30a. _____

(b) What is the Ford specification for this situation?

30b. _____

(c) Does the taper for this cylinder exceed the taper service limits as set forth in the specifications for a Ford?

30c. _____

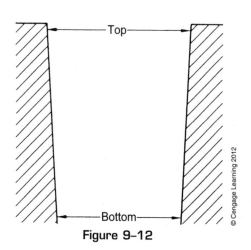

Figure 9–12

31–33. For Problems 31 through 33, find the taper for each cylinder in Figure 9–13. Look in Appendix A to determine the specifications and decide if the taper is within specifications.

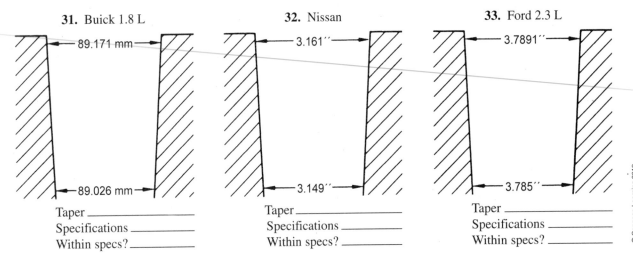

31. Buick 1.8 L

89.171 mm

89.026 mm

Taper _____

Specifications _____

Within specs? _____

32. Nissan

3.161″

3.149″

Taper _____

Specifications _____

Within specs? _____

33. Ford 2.3 L

3.7891″

3.785″

Taper _____

Specifications _____

Within specs? _____

© Cengage Learning 2012

Figure 9–13

34. We can determine the force that is acting on the top of each piston by multiplying the area of the top of the piston by the force acting on that area. If the top of the piston has an area of 9.5 in.^{2} and the cylinder combustion pressure is 500 pounds per square inch (lb/in.^{2} or psi), how much force is exerted by the piston through the connecting rod to the crankshaft?

34. _____

35. Proper piston fit within the cylinder is important. Piston seizure results if the piston is too tight. Poor performance and excessive oil consumption result from a piston that is fitted too loosely. When measuring the piston fit on a Ford, we find the bore to be 3.7795 in. and the piston to be 3.7758 in. This is a turbocharged engine.

    (a) What is the piston-to-bore clearance?

35a. _____

    (b) Is this clearance within specifications?

35b. _____

36. Will the pistons on the Ford in Problem 35 need to be replaced or can they be reused?

36. _____

37. Piston pins must be fitted correctly to prevent engine noises and/or damage. On a Buick, the piston pins measure 22.9709 mm and the pinholes in the piston measure 22.9930 mm.

    (a) What is the piston pin clearance?

37a. _____

    (b) Must new oversized pins be installed on this Buick?

37b. _____

38. Main bearing journal number 5 on a Buick Skyhawk measures 63.336 mm at one location and at 90 degrees (°) opposite measures 63.361 mm.

    (a) How much out-of-round does this journal have?

38a. _____

    (b) Is this journal too far out of round for further use?

38b. _____

39. The inside diameter of the rod with bearing is 50.762 mm. The outside diameter of the corresponding rod journal is 50.730 mm.

    (a) What is the rod-bearing clearance on this Buick?

    (b) What is the Buick specification for rod bearing clearance?

    (c) Is the rod-bearing clearance within specifications?

39a. _____

39b. _____

39c. _____

40. Each cylinder on an 8-cylinder engine has both an intake valve and an exhaust valve. If all of the valves must be replaced during a rebuild, how many valves do you need to order?

40. _____

41. To determine the actual valve lift on an engine that uses pushrods, the camshaft lobe lift must be multiplied by the rocker arm ratio. The rocker arm ratio can usually be found in the engine specifications. What is the actual valve opening/lift in an engine that has a lobe of 0.2437 in. and a rocker arm ratio of 1.64?

41. _____

42. Using the specifications given for a Buick in Appendix A, determine the valve opening/lift for that engine.

42. _____

The compression ratio of an engine is determined by comparing the volume of the cylinder when the piston is at the bottom of its stroke (Figure 9–14a), with the volume when the piston is at the top of the stroke (Figure 9–14b). If volume A is 48 in.^{3} and volume B is 6 in.^{3}, then the compression ratio is 48 to 6, or (dividing by 6) 8 to 1. The ratio 48 to 6 is often written as 48:6, and 8 to 1 can be written as 8:1. As you can see, the type of volume is not important. The volume could be in cubic inches, cubic centimeters, or liters. Only the numerical comparison of the two volumes is important, and it does not have any units.

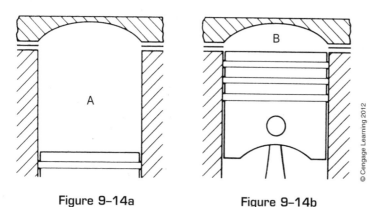

      Figure 9–14a   Figure 9–14b

© Cengage Learning 2012

43. Find the compression ratio of an engine with volume A listed as 360 in.^{3} and volume B as 45 in.^{3}.

43. _____

44. What is the compression ratio of an engine with volume A listed as 2277 cm^3 and volume B as 253 cm^3?

44. _____

Use the graph in Figure 9–15 for Problems 45 through 50.

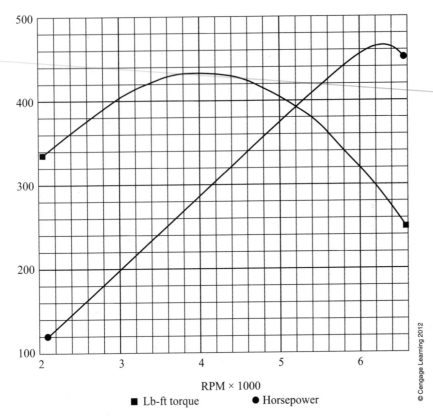

RPM × 1000

■ Lb-ft torque   ● Horsepower

© Cengage Learning 2012

Figure 9–15

45. (a) At what engine speed (rpm) is peak horsepower reached?

    (b) What is the peak horsepower?

46. (a) At what engine speed is peak torque reached?

    (b) What is the peak torque?

47. What is the engine speed difference between peak horse power and peak torque?

48. At what engine speed are both horsepower and torque the same?

49. What is the torque loss, or drop, between peak torque speed and peak horse-power speed?

50. What is the horsepower gain between peak torque speed and peak horsepower speed?

45a. _____

45b. _____

46a. _____

46b. _____

47. _____

48. _____

49. _____

50. _____

51. Figure 9–16 contains compression readings for six different engines. Determine if each engine compression is within recommendations. Indicate that compression is within range by writing a "Yes" in the space provided. If the engine's compression is not within range, write a "No" in the space provided.

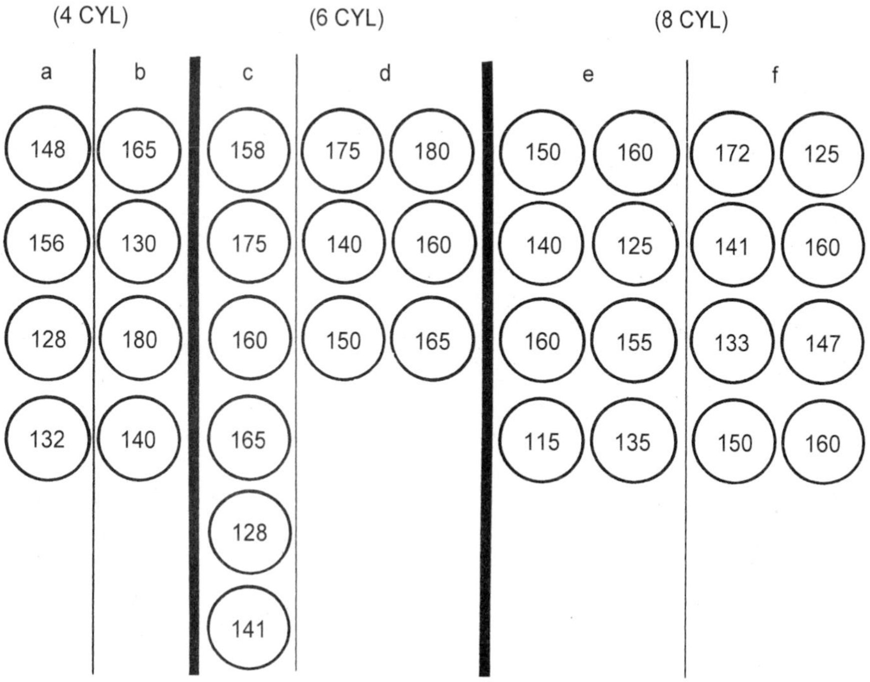

(4 CYL)		(6 CYL)		(8 CYL)	
a	b	c	d	e	f

51a. _____

51b. _____

51c. _____

51d. _____

51e. _____

51f. _____

(4 CYL)
a: 148, 156, 128, 132
b: 165, 130, 180, 140

(6 CYL)
c: 158, 175, 160, 165, 128, 141
d: 175, 140, 150 / 180, 160, 165

(8 CYL)
e: 150, 140, 160, 115 / 160, 125, 155, 135
f: 172, 141, 133, 150 / 125, 160, 147, 160

© Cengage Learning 2012

Figure 9–16

52. A race engine is having the pistons replaced. The current rotating assembly has been blueprinted and balanced with a reciprocating weight of 640 grams. Any new replacement parts must match the weights of the original parts, ± 5 grams, or the crankshaft must be rebalanced. The new reciprocating parts have the following individual weights:

Piston = 435 grams
Piston pin = 145 grams
Piston pin locks = 3 grams
Piston rings = 102 grams

(a) What is the total weight of these parts?

(b) Is the answer to part (a) within ± 5 grams of the original?

(c) Must the crankshaft be rebalanced?

52a. _____

52b. _____

52c. _____

53. Combustion cycle/valve timing diagrams are used to graphically represent the relationship between the valves and the piston strokes. The valve timing diagram in Figure 9–17 shows that the intake valve opens at 18° before top dead center (BTDC) and closes at 137° BTDC. The exhaust valve opens at 136° after top dead center (ATDC) and closes at 18° ATDC. In the spaces provided note the number of degrees that both the intake and exhaust valve remain open (this is called the *valve duration*). Also determine the number of degrees of valve overlap, the time when both valves are open at the same time.

53. Intake _____

Exhaust _____

Overlap _____

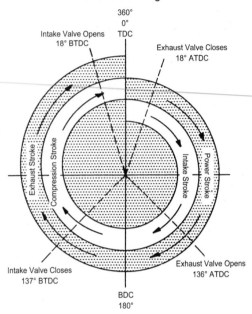

Figure 9–17

54. Complete Figure 9–18 by filling it in similar to Figure 9–17. Use the valve timing specifications provided below.

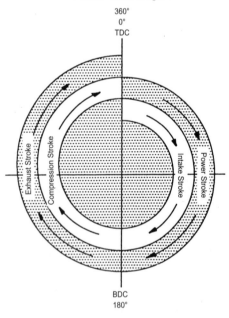

Figure 9–18

	Opens	Closes
Intake	8° BTDC	22° ABDC
Exhaust	20° BBDC	13° ATDC

54. Intake _____

Exhaust _____

Overlap _____

55. Determine the cubic inch displacement of the following engines.

Engine Type	Bore (in.)	Stroke (in.)	
(a) I-4 OHC	3.516	3.126	55a. _____
(b) V-8	4.054	3.819	55b. _____
(c) I-4 SOHC	3.622	3.336	55c. _____
(d) V-6	3.780	3.189	55d. _____

56. Determine the cubic centimeter displacement of the following engines.

Engine Type	Bore (mm)	Stroke (mm)	
(a) V-6	96	81	56a. _____
(b) V-8	101.63	106.2	56b. _____
(c) I-4 OHC	82	88	56c. _____
(d) I-5 OHC	96.075	79.400	56d. _____

57. Determine the displacement of the following engines in liters.

Engine Type	Bore (mm)	Stroke (mm)	
(a) I-4	75.019	89.992	57a. _____
(b) V-8	101.60	88.90	57b. _____
(c) V-6	86.030	86.1	57c. _____
(d) I-5	72.910	87.29	57d. _____

58. Determine the compression ratio for each set of engine data provided. Notice that the bore, stroke, deck height, and gasket thickness are all given in inches.

Bore (in.)	Stroke (in.)	Deck Height (in.)	Gasket Thickness (in.)	Combustion Chamber Volume (cm^3)	
(a) 4.00	4.00	0.015	0.033	105	58a. _____
(b) 4.030	3.50	0.018	0.038	65	58b. _____
(c) 3.50	3.48	0.009	0.018	63	58c. _____
(d) 3.780	3.126	0.020	0.035	58	58d. _____

59. Determine engine horsepower from the following specifications.

Torque	RPM	
(a) 130	3200	59a. _____
(b) 155	5200	59b. _____
(c) 275	3000	59c. _____
(d) 340	3600	59d. _____

60. Because automotive engines operate on a 4-stroke cycle, it takes two complete crankshaft revolutions (720° of crankshaft revolution) to complete an operational cycle of the 4 strokes (see Figure 9–19). During two complete crankshaft revolutions, all cylinders in the engine will fire. A 4-cylinder engine fires every 180 degrees (720° ÷ 4 = 180°).

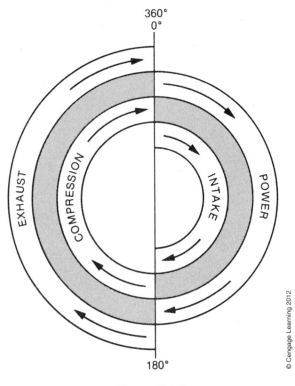

Figure 9–19

(a) What is the firing interval for an inline 6-cylinder engine?

60a. _____

(b) What is the firing interval for a V-8 engine?

60b. _____

(c) What is the firing interval for a 5-cylinder engine?

60c. _____

61. The following automobile is having some engine work completed, and to be sure it is running correctly a few other fairly routine maintenance procedures have been included. Assume a labor rate of $79.50/hour and a state sales tax of 7.5%. Complete and total the repair order in Figure 9–20.

61. _____

2007 MKZ, 3.5 L, AWD

Quan	Part No.	Name of Part	Sales Amount	
		Total Parts		

REPAIR ORDER

Name _J.J. Petrino_

Address _3 Corporate Dr_

Phone No. _555-0190_ **Date** _____

Speedometer Reading _132,856_ **VIN No.** _82167_

Year, Make, and Model	License No. and State	Engine
07 MKZ	MBA - 007	3.5 L AWD

Operation Number	SERVICE INSTRUCTIONS	TIME	AMOUNT
	Engine Oil Leak Diagnosis		
	R & R Oil Filter & Change Oil		
	R & R Serpentine Belt Tensioner		
	R & R Front Engine Mount		
	R & R Transaxle Mount		
	R & R Both Valve Cover Gaskets		
	R & R Front Crank Seal		

	Accessories	Amount	Total Labor	
			Total Parts	
@			Materials	
@			Accessories	
6	Qts. Oil @ 7.32		Tires, Tubes	
Lbs. Grease @			Outside Work	
Total Materials	Total Accessories		Total	

I hereby authorize the above repair work to be done along with the necessary material, and hereby grant you and/or your employees permission to operate the car, truck or vehicle herein described on streets, highways or elsewhere for the purpose of testing and/or inspection. An express mechanic's lien is hereby acknowledged on above car, truck or vehicle to secure the amount repairs thereto.

NOT RESPONSIBLE FOR LOSS OR DAMAGE TO CARS OR ARTICLES LEFT IN CARS IN CASE OF FIRE, THEFT OR ANY OTHER CAUSE BEYOND OUR CONTROL.

WORK AUTHORIZED BY _J.J. Petrino_ _____ DATE _____

Total	
Tax	
TOTAL AMOUNT	

© Cengage Learning 2012

Figure 9–20

10

AUTOMOBILE ENGINE SYSTEMS

Objectives: After studying this chapter, you should be able to:

- Use the formulas for horsepower loss, torque, cubic-inch displacement, theoretical air capacity, and volume efficiency.
- Determine engine fluid capacity.
- Calculate horsepower loss related to elevation.
- Calculate exhaust manifold tube diameter and length.
- Determine proper carburetor size for an engine.
- Calculate engine volumetric efficiency.

This chapter focuses on some of the numerical relationships that are involved in the lubrication, cooling, and fuel systems of an engine. Without a fuel system, the engine would not operate. Without the lubrication and cooling systems, it would not operate very long. Improper servicing of these systems can result in unreliable operation and/or engine damage. The following problems include situations within the automotive service trade as well as those of day-to-day operation.

■ EXAMPLE 10–1

A driver put 17,500 miles on a car during a year. If the oil was changed every 2500 miles, how many times was the oil changed during the year?

Solution

Car driven 17,500 miles
Oil changed every 2500 miles
Number of oil changes was 17,500 ÷ 2500.
17,500 ÷ 2500 = 7

The oil was changed 7 times.

■ EXAMPLE 10–2

Each oil change for the car in Example 10–1 required 4 quarts (qt) of oil. How many quarts were needed?

Solution

4 qt used each oil change.
7 oil changes
Total quarts used was 4 × 7 qt.
4 × 7 = 28

Therefore, 28 qt of oil were needed.

■ EXAMPLE 10–3

A case of oil contains 24 qt. How many cases of oil were needed for the car in Example 10–1?

Solution

One case contains 24 qt.
28 qt were needed.
Number of cases is 28 qt ÷ 24 qt/case.
28 ÷ 24 = 1.17

A total of 1.17 cases were needed.

■ EXAMPLE 10–4

If the oil filter was changed every other time the oil was changed, how many oil filters were needed for the car in Example 10–1?

Solution

There were 7 oil changes.
The oil filter was changed every 2 oil changes.
The number of filters needed was 7 ÷ 2.
7 ÷ 2 = 3.5.

The car needed a total of 3 oil filters.

There are 2 pints (pt) in 1 quart and 4 quarts in 1 gallon (gal). Use these measurements to help solve the next example.

■ EXAMPLE 10–5

A Ford has the following capacities:

Cooling system: 14 qt
Engine crankcase: 5 qt
Front axle: 4 pt
Rear axle: 6.5 pt
Fuel tank: 21 gal
Standard transmission: 7 pt
Transfer case: 6.5 pt
Power steering: 1 qt

How many pints of fluids are there?

Solution

All of the capacities given above should be given in pints. If the capacity is already in pints, then no work needs to be done. If the capacity is not in pints, then we have to determine how many pints of fluid can be held for each engine system.

Cooling system: 14 qt × 2 pt/qt = 28 pt
Engine crankcase: 5 qt × 2 pt/qt = 10 pt
Front axle: 4 pt
Rear axle: 6.5 pt
Fuel tank: 21 gal × 4 qt/gal × 2 pt/qt = 168 pt
Standard transmission: 7 pt
Transfer case: 6.5 pt
Power steering: 1 qt × 2 pt/qt = 2 pt

To find the total, add these pint measurements: 28 + 10 + 4 + 6.5 + 168 + 7 + 6.5 + 2 = 232. There are 232 pints of fluid.

■ EXAMPLE 10–6

How many quarts of fluid are there?

Solution

Since there are 2 pints in 1 quart, each pint is the same as ½ quart. There are 232 pints. Therefore, there are 232 pints × ½ qt/pt = 116 quarts.

■ EXAMPLE 10–7

Listed below are various fluid capacities for a vehicle. Convert each fluid capacity to liters and round the answers to the nearest tenth liter.

Liters

Cooling system = 16 qt _____
Power steering = 3.5 pt _____
Fuel capacity = 18.5 gal _____

You should look up pints, quarts, and gallons in the "Customary to Metric" table (Table 4–3) to determine the necessary conversion factor for each.

Solution

Quarts are a volume measure, so look in the "Volume" section of the right-hand column of Table 4–3. Here you will see

qt   L   0.9464

This means that we need to multiply the number of quarts in the cooling system (16) by 0.9464 to determine the number of liters.

$$16 \times 0.9464 = 15.1424$$

The cooling system holds about 15.1 liters.
  Similar work leads to the following results for the power steering and the fuel capacity.

Power steering = 3.5 pt × 0.4732 = 1.6562
Fuel capacity = 18.5 gal × 3.7854 = 70.0299

Thus, we see that there is about 1.7 L of power steering fluid and the fuel capacity for this vehicle is about 70.0 L.

■ EXAMPLE 10–8

An automatic transmission holds 20 pt of fluid. When the filter is changed, 25% or ¼ of the total fluid is lost. How many quarts are needed to fill the transmission after servicing?

Solution

The transmission holds 20 pints. A filter change means ¼ × 20 = 5 pt are lost.

$$5 \div 2 = 2\frac{1}{2} \text{ qt}$$

Higher altitudes have a negative effect on engine output. If the engine is not equipped with a turbocharger or supercharger, the horsepower loss is approximately 3% per 1000 feet in elevation gain from sea level.

This can be calculated using the following formula:

$$\text{Horsepower Loss} = \frac{\text{Elevation in Feet}}{1000} \times 0.03 \times \text{Sea Level Horsepower}$$

■ EXAMPLE 10–9

An automobile is driven over a mountain pass at 8325 feet. The engine is rated at 255 hp at 5200 rpm. What is the horsepower loss due to elevation?

$$\text{Horsepower Loss} = \frac{\text{Elevation in Feet}}{1000} \times 0.03 \times \text{Sea Level Horsepower}$$

$$= \frac{8325}{1000} \times 0.03 \times 255$$

$$= 8.325 \times 0.03 \times 255$$

$$= 63.69$$

This automobile loses about 63.69 hp when it is driven over this mountain pass.

This automobile's final horsepower at 8325 feet is its original horsepower minus its loss in horsepower, or

$$\text{Final horsepower} = 255 - 63.69$$

$$= 191.31$$

This automobile's final horsepower at 8325 feet is 191.31 hp.

A reconfiguration of the horsepower formula allows you to solve for torque. This means that you can determine engine torque from horsepower specifications. This formula is:

$$\text{Torque} = \frac{5252 \times \text{Horsepower}}{\text{rpm}}$$

Torque is measured in lb-ft.

■ EXAMPLE 10–10

If an engine is rated at 502 hp at 6250 rpm, what is the torque output?

Solution

$$\text{Torque} = \frac{5252 \times \text{Horsepower}}{\text{rpm}}$$

$$= \frac{5252 \times 502}{6250}$$

$$\approx 421.8 \approx 422$$

The torque output of this engine is about 422 lb-ft.

One method to improve engine breathing and performance is to custom design the exhaust tubing from the engine. The formula for determining proper exhaust header tubing diameter is:

$$\text{Diameter} = \sqrt{\frac{\text{CID} \times 1900}{\text{Length} \times \text{rpm}}}$$

where "CID" is the cubic inch displacement of the engine and "length" is the exhaust header length in inches. The number 1900 is a constant; it never changes.

■ EXAMPLE 10–11

An exhaust system for a high-performance engine is being designed for proper fit and performance. Determine the correct diameter exhaust tubing for this application. The engine cubic inch displacement is 355, the tubing length is 24 inches, and the maximum engine speed will be 6500 rpm.

Solution

Use the following formula:

$$\text{Diameter} = \sqrt{\frac{\text{CID} \times 1900}{\text{Length} \times \text{rpm}}}$$

with CID = 355, length = 24 in., and rpm = 6500.

Plugging in these values, we get

$$\text{Diameter} = \sqrt{\frac{355 \times 1900}{24 \times 6500}}$$

$$= \sqrt{\frac{674,500}{156,000}}$$

$$\approx \sqrt{4.3237}$$

At this point you need to use the square root key, $\boxed{\sqrt{}}$, on your calculator to find that $\sqrt{4.3237} \approx 2.079$. So, the diameter is about 2.07 inches. This is commonly listed as 2 inches.

■ EXAMPLE 10–12

You are building a 355 CID engine with a redline of 8400 rpm. Exhaust header pipe diameter is 1⅞ inches. What should be the length of the exhaust header?

Solution

We will first convert $1\frac{7}{8}$ inches to its decimal equivalent of 1.875 inches.

$$\text{Length} = \frac{\text{CID} \times 1900}{\text{Diameter}^2 \times \text{rpm}}$$

$$= \frac{355 \times 1900}{1.875^2 \times 8400}$$

$$= \frac{674{,}500}{29{,}531.25}$$

$$\approx 22.84$$

The length of the exhaust pipe should be about 22.84 inches.

An engine is essentially a pneumatic device. It draws in large amounts of air to be mixed with fuel. The theoretical air capacity of an engine is determined using the following formula.

$$\text{Theoretical air capacity} = \frac{\text{rpm} \times \text{Displacement}}{2}$$

If engine displacement is given in cubic inches, the air capacity number becomes very large. To make it more manageable you should refer to it in terms of cubic feet per minute (CFM). To convert cubic inches to cubic feet, divide the cubic inch number by 1728, since there are 1728 cubic inches per cubic foot. The above formula then becomes

$$\text{Theoretical air capacity (CFM)} = \frac{\text{rpm} \times \text{Displacement}}{2 \times 1728}$$

$$\text{Theoretical air capacity (CFM)} = \frac{\text{rpm} \times \text{Displacement}}{3456}$$

■ EXAMPLE 10–13

Determine the air capacity or air flow in an engine of 350 cubic inch displacement operating at 5700 rpm.

Solution

$$\text{Air capacity (CFM)} = \frac{\text{rpm} \times \text{Displacement}}{3456}$$

$$= \frac{5700 \times 350}{3456}$$

$$\approx 577.26$$

The air flow in this engine is about 577.26 CFM.

As the air–fuel mixture burns in the combustion area it produces large amounts of gas, which provides power and must then be removed from the cylinder. The power output depends, to a large extent, on how well the engine breathes. This capability is described by the term *volumetric efficiency* (VE).

Volumetric efficiency is given as a percentage. It is a comparison of actual air flow to the theoretical air flow at a given speed. This comparison is usually made in either liters per minute (LPM) or CFM. To determine volumetric efficiency, you must have access to actual air-flow data while an engine is in operation. The formula for determining volumetric efficiency is:

$$\text{VE} = \frac{\text{Actual air capacity (CFM)} \times 100}{\text{Theoretical air capacity (CFM)}}$$

■ EXAMPLE 10–14

If a dynamometer sensor measures an air flow of 478 CFM at 5500 rpm in a 350 CID engine, what is the volumetric efficiency?

Solution

The actual air flow is 478 CFM. We need to determine the theoretical air flow.

$$\text{Theoretical air capacity (CFM)} = \frac{\text{rpm} \times \text{Displacement}}{3456}$$

$$= \frac{5500 \times 350}{3456}$$

$$= 557$$

Now that we know both the actual and the theoretical air flow, we can determine the volumetric efficiency.

$$\text{VE} = \frac{\text{Actual air capacity (CFM)} \times 100}{\text{Theoretical air capacity (CFM)}}$$

$$= \frac{478 \times 100}{577}$$

$$\approx 82.8 \approx 83$$

The volumetric efficiency is about 83%.

The volumetric efficiency of engines varies with their application and operating speed. It can range from 70% to over 100% with specialized tuning.

To determine proper carburetor size, assume volumetric efficiency to be 85% for most street applications and 110% for all-out racing engines. Most American carburetors are rated in CFM. The formulas for determining proper carburetor size are:

$$\text{Street carburetor CFM} = \frac{\text{rpm} \times \text{Displacement} \times 0.85}{3456}$$

$$\text{Racing carburetor CFM} = \frac{\text{rpm} \times \text{Displacement} \times 1.10}{3456}$$

■ EXAMPLE 10–15

You have an automobile you use for everyday driving. The 355 CID engine has been rebuilt and modified slightly for occasional racing at the drag strip. If the engine is redlined at 6500 rpm, what size carburetor should you install? Carburetors are sold in 50-CFM increments.

Solution

$$\text{Street carburetor CFM} = \frac{\text{rpm} \times \text{Displacement} \times 0.85}{3456}$$

$$= \frac{6500 \times 355 \times 0.85}{3456}$$

$$= 567.53$$

Purchase the next larger size, or a 600 CFM carburetor.

NAME _____ DATE _____ SCORE _____

1. A driver put 27,000 miles on a car in one year.

   (a) If the oil was changed every 3000 miles, how many times was the oil changed during the year?

   1a. _____

   (b) How many quarts of oil were needed if each oil change required 5 quarts of oil?

   1b. _____

   (c) How many cases of oil was this?

   1c. _____

2. Another driver put 48,000 km on a car in a year. The oil in this car was changed every 4000 km.

   (a) How many times was the oil in this car changed during the year?

   2a. _____

   (b) Each oil change required 6 liters of oil. How many liters of oil were needed during the year?

   2b. _____

   (c) If a case of oil contains 24 liters of oil, how many cases were used?

   2c. _____

   (d) The oil filter was changed every 8000 km. How many oil filters were needed?

   2d. _____

3. A Ford has the following capacities:

   Cooling system: 16.5 qt
   Engine crankcase: 5 qt
   Drive axle: 4 pt
   Fuel tank: 20 gal
   Standard transmission: 2 pt
   Power steering: 1.5 pt

   (a) How many pints of fluid are there?

   3a. _____

   (b) How many quarts of fluid is this?

   3b. _____

   (c) How many gallons is this?

   3c. _____

4–6. For each of the cars listed, determine the total number of pints, quarts, and gallons of fluid if the measurements are given in customary measurements. If the measurements are given in metric measurements, determine the total number of liters.

	4. DODGE	5. NISSAN	6. RENAULT
Cooling system	9.0 qt	4.7 L	4.8 qt
Engine crankcase	4.0 qt	3.9 L	3.25 qt
Fuel tank	14.0 gal	50 L	12.5 gal
Standard transmission	2.0 qt	2.7 L	3.6 qt
Power steering	2.5 pt	1.0 L	2.0 pt

4. _____ pt
   _____ qt
   _____ gal

5. _____ L

6. _____ pt
   _____ qt
   _____ gal

7. **(a)** How many gallons of fluid are there in the cars in Example 10–5?

7a. _____

 (b) How many liters of fluid are in the car in Example 10–2?

7b. _____

 (c) How many liters of fluid are in the car in Example 10–6?

7c. _____

8. You are going on a 4200-mile trip and your automobile uses a quart of oil for every 1050 miles traveled. How many quarts of oil must you add during the trip to arrive home with the oil level on the "full" mark?

8. _____

9. Oil is priced at $2.68 per quart. However, if you buy a case, the price is $69.64.

 (a) How much money do you save by buying by the case rather than by the quart?

9a. _____

 (b) How much money do you save on each quart?

9b. _____

10. The local store has a special sale on engine oil bought by the case and has reduced the price of a quart of oil by $0.26. The store advertisements state that an additional $0.17 per quart can be saved by sending in for a manufacturer's rebate. If the original price is $68.64 per case, what is the new cost per quart?

10. _____

11. You are going on a 9600-km trip and the automobile you are driving uses a quart of oil for every 1200 km traveled. How many quarts of oil must be added during this trip so that you arrive home with the oil level on the "full" mark?

11. _____

12. During an oil change at a service station, 6 quarts of oil, an oil filter, and a can of engine oil supplement were used. The oil was $2.48 per quart, the filter $5.69, and the supplement $5.16.

 (a) What was the cost of all the materials used for this oil change?

12a. _____

 (b) If the sales tax rate is 4%, what was the amount of tax on these items?

12b. _____

 (c) What was the total cost of the oil change?

12c. _____

13. In an effort to save money, the person in Problem 12 bought 6 qt of oil at a discount store for $1.98 each, the oil filter for $2.33, and the same type engine oil supplement for $3.58 and changed the oil at home.

 (a) What is the cost of all of these materials?

13a. _____

 (b) How much sales tax must be included?

13b. _____

 (c) What was the total cost?

13c. _____

 (d) How much money was saved compared to the oil change done by the service station in Problem 12?

13d. _____

14. The engine cooling system has the following capacities:

 Engine: 7.5 quarts
 Radiator: 3.25 quarts
 Heater core: 0.75 quarts
 Radiator and heater hoses: 1.5 quarts

 (a) What is the total capacity of the entire coolant system?

14a. _____

 (b) For recommended protection of the system, a 50% solution of antifreeze should be used. How much antifreeze is needed to give a final antifreeze strength of 50%?

14b. _____

15. Using the graph (Figure 10–1), what freeze protection is obtained by using a 50% solution of antifreeze?

15. _____

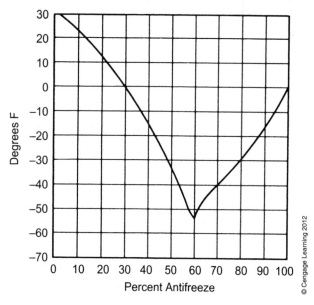

Figure 10–1

16. What is the maximum level of antifreeze protection available according to Figure 10–1?

17. What would the freeze protection be if 100% pure antifreeze were used? (See Figure 10–1.)

18. What freeze protection would you get if a 12-qt system contained 2 qt of antifreeze and the rest water? (See Figure 10–1.)

19. You want to change the coolant in your vehicle. The owner's manual lists the capacity of the cooling system as 9 quarts, including the heater and coolant reserve system.

    (a) How many quarts of antifreeze are needed to provide a 50% solution of antifreeze?

    (b) How many quarts of antifreeze are needed to provide the maximum protection?

    (c) How many quarts are needed to provide protection down to $-25°F$?

20. A vehicle is getting 20 miles per gallon (miles/gal) with the engine operating at approximately 180°F. Someone tells you that for each 10° increase in engine operating temperature above 160° the fuel economy increases 2.5%. If the 180° thermostat is removed and replaced with a 200° thermostat so that the engine operates at approximately 200°, what increase in fuel economy might the driver expect in miles per gallon?

21. A vehicle was getting 25 miles/gal using a direct-drive radiator (coolant) fan. This fan was removed and an electric fan was installed that only operated at times such as when the vehicle was stopped in a traffic jam on a hot day. The driver noticed a 2.5-miles/gal improvement in fuel economy since the new fan was installed. How many gallons of gasoline are saved by this new fan during a 5500-mile trip?

22. How many miles does the driver in Problem 21 have to drive before the fan has paid for itself? The fan cost $137.82 and the price of gasoline is $4.12 per gallon.

16. _____

17. _____

18. _____

19a. _____

19b. _____

19c. _____

20. _____

21. _____

22. _____

23. One gallon of gasoline weighs approximately 7.5 pounds. If a vehicle has a 13-gal fuel tank, how many pounds of fuel is the vehicle carrying when the tank is full?

23. _____

24. A vehicle gets 25 miles/gal of gasoline and has a full tank of 20 gal. (Use 7.5 pounds as the weight of 1 gallon of fuel.)

    (a) How many pounds of fuel are left in the tank after traveling 375 miles?

24a. _____

    (b) How many pounds were used to drive the 375 miles?

24b. _____

25. How many pounds of gasoline were needed for the vehicle in Problem 24 to travel 225 miles?

25. _____

26. A fuel tank has a capacity of 160 pints. The vehicle gets 5 miles per quart.

    (a) How far can a driver expect to go on a full tank of fuel?

26a. _____

    (b) What is the fuel economy in miles per gallon?

26b. _____

27. A fuel tank has a capacity of 75 liters. If the car gets 5 miles per liter, how many miles can be driven on a tank of gas?

27. _____

28. Convert the following fluid capacities to liters.

    (a) Cooling system: 9.9 qt

28a. _____

    (b) Engine crankcase: 5.5 qt

28b. _____

    (c) Fuel capacity: 22 gal

28c. _____

    (d) Rear axle: 4 pt

28d. _____

29. Convert the following fluid capacities to gallons, quarts, or pints, whichever is appropriate based on its use in Problem 28.

    (a) Cooling system: 13.3 L

29a. _____

    (b) Engine crankcase: 4.3 L

29b. _____

    (c) Fuel capacity: 70 L

29c. _____

    (d) Rear axle: 1.5 L

29d. _____

30. Which fuel tank has the greater capacity, tank A at 90 L or tank B at 25 gal?

30. _____

31. A driver puts 17 gallons of diesel fuel at $4.18 per gallon in a car. Because the fuel is paid for in cash, a discount of 3 cents per gallon is received. What amount of money is paid for this fuel?

31. _____

32. A driver was trying to save money and bought a diesel-powered automobile. The diesel gets 32 miles per gallon, compared to a gasoline-powered car that gets 24 miles per gallon. The diesel fuel costs $4.32 per gallon and the gasoline costs $3.78 per gallon. The diesel engine option on the new car costs $2018 above the cost of a gasoline-powered vehicle. How many miles must the driver travel before the extra cost of the diesel is paid off and the driver begins to save money?

32. _____

33. An automotive diesel fuel-injection pump pumps approximately 300 liters of fuel per hour. If 4.5% of the fuel being pumped is burned in the engine, how many liters are being returned to the fuel tank each hour?

33. _____

34. A diesel fuel-injection nozzle opening pressure is listed as 200 bars. If a bar is equal to 14.5 lb/in.^{2}, what is the opening pressure of the injection nozzle in pounds per square inch?

34. _____

35. To test for injection nozzle leakage, the pressure must be brought to 15 bars below the valve opening pressure. Referring to the nozzle in Problem 34, what is the proper pressure in pounds per square inch for a leakage test?

35. _____

36. A diesel fuel-injection pump is calibrated to deliver 40 mm^3 of fuel per stroke at 1500 rpm during full-throttle operation. This fuel delivery is for one cylinder only.

(a) If the pump is operated for exactly 1000 strokes, how many cubic centimeters of fuel are delivered during the period of operation?

36a. _____

(b) What is the total delivery volume from the pump for all 8 cylinders?

36b. _____

37. During the assembly of a high-performance engine the rebuilder installed a high-volume oil pump, which had an output 20% greater than the original pump. If the original pump had a capacity of 5 gallons per minute (GPM), what is the GPM output of the new pump?

37. _____

38. The engine cooling system usually operates under pressure. The radiator cap is the component that seals and regulates this pressure. The boiling point of the cooling system is raised 3°F for each pound per square inch (psi) of pressure in the system. What is the maximum boiling point if the system is fitted with a 15-psi cap? Assume the unpressurized coolant has a 200°F boiling point.

38. _____

39. An engine is tested with a dynamometer and produces 385 horsepower. Following the dynamometer test, a tuned intake manifold is installed, which is advertised to increase horsepower by 12%.

(a) If the advertisement is correct, what horsepower gain could one expect?

39a. _____

(b) What would the total horsepower then be?

39b. _____

40. You have just purchased a new automobile with a horsepower rating of 215 at 4200 rpm and are planning a trip through some of the Western states. During the trip you will be driving over various mountain passes. Determine the final horsepower and torque at the top of each pass.

Name of Pass	Altitude (ft)
(a) Sonora Pass	9,628
(b) Deadman's Pass	8,036
(c) Vail Pass	10,666
(d) Rabbit Ear's Pass	9,426
(e) Lizard Head Pass	10,222
(f) Independence Pass	12,095
(g) Devil's Gate	7,519

40a. _____ HP
_____ T
40b. _____ HP
_____ T
40c. _____ HP
_____ T
40d. _____ HP
_____ T
40e. _____ HP
_____ T
40f. _____ HP
_____ T
40g. _____ HP
_____ T

41. Determine the proper exhaust header tubing diameter from the following information.

CID	Length of Exhaust Header	RPM
(a) 377	20	5800
(b) 454	27	6500
(c) 427	32	6000
(d) 350	30	7200

41a. _____

41b. _____

41c. _____

41d. _____

42. Determine the proper length of an exhaust header pipe from the following information.

CID	Diameter of Exhaust Tubing	RPM
(a) 355	2	6500
(b) 377	1¾	6200
(c) 428	2	5500
(d) 327	1⅞	7000

42a. _____

42b. _____

42c. _____

42d. _____

43. Determine the volumetric efficiency for these engines. (Round your answers to the nearest tenth of a percent.)

CID	RPM	Measured Air Flow (CFM)
(a) 302	4450	305
(b) 243	3800	239
(c) 172	5100	221
(d) 488	3200	478

43a. _____

43b. _____

43c. _____

43d. _____

44. Determine the proper size of carburetors to purchase for the following applications. Carburetors are sold in 50-CFM increments. You should purchase a carburetor rated closest *above* your actual calculation. In (i) give your calculated carburetor size; in (ii) give the recommended purchase size.

	Application	CID	Max RPM
(a)	Street	351	5700
(b)	Racing	383	6700
(c)	Street	302	6000
(d)	Racing	355	8400

44a. (i) _____
(ii) _____
44b. (i) _____
(ii) _____
44c. (i) _____
(ii) _____
44d. (i) _____
(ii) _____

45. Engine variable valve timing (VVT) is a new strategy used to make engines more fuel efficient. One manufacturer suggests that this technology can result in an increase of approximately 4.5% in fuel economy. If a 2010-model car using traditional engine design achieves an average fuel economy rating of 31.6 mpg, what might be expected from a 2011 model of the same engine using VVT technology?

45. _____

46. It has been determined that each gallon of gasoline consumed by motor vehicles produces approximately 19.4 lb (or 8.8 kg) of CO_2. It has also been determined that the average automobile in the United States travels approximately 12,000 miles per year. If the average car gets 32.2 mpg, how many pounds of CO_2 are produced per year? How many kilograms per year?

46. _____ lb
_____ kg

47. If 60% of the fuel energy is lost to engine and exhaust heat, 12% is lost to deceleration and idling, 10% is lost to rolling resistance, and 5% is used for accessory operation, what percentage remains to drive the vehicle down the road?

47. _____

48. This automobile is undergoing a variety of service procedures related to its engine operating systems. The labor rate is $89.95/hour and the sales tax rate is 4%. Complete and total the repair order in Figure 10–2.

2008 Milan
2.3 L engine
FWD automatic transmission
California/Green State certification

48. _____

Quan	Part No.	Name of Part	Sales Amount	
	Total Parts			
1	Battery		85	29
		@		
6	Qts. Anti Freeze	@ 4.37		
	Qts. Oil	@		
	Lbs. Grease	@		
	Total Materials			

REPAIR ORDER

Name Sonja Chen

Address 1188 Marco Blvd.

Phone No. 555-2718 **Date** _____

Speedometer Reading 65,382 **VIN No.** _____

Year, Make, and Model	License No. and State	Engine
08 Milan	FWD Automatic Calif/Green	2.3 L

Operation Number	SERVICE INSTRUCTIONS	TIME	AMOUNT
	R & R Vapor Cannister Purge Control Valve		
	R & R Knock Sensor		
	Perform Cooling Sys. Pressure Test		
	R & R Water Pump		
	Perform Battery Test		
	R & R Battery		
	Perform Fuel Pump Test		
	R & R Fuel Pump		
	R & R Exhaust Manifold Gasket		

Accessories	Amount		Total Labor	
			Total Parts	
			Materials	
			Accessories	
			Tires, Tubes	
Total Accessories			Outside Work	
			Total	
			Tax	
			TOTAL AMOUNT	

I hereby authorize the above repair work to be done along with the necessary material, and hereby grant you and/or your employees permission to operate the car, truck or vehicle herein described on streets, highways or elsewhere for the purpose of testing and/or inspection. An express mechanic's lien is hereby acknowledged on above car, truck or vehicle to secure the amount repairs thereto.

NOT RESPONSIBLE FOR LOSS OR DAMAGE TO CARS OR ARTICLES LEFT IN CARS IN CASE OF FIRE, THEFT OR ANY OTHER CAUSE BEYOND OUR CONTROL.

WORK AUTHORIZED BY _Sonja Chen_ _____ DATE

© Cengage Learning 2012

Figure 10–2

AUTOMOBILE ELECTRICAL SYSTEMS

Objectives: After studying this chapter, you should be able to:

- Explain the relationship between current flow, resistance, and voltage.
- Use Ohm's law to calculate current flow, resistance, and voltage.
- Measure battery electrolyte specific gravity.
- Use Ohm's law to determine resistance, voltage, and current flow of a circuit.
- Describe battery ratings.
- Determine engine firing intervals.
- Determine proper wire size and length for a given current flow.
- Use circuit voltage drops to determine circuit resistance.

As federal regulations on automotive emissions and fuel economy increase, the role of vehicular electrical systems becomes increasingly more important. In the near future, virtually the entire process of vehicle operation will be monitored and/or controlled through electrical circuits and electronic devices of one type or another.

Chapter 11 shows the relationship among current flow, resistance, and voltage in a mathematical setting that is related to basic electrical circuits, the battery, the starter and alternator, and the ignition system.

The relationship between voltage, measured in volts; resistance, measured in ohms; and current flow, measured in amperes (amps), is given in a formula known as Ohm's law. Ohm's law is a mathematical formula that is used to calculate an unknown value in an electrical circuit when two other values are known. Ohm's law may be expressed as shown in Figure 11–1, where E = electromotive force (volts), I = current flow (amps), and R = resistance (ohms).

To use Ohm's law, cover the unit in Figure 11–1 that you wish to determine and proceed as indicated below.

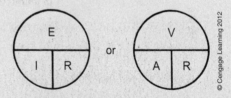

Figure 11–1

© Cengage Learning 2012

To determine voltage, cover the E. This leaves I, current flow in amps, times R, resistance in ohms ($E = I \times R$).

To determine amps, cover the I and divide the voltage E by the resistance R ($I = E/R$).

To determine the resistance, cover the R and divide the voltage E by the amps I ($R = E/I$).

■ EXAMPLE 11–1

Determine the voltage that is required to pass a current through a circuit with a current flow of 9 amps and 1.5 ohms resistance.

Solution

In this problem you are trying to find the voltage, *V*. Cover the *V* in Figure 11–1 with your finger (see Figure 11–2). The two remaining letters, *A* and *R*, are next to each other. This indicates that *A* and *R* should be multiplied.

$$A = 9 \text{ amps}$$
$$R = 1.5 \text{ ohms}$$
$$V = A \times R = 9 \times 1.5$$
$$= 13.5$$

Therefore, the voltage in this circuit is 13.5 volts.

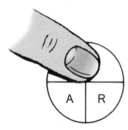

Figure 11–2

■ EXAMPLE 11–2

If an electrical circuit is operating at 14 volts and has a resistance of 2 ohms, what is the current flow?

Solution

You are trying to find the current flow in amps. Cover the *A* in Figure 11–1 with your finger as shown in Figure 11–3. The two letters that are not covered, *V* and *R*, are in the position shown in Figure 11–3. This relationship, V/R, indicates that *V* must be divided by *R*.

$$V = 14 \text{ volts}$$
$$R = 2 \text{ ohms}$$
$$A = \frac{V}{R} = \frac{14}{2}$$
$$= 7$$

The current flow is 7 amps.

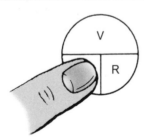

Figure 11–3

The symbol *A* represents amperes or amps, *V* stands for volts, and Ω is used for ohms.

Electrical power is measured in watts. The symbol *W* is used for watts. One watt is equal to 1 A of current flow under the electromotive force or pressure of 1 V.

Wattage is determined by multiplying the current flow in amperes by the circuit electromotive force in volts.

Watts of electrical power can also be compared to horsepower (hp). One horsepower is equal to 746 W (1 hp = 746 W). To solve a problem that asks for a horsepower equivalent of a given wattage, simply divide the wattage by 746.

■ EXAMPLE 11–3

What is the horsepower equivalent of 1492 W?

Solution

$$\text{Wattage} = \text{Horsepower} \times 746$$
$$1492 \text{ W} = \text{hp} \times 746$$
$$1492 \div 746 = \text{hp}$$
$$1492 \div 746 = 2$$
$$\text{Therefore, } 1492 \text{ W} = 2 \text{ hp}$$

A series electrical circuit is one in which the current has only one path to flow (see Figure 11–4). This means that the current flow is the same in all parts of the circuit. Total resistance in such a circuit is equal to the sum of the individual resistances. To find the total resistance in a series circuit, simply add the individual resistances.

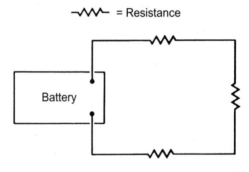

Figure 11–4

■ EXAMPLE 11–4

What is the total resistance in the circuit in Figure 11–5?

Solution

$$1.3 \, \Omega + 0.6 \, \Omega + 1.5 \, \Omega = 3.4 \, \Omega$$

The total resistance is 3.4 Ω.

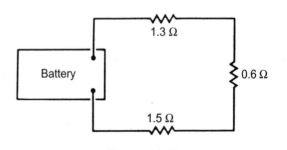

Figure 11–5

A parallel electrical circuit is one that has more than one path for current flow (see Figures 11–6 and 11–7). The current in Figure 11–6 has two paths that it can use to get from *A* to *B*. The current in Figure 11–7 can follow any of four paths from *C* to *D*. Total current flow is equal to the sum of the amount of current flow in each path. Total resistance in a parallel circuit is always less than the lowest resistance in the circuit and may be calculated by using the formula

$$R_{total} = \cfrac{1}{\cfrac{1}{R_1} + \cfrac{1}{R_2} + \cfrac{1}{R_3} + \cfrac{1}{R_4}}$$

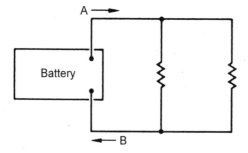

Figure 11–6

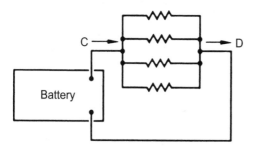

Figure 11–7

■ EXAMPLE 11–5

Determine the total resistance in the circuit in Figure 11–8.

Solution

$$R_{total} = \cfrac{1}{\cfrac{1}{R_1} + \cfrac{1}{R_2} + \cfrac{1}{R_3} + \cfrac{1}{R_4}}$$

$$= \cfrac{1}{\cfrac{1}{16} + \cfrac{1}{4} + \cfrac{1}{8} + \cfrac{1}{16}}$$

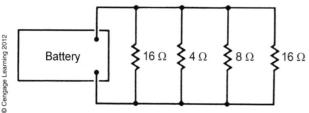

Figure 11–8

A common denominator is 16. So, R_{total} then becomes

$$R_t = \cfrac{1}{\cfrac{1}{16} + \cfrac{4}{16} + \cfrac{2}{16} + \cfrac{1}{16}} = \cfrac{1}{\cfrac{8}{16}}$$

$$= 1 \div \frac{8}{16}$$

$$= 1 \times \frac{16}{8} = \frac{16}{8} = 2$$

Thus, $R_t = 2$.
The total resistance in this circuit is 2 ohms or 2 Ω.

The strength of an electromagnetic coil is expressed in its ampere-turn rating. If a current of 3 amps flows through a coil having 200 turns, the coil is said to have a 600 ampere-turn rating (3 amperes × 200 turns = 600 ampere-turns).

■ EXAMPLE 11–6

Determine the ampere-turn rating of a coil of 425 turns that is passed through a current of 4 amps.

Solution

$$425 \text{ turns} \times 4 \text{ A}$$
$$425 \times 4 = 1700$$

The rating is 1700 ampere-turns.

Total resistance in a series circuit is always equal to the sum of the individual resistances. Also, the source voltage in a series circuit is always equal to the sum of the individual voltage drops. A voltage drop is the amount of voltage "used up" to make current flow through a resistance.

■ EXAMPLE 11–7

In the series circuit in Figure 11–9, you can see three resistances: *A* has 12 ohms, *B* has 4 ohms, and *C* has 8 ohms. Determine the voltage drops in this circuit.

Solution

If resistances *A*, *B*, and *C* represent the total resistance in the circuit, or 100%, you need to assign values, or

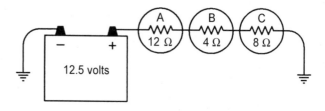

Figure 11–9

percentages, to each resistance. The sum of the given resistances is $12 + 4 + 8 = 24$. Use this sum, 24, as the common denominator to determine the fractional part and percentage that each resistance is of the total resistance. Resistance A then becomes $^{12}/_{24}$ or $^{1}/_{2}$, B becomes $^{4}/_{24}$ or $^{1}/_{6}$, and C becomes $^{8}/_{24}$ or $^{1}/_{3}$.

Next, convert these fractions into percentages. To do this, divide the numerator by the denominator and multiply the answer by 100.

$$A = \frac{1}{2} \times 100 = 50\%$$

$$B = \frac{1}{6} \times 100 = 16.7\%$$

$$C = \frac{1}{3} \times 100 = 33.3\%$$

To determine the actual voltage drops in this circuit, multiply the source voltage by the percentages. Make sure to change the percentages to decimals.

$$A = 12.5 \times 0.50 = 6.25 \text{ volts}$$
$$B = 12.5 \times 0.167 = 2.09 \text{ volts}$$
$$C = 12.5 \times 0.333 = 4.16 \text{ volts}$$

By adding the individual drops you can see that the sum of the individual voltage drops does indeed equal the source voltage.

$$
\begin{array}{r}
6.25 \text{ volts} \\
2.09 \text{ volts} \\
\underline{4.16 \text{ volts}} \\
12.5 \quad \text{volts}
\end{array}
$$

NAME _____ DATE _____ SCORE _____

1. Determine the voltage that is required to pass a given amount of current through a given amount of resistance in the examples below.

    (a) Ignition circuit with a current flow of 3 A through 4.5 Ω   1a. _____

    (b) Windshield wiper motor with a current flow of 15 A through 0.9 Ω of resistance   1b. _____

    (c) Starter motor with 200 A flowing through 0.05 Ω of resistance   1c. _____

2. Determine the current flow through the following components in an automotive electrical circuit that is operating on 14 V.

    (a) Headlights with 1 Ω resistance   2a. _____

    (b) Compressor clutch with 2.5 Ω resistance   2b. _____

    (c) Dome light with 8 Ω resistance   2c. _____

3. Determine the resistance in the following circuits that are operating on 14 V.

    (a) Power seats with a current flow of 18 A   3a. _____

    (b) Dash lights with a current flow of 2 A   3b. _____

    (c) Brake lights with a current flow of 8 A   3c. _____

4. Determine the missing quantity of amps, volts, or ohms in each part.

    (a) What is the resistance of an ignition circuit that has a current flow of 4 A and is operating on 14 V?   4a. _____

    (b) What is the voltage on power seats with a current force of 18 A and 0.75 Ω resistance?   4b. _____

    (c) What is the resistance of a starter motor with 200 A operating on 10.2 V?   4c. _____

    (d) What is the current flow through a compressor clutch operating on 14 V with 1.4 Ω resistance?   4d. _____

5. Determine the wattage in a 14-V circuit that has a current flow of 6 A.   5. _____

6. How much electrical power goes into a starter motor on an automotive diesel during startup on a cold morning if the system draws 600 A at 10 V?   6. _____

7. What is the approximate horsepower equivalent to the wattage in Problem 6?   7. _____

8. What is the approximate horsepower drain placed on an engine by an alternator that is generating 54 A at 14 V? Disregard any frictional losses.   8. _____

9. (a) What is the total resistance in the circuit in Figure 11–10?   9a. _____

    (b) Using Ohm's formula, what is the current flow if the circuit voltage is 12 V?   9b. _____

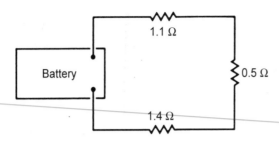

Figure 11–10

10. **(a)** Determine the total resistance of the circuit in Figure 11–11.

    (b) There are 5 A flowing through the circuit. What voltage is necessary to "push" the current through this circuit?

10a. _____

10b. _____

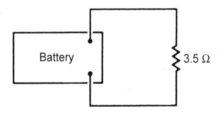

Figure 11–11

11. What is the total resistance of the circuit in Figure 11–12?

11. _____

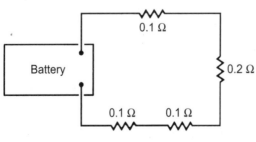

Figure 11–12

12. Determine the number of watts consumed by the circuit shown in Figure 11–13. Current flow is 28 A.

12. _____

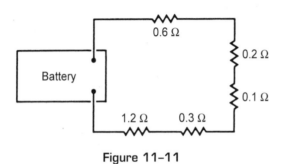

Figure 11–13

13. Determine the total resistance in the parallel circuit in Figure 11–14.

13. _____

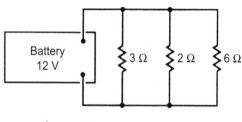

Figure 11–14

14. (a) Determine the total resistance in the parallel circuit in Figure 11–15.

14a. _____

(b) What amount of current is flowing in this circuit?

14b. _____

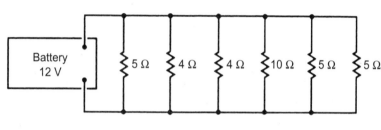

Figure 11–15

15. Determine the ampere-turn rating of a coil of 375 turns that is passing a current of 2 A.

15. _____

16. Electromagnet *A* has 275 turns with 6 A flowing and electromagnet *B* has 413 turns with 4 A flowing.

(a) What is the rating of magnetic coil *A*?

16a. _____

(b) What is the rating of magnetic coil *B*?

16b. _____

(c) Which coil has the higher rating?

16c. _____

17. The procedure for a battery capacity test requires that a load equal to 3 times the battery ampere-hour rating be applied for 15 sec. If the battery ampere-hour rating is 68, what is the amp load that is to be applied during the capacity test?

17. _____

18. A specific gravity reading on a battery of 0.001 is called a *point*. A specific gravity reading of 0.025 is 25 points. Many service manuals specify that a battery should be replaced if there is a 50-point or more difference between the specific gravity reading of the battery cells. The following are specific gravity readings that were taken from the battery in Figure 11–16: cell 1, 1.165; cell 2, 1.145; cell 3, 1.175; cell 4, 1.185; cell 5, 1.160; and cell 6, 1.140.

(a) What is the highest reading?

18a. _____

(b) What is the lowest reading?

18b. _____

(c) What is the difference between the highest reading and the lowest reading?

18c. _____

(d) Should the battery be replaced?

18d. _____

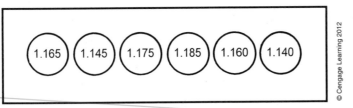

© Cengage Learning 2012

Figure 11–16

19. A battery has a reserve capacity of 150 minutes. How many hours and minutes is this?

19. _____

20. An alternator has an output of 40 A and is diagnosed as being defective as its rated output is 60 A. After being repaired, tests reveal the output to be 60 A—up to its rated output. What percent increase was there from the output of the defective alternator to the output after it was repaired?

20. _____

21. Before doing a tune-up, the electrical requirements for the secondary ignition circuit (spark plugs, distributor cap, and rotor, etc.) were 17,000 volts (V) or 17 kilovolts (17 kV). After the tune-up, the voltage requirements were found to be 7 kV. What percent decrease in secondary voltage occurred after the tune-up?

21. _____

22. An ignition system normally has a peak operating voltage of 14 kV. If the system has a maximum capability of 21 kV,

    (a) How many kilovolts are in reserve?

22a. _____

    (b) What percent of total capable output is in reserve?

22b. _____

Electrical resistance is measured in ohms (Ω). A reading of zero (0) indicates no resistance. A reading of 3 Ω or 14 Ω is considered higher resistance. The higher the reading, the higher the resistance.

23. Secondary or spark plug cables must be checked from time to time for excessive resistance. Accepted resistance for secondary cables is approximately 2000 Ω per inch. A cable is measured to have 21,000 Ω of resistance and is 1 foot long.

    (a) What is an acceptable resistance value for the cable

23a. _____

    (b) Is the cable within specification limits?

23b. _____

24. An imported car lists a 1500-Ω-per-centimeter specification for secondary cable resistances. What is the maximum resistance allowed for a 14″ wire?

24. _____

25. The starter draws 360 A at 10 V and takes 15 sec to start the engine. Once started, the charging system produces 80 A at 13.5 V.

    (a) How many watts does the starting system use to start the engine?

25a. _____

    (b) How many watts of electrical energy does the charging system produce?

25b. _____

    (c) Compared to the time needed to start the engine, approximately how long does it take to recharge the battery? Assume that all the charging system output is used to recharge the battery.

25c. _____

26. The following systems and/or components draw the listed amounts of current.

    Air conditioner: 18 amps
    Windshield wiper motor: 15 amps
    Radio/stereo system: 6 amps
    Headlights: 17 amps
    Ignition system: 7 amps

(a) How much current is consumed by these units, assuming they are all used at the same time?

26a. _____

(b) Would a 60-A alternator supply sufficient current to meet the above demands?

26b. _____

27. Katie usually keeps her car for a long time and is thinking about buying a new car that would provide long-term, high-fuel-mileage operation. Her current car is 18 years old with over 200,000 miles on it. She is considering either a hybrid or a diesel-powered car.

The advantage of hybrid vehicles is that they provide relatively high fuel economy during the time they are operating on electrical energy from the onboard battery pack. The disadvantage, especially for a person like Katie who usually keeps her car for a long time, is the cost of battery replacement. Battery life would not likely last for 18 years or over 200,000 miles, the condition of her current car.

The advantage of a diesel vehicle is the lack of high-cost battery replacement. The disadvantage is the slightly higher cost per gallon of the diesel fuel and the slightly lower gas mileage ratings, depending on the type of driving.

The purchase price of the two different vehicles is very similar, depending on equipment options.

The current hybrid battery replacement cost is $4814.81, including labor. The current national average cost of gasoline is $2.89 and diesel fuel is $3.12. Assume that as time goes by these prices will adjust proportionally.

The combined city/highway miles per gallon rating of the hybrid is 48 mpg, and the combined rating of the diesel is 42 mpg.

In this exercise the cost of gasoline is $2.89, diesel fuel is $3.12, and the cost of hybrid battery replacement is $4814.81. If Katie drives 225,000 miles and the hybrid car requires a battery replacement at 200,000 miles, determine the following:

(a) What is the cost of fuel for driving the hybrid 225,000 miles?

27a. _____

(b) What is the combined cost of fuel and battery replacement for driving 225,000 miles?

27b. _____

(c) What is the cost of fuel for driving the diesel 225,000 miles?

27c. _____

(d) Which vehicle has the lower operating cost for 225,000 miles?

27d. _____

(e) What is the cost advantage?

27e. _____

28. The cold-cranking rating of a battery should match or exceed the cubic-inch displacement of the engine with which it is being used. If an engine has a displacement of 1.7 liters, what size battery should be used for this car in terms of a cold-cranking rating number?

28. _____

29. The reserve capacity rating of a battery refers to how many minutes the automobile will operate if the charging system is disabled before the battery is so discharged that the engine stalls.

(a) If a battery has a reserve capacity of 135 minutes, how many hours and minutes can the car be operated safely?

29a. _____

(b) How far can the car go at 45 mph?

29b. _____

(c) How far can it go at 60 mph?

29c. _____

(d) Can a drive of 115 miles at 55 mph be made safely?

29d. _____

30. When the specific gravity of the battery electrolyte is determined, any variation from a test temperature of 80°F must be compensated for. Figure 11–17 illustrates that relationship.

As an example, if a reading of 1.245 is taken at 50°F, it must be adjusted or compensated to 80°F. If we locate 50° on the chart we can see that a specific gravity correction of −0.012 must be made to the original reading.

Original reading	1.245
Temperature correction	− 0.012
Corrected reading	1.233

In the spaces provided, note the corrected readings for the specific gravities given.

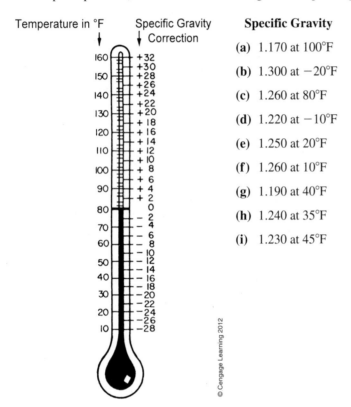

Specific Gravity

(a) 1.170 at 100°F   30a. _____

(b) 1.300 at −20°F   30b. _____

(c) 1.260 at 80°F   30c. _____

(d) 1.220 at −10°F   30d. _____

(e) 1.250 at 20°F   30e. _____

(f) 1.260 at 10°F   30f. _____

(g) 1.190 at 40°F   30g. _____

(h) 1.240 at 35°F   30h. _____

(i) 1.230 at 45°F   30i. _____

© Cengage Learning 2012

Figure 11–17
Note that the 12 on the thermometer actually indicates 0.012.

31. The procedure for performing a battery load test requires that the battery specific gravity reading be at least 1.220. Review the corrected readings in Problem 30 and indicate by a "Yes" or "No" if the specific gravity readings are high enough to permit a proper battery load test.

31a. _____

31b. _____

31c. _____

31d. _____

31e. _____

31f. _____

31g. _____

31h. _____

31i. _____

32. Table 11–1 can be used to determine the proper relationship among current flow, wire gauge number, and length of wire. For example, if to determine the proper gauge number of a 40′ wire required to carry 30 amps, first, find the amps (30) in the left column; second, cross-reference it with the length (40′) from the top horizontal listing; and third, locate the correct gauge number where the first two listings intersect. In this case, the correct gauge size is 10. Remember, the lower the wire gauge number, the larger the diameter of the wire.

In the spaces provided, list the proper gauge sizes for the following applications.

Circuit	Length	Amps		Wire Gauge
(a) Starter	5′	200		32a. _____
(b) Power convertible top	9′	30		32b. _____
(c) Head lights	5′	20		32c. _____
(d) Ignition switch feed	10′	40		32d. _____
(e) Brake lights	15′	20		32e. _____
(f) Tail lights	15′	10		32f. _____
(g) Horn	8′	20		32g. _____
(h) Dome light	9′	10		32h. _____

Table 11–1

Total Approx. Circuit Amperes 12 V	Wire Gauge (For Length in Feet)											
	3′	5′	7′	10′	15′	20′	25′	30′	40′	50′	75′	100′
1.0	18	18	18	18	18	18	18	18	18	18	18	18
1.5	18	18	18	18	18	18	18	18	18	18	18	18
2	18	18	18	18	18	18	18	18	18	18	16	16
3	18	18	18	18	18	18	18	18	18	18	14	14
4	18	18	18	18	18	18	18	18	16	16	12	12
5	18	18	18	18	18	18	18	18	16	14	12	12
6	18	18	18	18	18	18	16	16	16	14	12	10
7	18	18	18	18	18	18	16	16	14	14	10	10
8	18	18	18	18	18	16	16	16	14	12	10	10
10	18	18	18	18	16	16	16	14	12	12	10	10
11	18	18	18	18	16	16	14	14	12	12	10	8
12	18	18	18	18	16	16	14	14	12	12	10	8
15	18	18	18	18	14	14	12	12	12	10	8	8
18	18	18	16	16	14	14	12	12	10	10	8	8
20	18	18	16	16	14	12	10	10	10	10	8	6
22	18	18	16	16	12	12	10	10	10	8	6	6
24	18	18	16	16	12	12	10	10	10	8	6	6
30	18	16	16	14	10	10	10	10	10	6	4	4
40	18	16	14	12	10	10	8	8	6	6	4	2
50	16	14	12	12	10	10	8	8	6	6	2	2
100	12	12	10	10	6	6	4	4	4	2	1	1/0
150	10	10	8	8	4	4	2	2	2	1	2/0	2/0
200	10	8	8	6	4	4	2	2	1	1/0	4/2	4/0

© Cengage Learning 2012

33. Table 11–2 represents the relationship among wire length, current flow, and wire gauge number. Review Table 11–1 on wire gauge numbers and fill in the blank spaces in the table below with either "same," "increase," or "decrease."

For example, the top line asks, if the wire gauge number remains the same but the current flow is increased, what must be done to wire length to maintain a proper balance among length, current flow, and wire size? By reviewing the wire size table we can see that the wire length may be decreased.

Table 11–2

Wire Length	Current Flow (Amps)	Wire Gauge Number
Decrease	Increase	Same
Decrease	Same	
Same	Increase	
	Decrease	Same
Increase	Same	
Same	Decrease	

34. Determine the individual voltage drops in the circuit illustrated in Figure 11–18.

34. _____

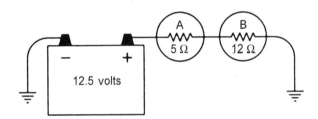

Figure 11–18

35. Determine the individual voltage drops in the circuit illustrated in Figure 11–19.

35. _____

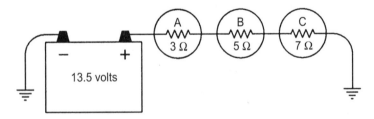

Figure 11–19

36. Determine the individual voltage drops in the circuit illustrated in Figure 11–20.

36. _____

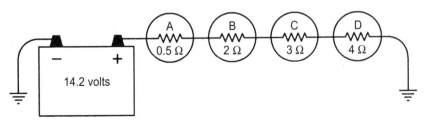

Figure 11–20

37. The starting motor circuit, illustrated in Figure 11–21, is being tested for excessive resistance. A technician conducts voltage drop test V-1, as shown, to evaluate the condition of the electrical circuit. Electrical specifications list a maximum allowable voltage drop for this entire insulated circuit as 0.5 V, with a maximum of 0.3 V for any single component. Test V-1, however, indicates a total voltage drop of 3.7 V. To pinpoint the problem, the technician conducts two additional tests. Test V-2 shows a voltage drop of 0.1 V and test V-3 shows a voltage drop of 0.2 V.

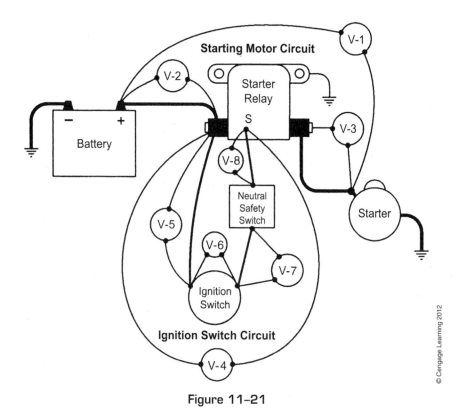

Figure 11–21

(a) What is the voltage drop across the starter relay?

37a. _____

(b) Which component in the starting motor circuit must be replaced?

37b. _____

(c) The technician also checks the ignition switch circuit for resistance using the same method as above. The maximum allowable voltage drop across the ignition switch insulated circuit is 0.5 V, with a maximum of 0.3 V for any single component. Voltage drop V-4 of the ignition switch circuit is measured at 3.8 V. Because it is over specifications, the technician proceeds with additional tests to locate the problem. If the voltage drop of V-5 is 0.1 V, voltage drop V-6 is 0.2 V, voltage drop V-7 is 0, and voltage drop V-8 is 0.1 V, what is the voltage drop of the neutral safety switch?

37c. _____

(d) Which component in the ignition switch circuit must be replaced?

37d. _____

38. Bill needs a replacement headlight switch. The dealership price is $86.20, while the salvage yard price is $8.99 if he removes it himself. For an additional $1.36 the switch is covered by a warranty. The state sales tax is 8.3%.

    (a) If Bill purchases the headlight switch from the salvage yard, what will be the final cost?

38a. _____

    (b) How much money would Bill save by getting the switch from the salvage yard rather than the dealership?

38b. _____

39. Fred bought a replacement audio system for his car on eBay for $845.00. The shipping cost was $26.85, and the insurance was an additional $8.37. He took it to an audio shop for installation. They took 3.7 hours to install the main module, the speakers, and the woofers. Their labor rate was $65.00/hour. Fred in turn sold his current audio system for $150.00. What was the final cost to Fred for this new system?

39. _____

40. After a hybrid car was driven 113,586 miles in 7 years, the battery failed. The new car dealer estimated the cost of a replacement battery at $3872.21. Shocked at the price the owner went online and located a do-it-yourself battery repair service. This company provided an instructional video, an instruction manual, and a toll-free number for $49.95. Using these materials and a few basic tools, the owner discovered some defective and corroded cells that had to be replaced and ordered three replacement cells at $49.95 each. He took these cells to an electrical technician, who charged $576.00 to make the repair.

    (a) What was the total cost of the repair?

40a. _____

    (b) What is the difference between the dealer estimate and the actual cost of the repair?

40b. _____

41. A major obstacle to producing fuel cell-powered vehicles is the high cost of platinum. If the fuel cell stack requires 14 grams of platinum and the unit cost of platinum is $1460 per ounce, what is the cost for the platinum for a single car? There are 28.35 grams per ounce.

41. _____

42. This automobile has come into the garage for electrical work. The labor rate is $89.50/hour and the sales tax rate is 6%. You will not find spark plugs listed in the *Parts and Time Guide*. Four XP5365 spark plugs will be used in this engine, at $6.32 per plug. Complete and total the repair order in Figure 11–22.

42. _____

2009 Fusion 2.3L

Quan	Part No.	Name of Part	Sales Amount

REPAIR ORDER

Name Bradford Bullett

Address 714 Broom Field

Phone No. 555-2236   **Date**

Speedometer Reading   **VIN No.**

Year, Make, and Model	License No. and State	Engine
2009 Fusion		2.3 L

Operation Number	SERVICE INSTRUCTIONS	TIME	AMOUNT
	Compression Test		
	R & R Spark Plugs		
	R & R Both Wiper Blades		
	R & R Mass Air Flow Sensor		
	R & R Both Oxygen Sensors		
	Clean Battery Terminals		
	Align Headlamps		
	R & R Oil Pressure Sending Unit		
	R & R Hybrid Battery		

Total Parts

Accessories	Amount		
		Total Labor	
		Total Parts	
		Materials	
		Accessories	
		Tires, Tubes	
		Outside Work	

Quan		@	
4	Spark Plugs	@ 6.32	
		@	
	Qts. Oil	@	
	Lbs. Grease	@	

Total Materials   **Total Accessories**

Total	
Tax	
TOTAL AMOUNT	

I hereby authorize the above repair work to be done along with the necessary material, and hereby grant you and/or your employees permission to operate the car, truck or vehicle herein described on streets, highways or elsewhere for the purpose of testing and/or inspection. An express mechanic's lien is hereby acknowledged on above car, truck or vehicle to secure the amount repairs thereto.

NOT RESPONSIBLE FOR LOSS OR DAMAGE TO CARS OR ARTICLES LEFT IN CARS IN CASE OF FIRE, THEFT OR ANY OTHER CAUSE BEYOND OUR CONTROL.

WORK AUTHORIZED BY _Bradford Bullett_ DATE

Figure 11–22

12

THE AUTOMOBILE DRIVE TRAIN

Objectives: After studying this chapter, you should be able to:

- Calculate torque, force, pressure, and gear ratio.
- Determine transmission gear ratios for external gears.
- Determine gear ratios for planetary gear sets.
- Determine total torque multiplication of the entire driveline.
- Describe the relationship between speed and torque.
- Explain the relationships of engine speed, transmission ratio, drive-axle ratio, and tire diameter to vehicle speed.

Engine power is transmitted through the drive train to the drive wheels of a vehicle. Many of the mechanical changes in today's automobile have occurred in the drive train after the introduction of the transaxle/front wheel drive and because of increased environmental concerns.

Regardless of the type of drive used, standard or automatic transmission, rear axle or front transaxle, proper servicing is essential. An important part of that servicing is the technician's identification of ratios and measurements in order to detect excessive wear. The problems in this chapter deal with these concerns as well as other matters that are related to the power train.

Torque is described as a twisting force and is measured in pound-feet (lb-ft) or in newton meters (N·m). Torque can be calculated by multiplying the force (in pounds or newtons) by the distance (in feet or meters) through which the force is exerted. (See Figure 12–1.)

A technician who is exerting a force of 10 pounds on a wrench 1 foot long is said to be exerting 10 lb-ft of torque. When the technician is working on small components, he or she may express the torque in pound-inches (lb-in.). The pound-feet of torque may be converted to pound-inches by multiplying by 12. In the example above, 10 lb-ft = 10 × 12 lb-in. or 120 lb-in.

Similarly, a technician who is exerting a force of 5 newtons on a wrench 0.3 meters long is said to be exerting 1.5 N·m of torque on the bolt.

■ EXAMPLE 12–1

How much torque is developed by a force of 35 pounds acting through a distance of 3 feet?

Solution

$$\text{Force} = 35 \text{ lb}$$
$$\text{Distance} = 3 \text{ ft}$$
$$\text{Torque} = \text{Force} \times \text{Distance}$$
$$= 35 \text{ lb} \times 3 \text{ ft}$$
$$35 \times 3 = 105$$

The torque is 105 lb-ft.

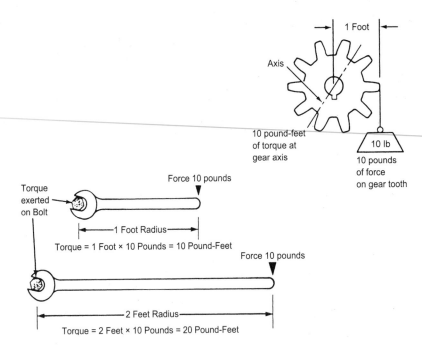

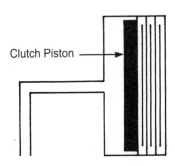

Figure 12-1

■ EXAMPLE 12-2

What is the torque developed by a force of 7 newtons acting through a distance of 1.3 meters?

Solution

Force = 7 N
Distance = 1.3 m
Torque = 7 N × 1.3 m
7 × 1.3 = 9.1

The torque is 9.1 N·m.

Clutches within an automatic transmission are applied by using oil pressure acting on a piston. The force that the piston exerts may be determined by multiplying the pressure by the area on which the pressure is acting.

The top of a piston, such as the clutch piston in Figure 12–2, is a circle. In order to determine the force a piston exerts, you must know the area of the piston.

Figure 12-2

To find the area of a circle (in this case, a piston), multiply the diameter of the piston by itself and then multiply this answer by 0.785. If A is the area of the piston and D is the piston's diameter, this can be written as $A = D^2 \times 0.785$ or $A = 0.785 \times D^2$.

■ EXAMPLE 12-3

Find the area of a piston with a diameter of 50 mm.

Solution

$$
\begin{aligned}
\text{Area} &= D^2 \times 0.785 \\
&= 50^2 \times 0.785 \\
&= 50 \times 50 \times 0.785 \\
&= 2500 \times 0.785 \\
&= 1962.5
\end{aligned}
$$

The area of this piston is 1962.5 mm^2.

■ EXAMPLE 12-4

If a piston has an area of 18 sq in. and the oil pressure is 160 pounds per square inch (psi), what force does the piston exert?

Solution

Area = 18 sq in.
Pressure = 160 psi
Force = Area × Pressure
= 18 sq in. × 160 psi
18 × 160 = 2880

The force is 2880 pounds.

■ EXAMPLE 12–5

A piston in an automatic transmission has an area of 125 cm^2. If the oil pressure on this piston is 975 kPa, what force is the piston exerting? (In the metric system, 1 kPa = 0.1 N/cm^2. Computations are easier if we change kPa to N/cm^2.)

Solution

$$1 \text{ kPa} = 0.1 \text{ N/cm}^2, \text{ so}$$
$$975 \text{ kPa} = 975 \times 0.1$$
$$= 97.5 \text{ N/cm}^2$$
$$\text{Force} = \text{Area} \times \text{Pressure}$$
$$= 125 \text{ cm}^2 \times 97.5 \text{ N/cm}^2$$
$$125 \times 97.5 = 12{,}187.5$$

The force is 12,187.5 N.

Total torque multiplication for the entire drive train in any given gear can be determined by multiplying the engine torque by both the transmission ratio for the given gear and the drive-axle ratio.

■ EXAMPLE 12–6

Determine the total torque multiplication for the entire drive train in first gear if the engine is developing 335 N·m of torque, the first gear has a ratio of 2.55:1, and the drive-axle ratio is 3.84:1.

Solution

Total torque multiplication

$$= 335 \text{ N·m} \times 3.84{:}1 \times 2.55{:}1$$
$$= 335 \text{ N·m} \times 3.84 \times 2.55$$
$$= 3280.32 \text{ N·m}$$

The total torque multiplication is 3280.32 newton meters.

(Remember, a ratio such as 3.84:1 is equivalent to $^{3.84}\!/_1$. But, $^{3.84}\!/_1 = 3.84$.)

■ EXAMPLE 12–7

Determine the gear ratio illustrated in Figure 12–3. The smaller gear with 10 teeth is the drive or input gear, and the larger gear with 20 teeth is the driven or output gear.

Solution

$$\text{Gear Ratio} = \frac{\text{Driven Gear}}{\text{Drive Gear}}$$
$$= \frac{20}{10}$$
$$= 2$$

Thus, the gear ratio is expressed as 2:1.

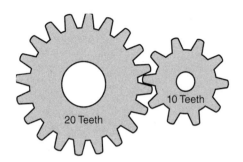

Figure 12–3

Automatic transmissions use planetary gear sets to provide several drive ratios. A simple planetary gear set is illustrated below in Figure 12–4. It has three basic parts: a sun gear (S) in the center; an internal, or ring, gear (I) around the outside; and planet pinion (P) gears located between the sun gear and the internal gear. The planet pinion gears are held together and supported by a planet carrier.

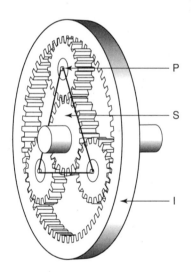

Figure 12–4

Gear ratio calculations for planetary gears are much different than those for external gears. This gear set usually has an input member, an output member, and a reaction or locked member. These three members have different functions for different ratios. Table 12–1 shows these different functions. The table also shows on the right the formulas used to determine the gear ratios for three operational modes, where R equals the gear ratio, N equals the number of gear teeth of a particular gear, and the subscript identifies the particular gear in question. The bottom formula shows N_i as a negative number because the gear motion is in reverse.

Table 12–1

	Input	Output	Locked	Ratio Formula
1st	S	P	I	$R = 1 + N_i/N_s$
2nd	I	P	S	$R = 1 + N_s/N_i$
Rev.	S	I	P	$R = -N_i/N_s$

■ EXAMPLE 12–8

Suppose that the gear tooth count is

Sun gear   = 28 teeth
Planet gears = 17 teeth
Internal gear = 62 teeth

Determine the first-gear ratio.

Solution

From Table 12–1 we see that the formula for the first-gear ratio is

$$R = 1 + \frac{N_i}{N_s}.$$

Here $N_s = 28$ and $N_i = 62$.

Thus, the desired ratio is

$$R = 1 + \frac{62}{28} = 1 + 2.21 = 3.21 : 1.$$

NAME _____ DATE _____ SCORE _____

1. **(a)** How many pound-feet of torque are developed by a force of 58 pounds
   acting through a distance of 2 feet?   1a. _____

   (b) How many pound-inches is this?   1b. _____

2. How many pound-feet of torque are developed by a force of 148 pounds acting
   through a distance of 6 inches?   2. _____

3. How much torque is developed by a force of 24 newtons acting through a distance of 0.75 meters?   3. _____

4. How much torque is developed by a force of 85 newtons acting through a distance of 0.8 meters?   4. _____

5. A standard transmission clutch is designed to handle 125% of the maximum
   engine torque. If a particular engine develops 360 pound-feet of torque and the
   clutch is designed to handle 125% of this figure, what is the maximum amount
   of torque that this clutch can handle?   5. _____

6. The clutch pedal leverage ratio is important because it reduces the clutch pedal
   effort that is required to operate the clutch. A ratio of 4:1 means that a pressure
   of 10 pounds of force on the pedal exerts a pressure of 40 pounds at the clutch
   pushrod, as shown in Figure 12–5. If a particular clutch has a leverage ratio of 7:1,
   what force is exerted at the clutch if a force of 23 pounds is applied on the pedal?   6. _____

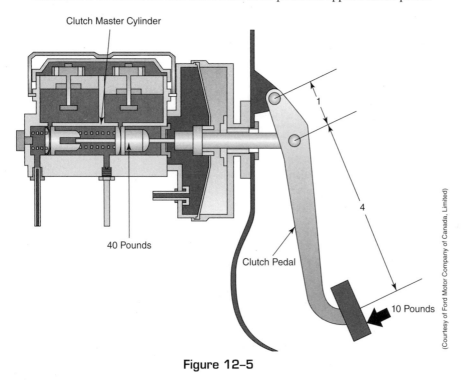

Clutch Master Cylinder

40 Pounds

Clutch Pedal

1

4

10 Pounds

(Courtesy of Ford Motor Company of Canada, Limited)

Figure 12–5

7. Some clutches have weights cast on the outer dimension of the release levers. At high speeds, these weights, acting through centrifugal force, add additional force to the clutch pressure plate to provide an increase in the torque capacity of the clutch. If the clutch plate force is 648 pounds at rest and increases 21% at higher speeds, what is the maximum force that the clutch exerts at high speeds?

7. _____

8. The clutch plate force on a centrifugally assisted clutch is 2950 newtons at rest and increases 23% at higher speeds.

   (a) What is the maximum number of newtons exerted by the clutch at higher speeds?

   8a. _____

   (b) Express the above answer in pounds.

   8b. _____

9. Determine the difference in angle between the transmission and the rear-axle differential housing. The transmission angle is ¾ degree and the differential angle is 1⅝ degrees.

   9. _____

10. The transmission angle is adjusted by using shims (spacers) under the transmission extension housing. If one shim changes the transmission angle ⅛ degree, how many shims must be added to change the transmission angle 1¼ degrees?

    10. _____

11. Engine torque may be increased or decreased at the transmission output shaft by using different gear ratios in the transmission. A ratio greater than 1 results in a torque multiplication or increase, and a ratio less than 1 results in a torque reduction. If an engine is developing 348 pound-feet of torque, what is the torque at the transmission output shaft with a transmission ratio of 2.5 : 1?

    11. _____

12. If engine torque is 348 pound-feet, what is the transmission output torque in each gear from the following three-speed transmission ratios?

    (a) First gear 2.9 : 1

    12a. _____

    (b) Second gear 1.6 : 1

    12b. _____

    (c) Third gear 1 : 1

    12c. _____

13. If engine torque is 142 N · m, what is the transmission output torque in each gear from the following four-speed transmission ratios?

    (a) First gear 2.53 : 1

    13a. _____

    (b) Second gear 1.87 : 1

    13b. _____

    (c) Third gear 1.35 : 1

    13c. _____

    (d) Fourth gear 1 : 1

    13d. _____

    (e) Reverse 2.88 : 1

    13e. _____

14. Recently, overdrive transmissions have been used to improve fuel economy. When a vehicle is using an overdrive unit, the engine speed is reduced and at the same time the same vehicle speed is maintained. To determine engine speed reduction when using an overdrive unit, multiply engine speed by the transmission overdrive ratio.

    (a) If the engine speed is 2800 rpm while the transmission is operating at a 1 : 1 ratio, how fast will the engine be operating if the transmission is shifted into a 0.7 : 1 overdrive?

    14a. _____

    (b) What is the engine speed difference between the two ratios?

    14b. _____

15. What percent was engine speed reduced in Problem 14?

    15. _____

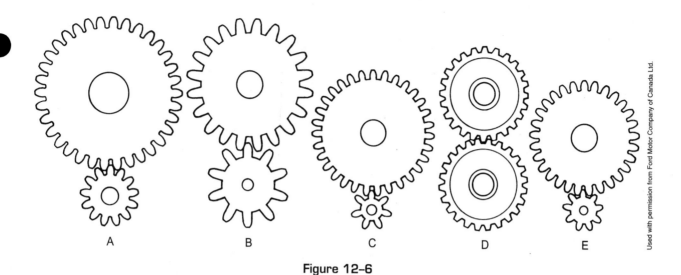

Figure 12–6

16. In the spaces provided, list the ratio for each gear set shown in Figure 12–6. Assume the bottom gear is the drive or input gear.

16a. _____

16b. _____

16c. _____

16d. _____

16e. _____

17. An engine is developing 225 pound-feet of torque. If the rear (drive)-axle ratio is 3.72 : 1, what is the final output torque for the entire drive train in each of the following transmission ratios?

   (a) First gear 2.63 : 1

   (b) Second gear 1.60 : 1

   (c) Third gear 1 : 1

   (d) Fourth gear 0.7 : 1

17a. _____

17b. _____

17c. _____

17d. _____

18. Vehicle A has a transmission high-gear ratio of 1 : 1 with a drive-axle ratio of 3.73 : 1. Vehicle B has a transmission high-gear ratio of 0.7 : 1 and a drive-axle ratio of 4.56 : 1.

   (a) What is the overall ratio of vehicle A?

   (b) What is the overall ratio of vehicle B?

   (c) If a lower overall ratio indicates better fuel economy, which of these two vehicles achieves the better fuel economy?

18a. _____

18b. _____

18c. _____

19. Automatic transmissions also have gear ratios. The transmission is coupled to the engine through a torque converter. Maximum torque multiplication of the converter is approximately 2.2 : 1. If the low-gear ratio is 2.5 : 1 and the drive-axle ratio is 2.89 : 1, what is the overall ratio of the power train in low gear during initial acceleration?

19. _____

20. Automatic transmissions have oil coolers to limit maximum oil temperature. If the transmission oil temperature prior to cooling is 110°C and 75°C after cooling, how many degrees were lost to the cooling system?

20. _____

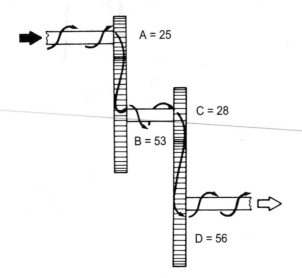

Figure 12-7

21. A multiple gear set is illustrated in Figure 12–7. Determine the A:B ratio, the C:D ratio, and the overall ratio for the complete gear set.

21. A:B _____

C:D _____

Total _____

22. A piston from an automatic transmission is illustrated in Figure 12–8. The shaded portion is the actual piston surface on which the oil pressure is acting. If the oil pressure is 170 psi, determine the force that is developed by this piston.

Outside diameter area

Inside diameter area

Total active piston area

Force developed by piston

22. _____

_____

_____

_____

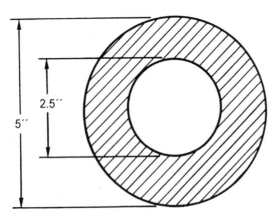

Figure 12-8

23. If an automatic transmission clutch piston has an area of 16 square inches and the oil pressure is 178 pounds per square inch, what force does the piston exert?

23. _____

24. If an automatic transmission clutch piston has an area of 100 cm^2 and the oil pressure is 1214 kPa, what force does the piston exert?

24. _____

25. When checking the transmission turbine shaft end play, we find the end play to be 0.076 inch. The specified end play is 0.003 to 0.024 inch. To reduce this end play, we must install a selective fit thrust washer. If washers come in five thicknesses, 0.020, 0.040, 0.060, 0.080, and 0.100 inch, which washer must be used to bring the end play within limits?

25. _____

26. The transmission input shaft end play on another car is 2.2 mm. The specified end play is 0.07–0.6 mm. Selective fit washers come in five metric thicknesses: 0.5, 1.0, 1.5, 2.0, and 2.5 mm. Which washer must be used to bring the end play within limits?

26. _____

27. A transmission oil-pressure check shows that the pressure is 68 pounds low. The pressure adjuster increases the pressure 16 pounds for each complete turn. Exactly how many turns of the adjuster are required to increase the pressure by 68 pounds?

27. _____

28. A transmission oil-pressure check on another car shows that the pressure is 3 kPa low. The pressure adjuster increases the pressure 0.8 kPa for each complete turn. How many turns are required to increase the pressure by 3 kPa?

28. _____

29. A transmission has an oil-pressure reading of 149 lb/in.^{2} when the shift selector is in Drive. The pressure increases to 245 lb/in.^{2} when the car is shifted into Reverse. How much is the pressure increased in Reverse?

29. _____

30. A transmission has an oil-pressure reading of 1027 kPa when the shift selector is in Drive and 1689 kPa when it is shifted into Reverse. How much is the pressure changed in Reverse?

30. _____

31. When a vehicle is cruising at highway speeds, the transmission oil pump pumps 92 liters per minute. How many liters are pumped during a half-hour trip?

31. _____

32. The ring gear on the transmission torque converter has 136 teeth, and the starter drive pinion gear has 8 teeth. What is the gear ratio between these two gears?

32. _____

33. Each gear selection in the transmission provides an overall transmission ratio (OTR). This overall transmission ratio can be calculated by using one of the following formulas:

$$OTR = \frac{\text{Front Counter Gear}}{\text{Clutch Shaft Gear}} \times \frac{\text{Last Gear}}{\text{Last Counter Gear in Use}}$$

or

$$OTR = \frac{\text{Front Driven Gear}}{\text{Front Drive Gear}} \times \frac{\text{Rear Driven Gear}}{\text{Rear Drive Gear}}$$

(a) Using the transmission illustrated in Figure 12–9, determine the OTR for each gear position. The number of gear teeth are listed for all required gears.

33a. 1st gear _____

2nd gear _____

3rd gear _____

4th gear _____

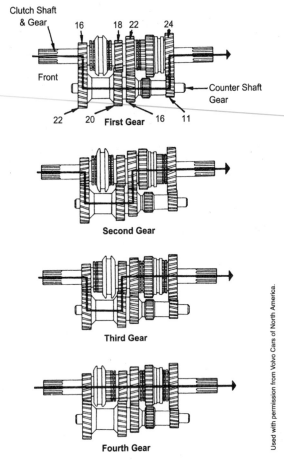

Figure 12–9

(b) If this vehicle is using a drive-axle ratio of 3.72 : 1, determine the total drive line ratio in each gear.

33b. 1st gear _____

2nd gear _____

3rd gear _____

4th gear _____

34. For racing applications, it is sometimes helpful to be able to determine the engine revolutions per minute under certain conditions. Engine revolutions per minute may be calculated by using the following formula:

$$\text{rpm} = \frac{168 \times R \times \text{mph}}{r}$$

where rpm = revolutions per minute of the engine, 168 = a constant factor, R = overall gear ratio, mph = vehicle ground speed in miles per hour, and r = rolling radius in inches of loaded-driving tire.

(a) Find the engine's rpm if $R = 2.87$, mph = 55, and $r = 12.5$ inches.

34a. _____

(b) What are the rpm for this same engine if it is traveling at 45 mph?

34b. _____

(c) What are the rpm at 75 mph?

34c. _____

35. Under certain conditions it may be helpful to calculate miles per hour at a given revolutions per minute. To do this, you should use the following formula:

$$\text{mph} = \frac{\text{rpm} \times r}{168 \times R}$$

where the letters mean the same as those in Problem 34.

(a) What are the mph for an engine going at 4000 rpm if $R = 4.11$ and $r = 12.5$ inches?

35a. _____

(b) What are the mph for this same engine if it is turning at 3500 rpm?

35b. _____

(c) What are the mph at 4500 rpm?

35c. _____

36. If a driver traveled 3786 kilometers in 48 hours, what was the average speed in kilometers per hour for the entire trip?

36. _____

As engine torque is passed through the transmission, it may increase, decrease, or remain the same, depending on the transmission ratio. (See Example 12–6.)

37. A customized sports car has these specifications:

Engine torque (lb-ft @ rpm) . . . . 412@4500

Gear ratios
First 2.97:1
Second 2.07:1
Third 1.43:1
Fourth 1.00:1
Fifth 0.80:1
Sixth 0.62:1
Drive-axle ratio 3.23:1
Tire diameter . . . . 25.66 inches

(a) When the engine is operating at 4500 rpm, how much torque is delivered to the rear wheels in each gear?

37a. 1st gear _____

2nd gear _____

3rd gear _____

4th gear _____

5th gear _____

6th gear _____

(b) How fast is the engine operating at 65 mph in sixth gear?

37b. _____

(c) What is the theoretical top speed of this automobile in each gear at the engine speeds indicated?

First @ 6250 rpm   **37c.** _____

Second @ 6250 rpm   _____

Third @ 6250 rpm   _____

Fourth @ 6250 rpm   _____

Fifth @ 5250 rpm   _____

Sixth @ 4250 rpm   _____

In an effort to make a motor home easier to operate under a wider variety of driving conditions, the owner installed a gear splitter on the back of the transmission. A gear splitter is a gear set which enables the driver to split or divide a given gear ratio into two different ratios. The splitter has two operational ratios, 1:1 and 0.78:1. As an example, if the transmission is operating in a 1:1 ratio, the splitter allows the driver to select either a 0.78:1 ratio or a 1:1 ratio.

38. An older motor home is equipped with a three-speed automatic transmission. The transmission ratios are 2.48:1 for first gear, 1.48:1 for second gear, and 1:1 for third. The drive-axle ratio is 4.88:1.

   (a) Determine the overall final drive ratio possibilities for each gear when the splitter is used. In (i), give the drive ratio if a 1:1 ratio is used, and in (ii), give the drive ratio if a 0.78:1 ratio is used.

   First gear   **38a.**(i) _____
   (ii) _____

   Second gear   (i) _____
   (ii) _____

   Third gear   (i) _____
   (ii) _____

   (b) What is the theoretical top speed of the motor home in each gear at the engine speeds indicated? The tire diameter is 31 inches.

   First gear @ 3500 rpm   **38b.**(i) _____
   (ii) _____

   Second gear @ 3500 rpm   (i) _____
   (ii) _____

   Third gear @ 3500 rpm   (i) _____
   3000 rpm   (ii) _____

39. In Example 12–8, the gear tooth count for planetary gears is given as

   Sun gear   = 28 teeth
   Planet gears = 17 teeth
   Internal gear = 62 teeth

   (a) Determine the second-gear ratio.   **39a.** _____

   (b) Determine the reverse-gear ratio.   **39b.** _____

40. The gear tooth count for the planetary gears are

   Sun gear   = 27 teeth
   Planet gears = 33 teeth
   Internal gear = 60 teeth

   (a) Determine the first-gear ratio.   **40a.** _____

   (b) Determine the second-gear ratio.   **40b.** _____

   (c) Determine the reverse-gear ratio.   **40c.** _____

41. Dual-clutch transmissions can achieve a 7–9% fuel economy advantage over the older four-speed automatic transmissions. A 2010 vehicle equipped with a four-speed automatic transmission averages 29.8 mpg. The new 2011 model vehicle is basically the same except for the new dual-clutch transmission. What fuel economy range might this vehicle expect to achieve?

41. _____

42. This automobile has come in for some drive train service as well as other miscellaneous work. The labor rate is $89.50/hour and the sales tax rate is 8.5%. Complete and total the repair order in Figure 12–10.

42. _____

2009 Fusion
3.0 L engine
Automatic transmission
AWD

Quan	Part No.	Name of Part	Sales Amount	
		Total Parts		

REPAIR ORDER

Name Shaquille Lincoln

Address Fareway Dr.

Phone No. 555-2645   **Date**

Speedometer Reading   **VIN No.**

Year, Make, and Model	License No. and State	Engine
09 Fusion	Automatic AWD, 5 Speed	3.0 L

Operation Number	SERVICE INSTRUCTIONS	TIME	AMOUNT
	Transaxle Diagnosis		
	R & I Transaxle		
	R & R Drive Plate		
	R & R Right Front Axle		
	R & R Differential Pinion Seal		
	R & R Viscous Coupler		
	Trans Fluid Change		
	R & R Trans Filter		

Accessories		Amount	Total Labor	
@			Total Parts	
@			Materials	
9 Qts. Oil ATF @ 6.16			Accessories	
Lbs. Grease @			Tires, Tubes	
Total Materials			Outside Work	
Total Accessories			Total	
			Tax	
			TOTAL AMOUNT	

I hereby authorize the above repair work to be done along with the necessary material, and hereby grant you and/or your employees permission to operate the car, truck or vehicle herein described on streets, highways or elsewhere for the purpose of testing and/or inspection. An express mechanic's lien is hereby acknowledged on above car, truck or vehicle to secure the amount repairs thereto.

NOT RESPONSIBLE FOR LOSS OR DAMAGE TO CARS OR ARTICLES LEFT IN CARS IN CASE OF FIRE, THEFT OR ANY OTHER CAUSE BEYOND OUR CONTROL.

WORK AUTHORIZED BY _Shaquille, Lincoln_ DATE

Figure 12–10

13

THE AUTOMOBILE CHASSIS

Objectives: After studying this chapter, you should be able to:

- Calculate the area of a circle and the volume of a cylinder in order to determine hydraulic force.
- Apply angular relationships when dealing with alignment angles and vehicle cornering.
- Explain the hydraulic relationship of pressure, area, and force in brake operation.
- Explain the relationship between tire diameter, gear ratio, and speedometer accuracy.
- Describe the chassis dynamics of cornering force and weight transfer.
- Use brake specifications to determine serviceability of worn brake components.
- Read and interpret alignment adjustment angles.

Most of the material that we have discussed up to this point has been concerned with starting and maintaining vehicle movement. This includes both short trips around town and longer interstate travel.

Chapter 13 focuses on the vehicle control systems that have an important role in vehicle comfort and safety. The following problems relate to the considerations of chassis design as well as to the mathematical situations that service technicians encounter in day-to-day activities.

The brake and power-steering systems depend on hydraulic action to operate. The hydraulic relationship is expressed in the formula Force = Pressure × Area. Hydraulic force (pounds or kilograms) is calculated by multiplying pressure (pounds per square inch, psi, or kilopascals, kPa) by area (square inches or square centimeters). Thus, the equation for force is $F = PA$.

You may calculate pressure by dividing force by area ($P = {}^F\!/_A$). You may determine the area by dividing the force by the pressure ($A = {}^F\!/_P$). You might want to use an aid such as Figure 13–1 to help you solve problems that involve area, force, and pressure. Use Figure 13–1 in the same way in which you used Figure 11–1 to solve Ohm's law problems.

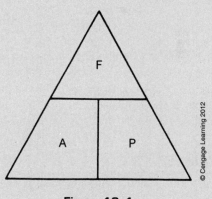

© Cengage Learning 2012

Figure 13–1

■ EXAMPLE 13–1

A piston is exerting a force of 8792 pounds. If the piston has an area of 14 in.^{2}, what is the pressure on the piston?

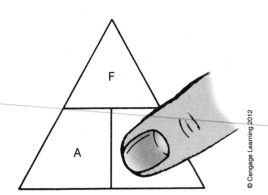

Figure 13–2

Solution

You are trying to find the pressure P. Cover the P in Figure 13–1 with your finger as shown in Figure 13–2. The two letters that are not covered, F and A, are in the position shown in Figure 13–2. The relationship, F/A, indicates that F must be divided by A.

$$F = 8792 \text{ lbs}$$
$$A = 14 \text{ in.}^2$$
$$P = \frac{F}{A} = \frac{8792}{14}$$
$$= 628$$

The pressure is 628 psi.

You might review Examples 12–4 and 12–5 for other problems that use pressure, area, and force.

Because most hydraulic components are circular in nature, it is important to know how to calculate the area of a circle such as the end of a piston. To find the area of a circle, multiply the diameter by itself and then multiply this answer by 0.7854 ($A = D^2 \times 0.7854$).

■ EXAMPLE 13–2

Find the area of a circle with a diameter of 4 cm.

Solution

$$\text{Area} = D^2 \times 0.7854$$
$$= 4^2 \times 0.7854$$
$$= 4 \times 4 \times 0.7854$$
$$= 16 \times 0.7854$$
$$= 12.57$$

The area is 12.57 cm^2.

If you know D^2 and want D, then take the square root of D^2, written $\sqrt{D^2}$.

■ EXAMPLE 13–3

The area of a circle is 38.465 cm^2. What is its diameter?

Solution

Again we use the formula Area $= D^2 \times 0.7854$. We have

$$38.465 = D^2 \times 0.7854$$
$$D^2 \times 0.7854 = 38.465.$$
$$D^2 = 38.465 \div 0.7854 = 49.$$
$$D^2 = 49, D = \sqrt{49} = 7.$$

So, the diameter of this circle is 7 cm.

■ EXAMPLE 13–4

If a force of 3140 pounds is created with a pressure of 1000 psi, determine the diameter of the piston being used in this cylinder.

Solution

$$F = 3140 \text{ lb}$$
$$P = 1000 \text{ psi}$$
$$A = \frac{F}{P}$$
$$A = \frac{3140}{1000}$$
$$= 3.14$$

We also know that

$$A = D^2 \times 0.7854$$
$$3.14 = D^2 \times 0.7854$$
$$\frac{3.14}{0.785} = D^2 \times \frac{0.7854}{0.7854}$$
$$4 = D^2$$
$$D = 2$$

So, the diameter of this piston is 2 inches.

The volume of a cylinder is calculated by multiplying the circular area by the length of the cylinder.

■ EXAMPLE 13–5

Find the volume of a cylinder 5 cm in diameter and 12 cm long.

Solution

To find the volume V, multiply the diameter D by itself, multiply that answer by 0.7854, and then multiply by the length L of the cylinder.

$$V = D^2 \times 0.7854 \times L$$
$$= 5^2 \times 0.7854 \times 12$$
$$= 5 \times 5 \times 0.7854 \times 12$$
$$= 25 \times 0.7854 \times 12$$
$$= 235.6$$

The volume of this cylinder is 235.6 cm^3.

To determine how much force is being applied to the piston in the master cylinder, multiply the brake-pedal effort that the driver applies by the brake-pedal lever ratio. The pressure that the master cylinder develops is transmitted to the wheel cylinders by hydraulic brake lines. This pressure, acting on the wheel cylinder pistons, causes a force that moves the brake shoes and mechanical braking system.

■ EXAMPLE 13–6

A driver puts a force of 360 newtons on a brake pedal. The brake-pedal ratio is 8 : 1 and the master cylinder diameter is 3 cm. The diameter of the front-wheel caliper piston is 3.2 cm. What is the force that the front-wheel caliper piston develops?

Solution

There are several steps to solving this problem. First, we must find the force that acts on the master cylinder piston as a result of the force that the driver exerts on the brake pedal times the brake-pedal ratio.

Force on master cylinder
$$= \text{(force on brake pedal)} \times \text{(brake-pedal lever ratio)}$$
$$= 360 \text{ N} \times 8$$
$$= 2880 \text{ N}$$

Second, we must find the area of the master cylinder piston.

$$\text{Area of master cylinder piston} = D^2 \times 0.7854$$
$$= 3^2 \times 0.7854$$
$$= 7.069 \text{ cm}^2$$

Third, we find the pressure developed within the master cylinder.

$$P = F \div A$$
$$= 2880 \text{ N} \div 7.069 \text{ cm}^2$$
$$= 407.41 \text{ N/cm}^2$$

Next, find the area of the front-wheel caliper piston. The diameter of the front-wheel caliper piston is 3.2 cm.

$$\text{Area} = D^2 \times 0.7854$$
$$= (3.2)^2 \times 0.7854$$
$$= 10.24 \times 0.7854$$
$$= 8.04 \text{ cm}^2$$

Finally, we find the force that results from the pressure within the master cylinder acting on a front-wheel caliper piston.

$$F = P \times A$$
$$= 407.41 \text{ N/cm}^2 \times 8.04 \text{ cm}^2$$
$$= 3275.6 \text{ N}$$

So, the answer to the problem is 3275.6 newtons.

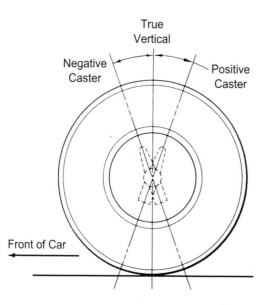

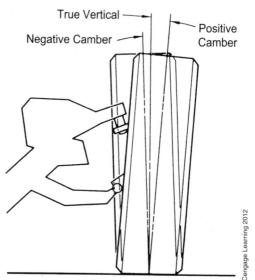

Figure 13–3

During a wheel alignment, the caster and camber angles of the front wheels are adjusted. These alignment angles are measured in degrees or parts of degrees from a true vertical. (See Figure 13–3.) The parts of degrees may be either in fractions or decimals.

The toe-in setting, which is also part of a wheel alignment, ensures that the front wheels are both rolling parallel to each other and in a straight-ahead position when driving down a straight road. (See Figure 13–4.) The toe adjustment may be measured in inches or degrees and is usually adjusted to be toed-in slightly when the vehicle is at rest. Readings before adjustments, however, may vary and measure either as toe-in or toe-out.

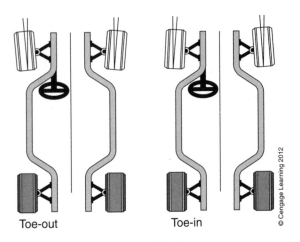

Toe-out   Toe-in

© Cengage Learning 2012

Figure 13–4

■ EXAMPLE 13–7

The camber reading of a wheel on a Buick is −.125°. How large an adjustment must be made to bring the wheel within specifications?

Solution

The specifications are in Appendix A. The specifications for the camber are +.60° ± .50°. This means that the camber can be between +.60° − .50° = +.10° and +.60° + .50° = +1.10°. The adjustment can be found by determining the difference between the specifications and the reading. Because the specification of +.10° is closer to the reading of −.125°, this will give the minimum adjustment needed.

$$\text{Adjustment} = .10° - (-.125°)$$
$$= .10° + .125°$$
$$= .225°$$

An adjustment of at least .225° must be made.

Tire size information is indicated on the side of each tire. A tire sized as P205/70R14 is designated for passenger car use, has a section width of 205 mm, has an aspect ratio (the height-to-width relationship) of 70%, has radial construction, and fits a 14-inch wheel rim.

To determine the actual diameter in inches, use the following formula:

$$\text{Tire Diameter} = \frac{\text{Section Width} \times \text{Aspect Ratio}}{1270} + \text{Rim Diameter}$$

■ EXAMPLE 13–8

Determine the actual diameter in inches of the P205/70R14 tire described above.

Solution

Here the section width is 205, the aspect ratio is 70, and the rim diameter is 14. We substitute these in the above formula.

$$\text{Tire Diameter} = \frac{\text{Section Width} \times \text{Aspect Ratio}}{1270} + \text{Rim Diameter}$$
$$= \frac{205 \times 70}{1270} + 14$$
$$= 11.3 + 14$$
$$= 25.3 \text{ inches}$$

Sometimes larger tires are installed on a vehicle to increase the ground clearance. Whenever a switch like this is done it also affects the ability of the vehicle to accelerate or climb steep grades by changing the effective drive ratio. This may be calculated by using the following formula:

$$\text{New Effective Drive Ratio} = \frac{\text{Old Tire Diameter}}{\text{New Tire Diameter}} \times \text{Drive-Axle Ratio}$$

■ EXAMPLE 13–9

A truck was equipped with LT215/75R15 tires. In an effort to gain added ground clearance the owner installs LT235/85R16 tires and matching rims. The truck has a drive-axle ratio of 3.54 : 1. What is the new effective drive ratio?

Solution

To solve this problem, first determine the diameter of the original tires and the new replacement tires.

$$\text{Tire Diameter} = \frac{\text{Section Width} \times \text{Aspect Ratio}}{1270}$$
$$+ \text{ Rim Diameter}$$

$$\begin{aligned}\text{Original} \atop \text{tire diameter} &= \frac{215 \times 75}{1270} + 15 \\ &= 27.7 \text{ inches}\end{aligned}$$

$$\begin{aligned}\text{New tire diameter} &= \frac{235 \times 85}{1270} + 16 \\ &\approx 31.7 \text{ inches}\end{aligned}$$

Now insert these tire diameter values into the new effective drive ratio formula.

$$\begin{aligned}{\text{New Effective} \atop \text{Drive Ratio}} &= \frac{\text{Old Tire Diameter}}{\text{New Tire Diameter}} \\ &\quad \times \text{Drive-Axle Ratio} \\ &= \frac{27.7}{31.7} \times 3.54 \\ &\approx 3.09\end{aligned}$$

The truck has a new effective drive ratio of 3.09.

If, for some reason, you need to determine which drive-axle ratio is needed to bring back original performance levels, use the following formula:

$$\begin{aligned}{\text{New} \atop \text{Drive} \atop \text{Ratio}} &= \frac{\text{New Tire Diameter}}{\text{Old Tire Diameter}} \times \text{Original Drive Ratio}\end{aligned}$$

■ EXAMPLE 13–10

After driving the truck in Example 13–9, with the new larger tires, the owner liked the look and the added ground clearance but was not satisfied with the performance that resulted from the new effective drive ratio. Determine what drive-axle ratio is needed to bring back the original performance level.

Solution

$$\begin{aligned}\text{New Drive Ratio} &= \frac{\text{New Tire Diameter}}{\text{Old Tire Diameter}} \\ &\quad \times \text{Drive-Axle Ratio} \\ &= \frac{31.7}{27.7} \times 3.54 \\ &\approx 4.05\end{aligned}$$

The truck needs a new drive ratio of 4.05 to return to its original performance level.

Whenever tire sizes are changed, the accuracy of the speedometer is affected. You can determine speedometer error by using the following formula:

$$\text{True Speed} = \frac{\text{New Tire Diameter}}{\text{Old Tire Diameter}} \times \text{Indicated Speed}$$

■ EXAMPLE 13–11

The speedometer of the vehicle in Example 13–10 indicates that it is going 55 mph. Determine the speedometer error.

Solution

For this vehicle, the new tire diameter is about 31.7 in., the old tire diameter is about 27.7 in., and the indicated speed is 55 mph.

$$\begin{aligned}\text{True Speed} &= \frac{\text{New Tire Diameter}}{\text{Old Tire Diameter}} \times \text{Indicated Speed} \\ &= \frac{31.7}{27.7} \times 55 \\ &\approx 62.9\end{aligned}$$

The vehicle's true speed is about 63 mph.

If you wish to determine the equivalent indicated speed that is equal to a true speed, use the following formula:

$$\begin{aligned}{\text{Equivalent} \atop \text{Indicated Speed}} &= \frac{\text{Old Tire Diameter}}{\text{New Tire Diameter}} \times \text{True Speed}\end{aligned}$$

■ EXAMPLE 13–12

What should the speedometer of the vehicle in Example 13–11 indicate if the vehicle is actually going 55 mph?

Solution

Here again, the new tire diameter is about 31.7 in., and the old tire diameter is about 27.7 in. The true speed is 55 mph.

$$\begin{aligned}{\text{Equivalent} \atop \text{Indicated} \atop \text{Speed}} &= \frac{\text{Old Tire Diameter}}{\text{New Tire Diameter}} \times \text{Indicated Speed} \\ &= \frac{27.7}{31.7} \times 55 \\ &\approx 48.1\end{aligned}$$

If this vehicle wants to have a true speed of 55 mph then it should travel so that the speedometer reads 48 mph.

Sometimes it becomes necessary to check the accuracy of your speedometer. Some highways have specially marked speedometer check distances. As an alternative you can use the mile markers found on interstate highways. Obviously you must carefully and accurately maintain a steady speed during the check. To perform the check you must time your speed over a distance of one mile. To determine your true speed use the following formula.

$$\text{True Speed} = \frac{3600}{\text{Seconds per Mile}}$$

■ EXAMPLE 13–13

If your timed mile took 68 seconds, determine the true speed.

Solution

$$\begin{aligned}
\text{True Speed} &= \frac{3600}{\text{Seconds per Mile}} \\
&= \frac{3600}{68} \\
&\approx 52.94
\end{aligned}$$

Your true speed was about 53 mph.

To determine percent error, use the following formula.

$$\text{Percent Error} = \frac{\text{True Speed} - \text{Indicated Speed}}{\text{True Speed}} \times 100$$

■ EXAMPLE 13–14

If you discover that at an indicated speed of 55 mph you are actually traveling at a true speed of 63 mph, what is the percent error?

Solution

$$\begin{aligned}
\text{Percent Error} &= \frac{\text{True Speed} - \text{Indicated Speed}}{\text{True Speed}} \times 100 \\
&= \frac{63 - 55}{63} \times 100 \\
&\approx 12.70
\end{aligned}$$

The error is about 12.7%

Cornering force, also referred to as lateral acceleration, is sometimes used to characterize vehicular performance around corners. It is usually expressed in terms of its relationship to gravity, or *g*s. One g is equal to the force of gravity, two *g*s to twice the force of gravity, and so on. It is unusual for a street vehicle to exceed one *g* of cornering force.

To calculate cornering force, use the following formulas, in which velocity is expressed in feet per second and the radius or diameter of the turn is given in feet.

$$\text{Cornering force} = \frac{\text{Velocity}^2}{32 \times \text{Radius}}$$

If the diameter, rather than the radius, of the turn is given, then use the relation

$$\text{Radius} = \frac{\text{Diameter}}{2}$$

or

$$\text{Cornering Force} = \frac{\text{Velocity}^2}{16 \times \text{Diameter}}$$

To convert velocity in miles per hour to velocity in feet per second (fps), use either of the following two formulas:

$$\begin{aligned}
\text{Velocity (fps)} &= \text{velocity (mph)} \times \frac{22}{15} \\
&= \text{velocity (mph)} \times 1.467
\end{aligned}$$

■ EXAMPLE 13–15

A vehicle maintains a steady speed of 62 mph around a circle that has a diameter of 525 feet. What is the cornering force generated?

Solution

We are given the diameter of the turning circle, so we use the formula

$$\text{Cornering Force} = \frac{\text{Velocity}^2}{16 \times \text{Diameter}}$$

The diameter is 525 ft.

We must change the velocity from miles per hour to feet per second.

$$\begin{aligned}
\text{Velocity (fps)} &= 62 \times 1.467 \\
&= 90.954
\end{aligned}$$

The velocity is about 90.954 fps.

Now we have all the information to determine the cornering force.

$$\begin{aligned}
\text{Cornering Force} &= \frac{\text{Velocity}^2}{16 \times \text{Diameter}} \\
&= \frac{90.954 \times 90.954}{16 \times 525} \\
&\approx 0.98
\end{aligned}$$

The cornering force is about 0.98 g.

Another method of determining cornering force is through the use of a skid pad. This is a flat safe area with a large circle marked on it. Across the circle is a marked start/finish line. Each time the vehicle passes the start line it is accelerated to a slightly faster maintained speed. This process continues until the vehicle loses traction. The previous lap speed is then considered its maximum. The formula for this calculation is:

$$\text{Cornering Force} = \frac{1.23 \times \text{Radius}}{\text{Time}^2}$$

where the radius is in feet and the time in seconds.

■ EXAMPLE 13–16

A vehicle completes a lap on a skid pad in 25 seconds. The pad has a radius of 425 feet. What is the cornering force generated?

Solution

$$\text{Cornering Force} = \frac{1.23 \times \text{Radius}}{\text{Time}^2}$$

$$= \frac{1.23 \times 425}{25 \times 25}$$

$$= \frac{522.75}{625}$$

$$= 0.8364$$

The cornering force is about 0.84 g.

NAME _____ DATE _____ SCORE _____

1. What is the area of the end of a piston that has a diameter 1.75 in.?

     1. _____

2. What is the area of the end of a piston that has a diameter 4.5 cm?

     2. _____

3. What is the volume of a wheel cylinder that is 1.25 inches in diameter and 3.25 inches long?

     3. _____

4. What is the volume of a wheel cylinder that is 3.2 cm in diameter and 8.5 cm long?

     4. _____

5. (a) How many cubic centimeters of brake fluid are needed to fill a master cylinder bore that has a 2.4-cm bore and is 17.5 cm long?

     5a. _____

  (b) How many milliliters (mL) is this?

     5b. _____

  (c) How many liters (L) is this?

     5c. _____

6. How many cubic inches of brake fluid are needed to fill a master cylinder bore that has a $^{15}/_{16}$-inch bore 7 inches long?

     6. _____

7. Cylinder A has a 1.250-inch bore and is 14.5 inches long; cylinder B has a 1⅞-inch bore and is 6⁹⁄₁₆ inches long.

  (a) What is the volume of cylinder A?

     7a. _____

  (b) What is the volume of cylinder B?

     7b. _____

  (c) What is the difference in volume between cylinder A and cylinder B?

     7c. _____

8. Cylinder C has a 3.3-cm bore and is 37 cm long; cylinder D has a 4.8-cm bore and is 17 cm long.

  (a) What is the volume of cylinder C?

     8a. _____

  (b) What is the volume of cylinder D?

     8b. _____

  (c) What is the difference in volume between cylinder C and cylinder D?

     8c. _____

9. The master cylinder is filled with brake fluid and has a piston in it with a diameter of 1 inch. The driver applies a force of 200 pounds to the piston. How much pressure is developed in this system? Assume there is no fluid flow from the master cylinder.

     9. _____

10. Determine the force that is developed by the front- and rear-wheel cylinder pistons in a drum-brake system as shown in Figure 13–5. The driver-pedal force is 80 pounds. The brake-pedal lever ratio is 7:1, and the master cylinder piston diameter is 1 inch. The front-wheel cylinder piston diameter is 1.250 inches, and the rear-wheel cylinder piston diameter is 0.875 inches.

  (a) What is the force acting on the master cylinder piston that results from the driver-pedal force and the brake-pedal lever ratio?

     10a. _____

  (b) What is the area of the master cylinder piston?

     10b. _____

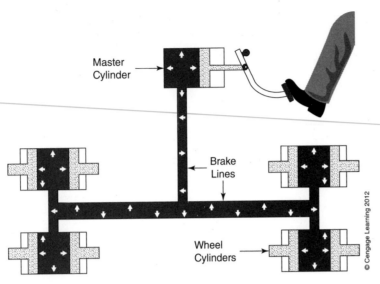

Figure 13–5

(c) How much pressure is developed within the master cylinder?

10c. _____

(d) What is the area of one front-wheel cylinder piston?

10d. _____

(e) How much force is developed by the system pressure acting on a front-wheel cylinder piston?

10e. _____

(f) What is the area of one rear-wheel cylinder piston?

10f. _____

(g) What is the force developed by the system pressure acting on the rear-wheel cylinder piston?

10g. _____

11. Use the system pressure that you found in Problem 10c to determine the force developed in a disc-brake system that uses a 2.5-inch diameter piston in one of the front brakes.

11. _____

12. A master cylinder is filled with brake fluid and has a piston in it with an area of 8 cm². The driver applies a force of 850 newtons to the piston. How much pressure is developed in this system? Assume there is no fluid flow from the master cylinder.

12. _____

13. Determine the force developed by the front- and rear-wheel cylinder pistons in a drum-brake system similar to the one described in Problem 10. The driver exerts a force of 150 newtons. The brake pedal ratio is 7.5:1, and the master cylinder diameter is 2.5 cm. The diameter of the front-wheel cylinder piston is 3 cm and the rear-wheel cylinder piston diameter is 2 cm.

(a) What is the force acting on the master cylinder piston that results from the driver-pedal force and the brake-pedal lever ratio?

13a. _____

(b) What is the area of the master cylinder piston?

13b. _____

(c) How much pressure is developed within the master cylinder?

13c. _____

(d) What is the area of one front-wheel cylinder piston?

13d. _____

(e) How much force is developed by the system pressure acting on a front-wheel cylinder piston?

13e. _____

(f) What is the area of one rear-wheel cylinder piston?

13f. _____

(g) What is the force developed by the system pressure acting on the rear-wheel cylinder piston?

13g. _____

14. Use the system pressure from Problem 13c to determine the force developed in a disc-brake system that uses a 7-cm diameter piston in one of the front disc brake units.

14. _____

15. During a panic stop, brake-system pressures may go as high as 1500 pounds per square inch. Using the specifications in Problem 10, what effort or force must the driver apply to the brake pedal to make the pressure go that high?

15. _____

16. A vehicle weighs 1530 kilograms. The front wheels support 65 percent of the total vehicle weight and the rear wheels support the rest.

    (a) How much weight is on the front wheels?

16a. _____

    (b) What percent of weight is on the rear wheels?

16b. _____

    (c) How much weight is on the rear wheels?

16c. _____

17. Because of the operating characteristics of disc brakes, the front brake-piston area must be larger than the front drum-brake pistons. The diameter of a drum-brake front-wheel cylinder piston is 1.119 inches. What is the *diameter* of a disc-brake piston if the *area* of the disc piston must be 5 times that of the drum piston?

17. _____

18. Brake-system pressure may reach 10,350 kPa during a panic stop. Use the specifications in Problem 13 to determine the force that a driver must exert to make the pressure go that high.

18. _____

19. The diameter of a drum-brake front-wheel cylinder piston is 2.8 cm. What is the diameter of a disc-brake piston if the area of the piston must be 5 times that of the drum piston?

19. _____

20. The car illustrated in Figure 13–6 has a weight distribution, while the car is in a steady state, of 57% in front and 43% in the rear. The total vehicle weight is 3600 lb. If the car brakes suddenly so that the weight distribution is 69% front and 31% rear, how much weight will be on the front tires and rear tires?

20. Front _____
    Rear _____

Total Weight 3600 Pounds

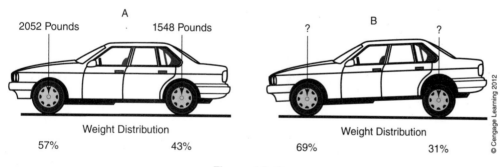

Figure 13–6

21. A vehicle has the following weight distribution: the front weighs 820 kg and the rear weighs 760 kg. Determine the percentages for front and rear weight distribution.

21. % Front _____
    % Rear _____

22. Disc-brake systems usually require higher hydraulic pressures than do drum systems—as much as 85 percent higher. If a drum system operates of an average pressure of 600 psi (lb/in.²), what is the average pressure in a disc-brake system if it operates at 85% higher pressure than the drum system?

22. _____

23. Suppose a drum system operates at an average pressure of 4140 kPa. What is the average pressure in a disc-brake system if it operates at 85% higher pressure than the drum system?

23. _____

24. A brake system was designed to provide 1 square inch of lining area for each 25 pounds of vehicle weight. If the vehicle weighs 3750 pounds, how many square inches of lining area are required?

24. _____

25. A brake drum was manufactured with an inside diameter of 11.500 inches. After the drum was used for 50,000 miles, it was measured and found to be 11.538 inches. During a brake job, the drums were resurfaced to their maximum of 11.560.

    (a) How much material had worn off during the 50,000 miles of driving?

25a. _____

    (b) How much material was removed during the resurfacing process?

25b. _____

26. A technician has measured several brake fluid lines for replacement purposes. A U.S. customary tape measure was used for the measurement. However, when the technician went for parts, he found out they only come in metric lengths.

In the spaces provided, list the closest metric line longer that the U.S. customary measurement. The metric lines come in increments of 10 cm.

The sizes shown in Figure 13–7 are diameters, not lengths.

26a. 6" _____

18" _____

7½" _____

14" _____

45" _____

70" _____

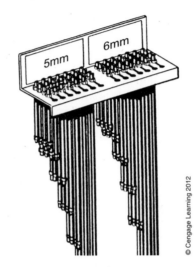

Figure 13–7

27. A customer with a Dodge complained about the brake pedal pulsating during braking. The technician removed the wheel and checked the run-out of the brake rotor as shown in Figure 13–8.

    (a) What was the dial indicator reading?

27a. _____

    (b) What are the brake specifications?

27b. _____

    (c) Is this run-out acceptable? Give your answer as either "Yes" or "No."

27c. _____

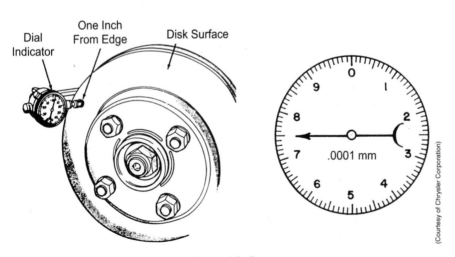

Figure 13–8

28. New tires are manufactured with a tread depth of approximately ⅜ inch. When the tires have worn to their design limit, the wear indicators or bars begin to show. Wear indicators begin to show when there is 1/16 inch of tread left. How much rubber has worn away during the design life of the tire?

28. _____

29. An F-78-15 tire is one that has an F-rated capacity, for which the height of the tire is 78 percent of the width, and that fits a 15-inch rim. If the cross-sectional width is 16.2 cm, how high is the tire?

29. _____

30. A technician is performing a wheel alignment on a Ford Mustang. The camber reading on one tire is +1¾°.

(a) What is the preferred reading from the specifications?

30a. _____

(b) What amount must the adjustment be changed to bring the wheel within specifications?

30b. _____

31. The caster angle is found to be +2¼° on a Nissan. This car has manual steering.

(a) What setting do the specifications recommend?

31a. _____

(b) What amount must the adjustment be changed to bring the wheel within specifications?

31b. _____

32. The toe measurement for the front wheels on a Dodge is 1⅜° toe-out.

(a) What setting is recommended in the specifications?

32a. _____

(b) How large an adjustment must be made to bring the toe measurement within specifications?

32b. _____

33. The camber setting on many vehicles is adjusted by adding or removing shims (spacers) between the frame and the pivot shaft. If the camber setting must be changed ⅜° and each shim changes the setting 1/16°, how many shims are needed?

33. _____

34. A camber reading is −¾° and must be changed to a preferred setting of +½°.

(a) How much must the camber be changed to meet specifications?

34a. _____

(b) If each shim changes the camber setting 1/16°, how many shims must be used to bring the camber within specifications?

34b. _____

35. A wheel bearing adjustment procedure is outlined in Figure 13–9. (See Chapter 15 for instruction on reading micrometers.)

(a) Step 2 asks you to back off the adjusting nut one-half turn. How many degrees is that?

35a. _____

(b) Step 3 asks that you tighten the adjusting nut one flat. How many degrees is that?

35b. _____

36. A Dodge brake rotor is being checked for parallelism in Figure 13–10. The readings on the micrometers are shown. Check this with the specifications to see if it meets specifications. List your answer as "Yes" or "No."

36. _____

37. The relationship between the turning of the steering wheel and the resulting movement of the wheels is referred to as the *steering ratio*. Each complete turn of the steering wheel is 360°. By dividing the total number of degrees that the steering wheel turns by the number of degrees that the wheels turn, we can determine the steering ratio. What is the steering ratio if the steering wheel makes 4⅔ complete turns from lock to lock and the front wheels turn 60°?

37. _____

38. Many convertible tops are powered with small hydraulic systems. If the force needed to raise the top is 275 pounds and the diameters of the two hydraulic cylinders are 1 inch, how much pressure is needed to raise the top?

38. _____

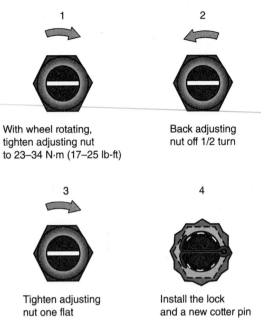

1

With wheel rotating,
tighten adjusting nut
to 23–34 N·m (17–25 lb-ft)

2

Back adjusting
nut off 1/2 turn

3

Tighten adjusting
nut one flat

4

Install the lock
and a new cotter pin

Figure 13–9

39. The power-steering pump pumps 24 liters per minute at an engine speed
of 3000 revolutions per minute. How much oil does it pump when the engine is
idling at 1000 revolutions per minute?

39. _____

40. An automobile is designed to crush ¾ inch upon impact for each mile per hour
over 10 miles per hour. How many inches are crushed in an accident that
occurs at 36 miles per hour?

40. _____

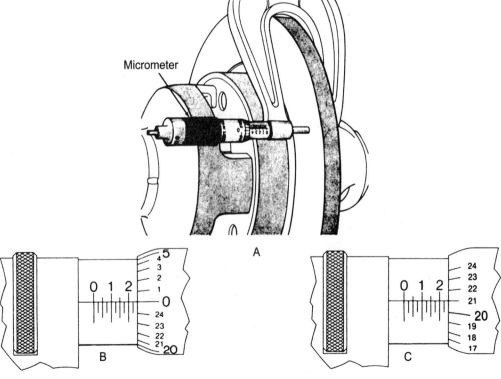

Micrometer

A

B

C

Figure 13–10

41. A proportioning valve is designed to control maximum rear-brake system hydraulic pressures during hard braking to prevent rear-wheel lockup. Reduction is sometimes as much as 20 percent. If the system pressure is 9490 kPa, how much pressure is applied to the rear brakes during a panic stop?

41. _____

42. Determine the diameter of the following tires. Round each answer to the nearest tenth.

    (a) P185/75R14

42a. _____

    (b) P165/80R13

42b. _____

    (c) P175/70R13

42c. _____

    (d) P225/70R15

42d. _____

    (e) LT235/70R15

42e. _____

    (f) LT235/85R16

42f. _____

    (g) P245/45R17

42g. _____

    (h) P275/40R17

42h. _____

    (i) P325/35R17

42i. _____

For Problems 43 through 56, round each answer to the nearest hundredth.

43. Determine the new effective drive ratios for the following:

	Old Tire Size	New Tire Size	Drive-Axle Ratio
(a)	P185/60R14	P195/60R14	3.73
(b)	P205/50R15	P225/50R15	4.10
(c)	P185/70R14	P195/70R14	3.08
(d)	P235/40R17	P315/35R17	3.54

43a. _____

43b. _____

43c. _____

43d. _____

44. Determine the new drive ratios for the following:

	Original Tires	New Tires	Old Drive-Axle Ratio
(a)	P205/50R15	P195/60R14	3.73
(b)	P205/50R15	P225/50R15	4.10
(c)	P185/70R14	P195/70R14	3.08
(d)	P235/40R17	P315/35R17	3.54

44a. _____

44b. _____

44c. _____

44d. _____

45. Determine the true speeds in the situations listed below. Assume the vehicles are traveling at an indicated speed of 60 mph.

	Old Tires	New Tires
(a)	P185/60R14	P195/60R14
(b)	P225/50R15	P205/50R15
(c)	P195/70R14	P185/70R14
(d)	P235/40R17	P315/35R17

45a. _____

45b. _____

45c. _____

45d. _____

46. Determine the equivalent indicated speed for the following situations. The true speed wanted is 65 mph.

	Old Tires	New Tires		
(a)	P195/60R14	P185/60R14	46a.	_____
(b)	P205/50R15	P225/50R15	46b.	_____
(c)	P185/70R14	P195/70R14	46c.	_____
(d)	P315/35R17	P235/40R17	46d.	_____

47. Determine the true speed during the following timed miles.

	Distance (miles)	Time		
(a)	1	55 s	47a.	_____
(b)	1	71 s	47b.	_____
(c)	2	36 s	47c.	_____
(d)	1.5	168 s	47d.	_____
(e)	2.5	2 min 35 s	47e.	_____

48. Determine the percent error involved with the following speeds.

	Indicated Speed (mph)	True Speed (mph)		
(a)	30	33	48a.	_____
(b)	45	49	48b.	_____
(c)	55	61	48c.	_____
(d)	65	72	48d.	_____

49. While building a race car, it is determined that the weight distribution will be as follows: LF = 27%, RF = 22%, LR = 28%, and RR = 23%. What percentage of the weight is on the:

49. Front _____
    Rear _____
    L/side _____
    R/side _____

50. If the front of the vehicle in Problem 49 weighs 1470 lb, determine the weight of each of the following parts.

50. Rear _____
    R/side _____
    L/side _____
    R/front _____
    L/front _____
    R/rear _____
    L/rear _____

51. Determine the cornering forces in the following situations:

	Speed (mph)	Track		
(a)	38	378 foot diameter	51a.	_____
(b)	49	209 foot radius	51b.	_____
(c)	58	225 foot radius	51c.	_____
(d)	65	578 foot diameter	51d.	_____

52. Determine the cornering forces with the following lap times and skid pad sizes.

	Lap Time	Pad Size
(a)	28.5 s	632 foot radius
(b)	26.5 s	375 foot radius
(c)	33.5 s	1575 foot diameter
(d)	0.5 min	1225 foot diameter

52a. _____

52b. _____

52c. _____

52d. _____

One of the popular new cars has the following specifications. Use this information for Problems 53 through 56.

Tire size	P245/45R17
Rear axle ratio	3.08 : 1
First speed transmission ratio	3.35 : 1
Second speed	1.99 : 1
Third speed	1.33 : 1
Fourth speed	1 : 1
Fifth speed	0.68 : 1
Weight distribution F/R%	57/43
Curb weight	3400 lbs
Fuel capacity	15.4 gal
Fuel mileage	19.1 mpg

53. What is the overall gear ratio in each gear?

(a) First gear

(b) Second gear

(c) Third gear

(d) Fourth gear

(e) Fifth gear

53a. _____

53b. _____

53c. _____

53d. _____

53e. _____

54. Determine vehicle speed in each gear at the RPM listed.

(a) First gear @ 5900 RPM

(b) Second gear @ 5900 RPM

(c) Third gear @ 5900 RPM

(d) Fourth gear @ 5750 RPM

(e) Fifth gear @ 4000 RPM

54a. _____

54b. _____

54c. _____

54d. _____

54e. _____

55. How much weight is on the:

(a) Front wheels

(b) Rear wheels

55a. _____

55b. _____

56. What is the theoretical driving range (in miles) of this vehicle?

56. _____

57. Following a relatively minor accident, José needed the steering-wheel airbag replaced. The new car dealer wanted $598.48 for a replacement airbag, and a salvage yard was asking $153.98. The warranty charge for the airbag was $5.98. The state sales tax was 4.8%

(a) If José purchased the airbag from the salvage yard, how much did it cost?

57a. _____

(b) How much money was saved by purchasing from a salvage yard rather than the new car dealer?

57b. _____

(c) José took the airbag to an independent garage to have it installed. The garage took 24 minutes to install the airbag. The labor rate was $75.00/hour. What was the labor charge for this installation?

57c. _____

(d) What was José's final cost for this airbag replacement?

57d. _____

58. New tire technology is anticipated to save 45 gallons of fuel per vehicle over the lifetime of the tires when compared to older tire technology. If each unused gallon of fossil fuel saves about 8 kg of CO_2 from entering the atmosphere, how many kg of CO_2 are saved from entering the atmosphere when new tire technology is used on a vehicle?

58. _____

59. Ramone needed to purchase four new tires and wanted to be sure the new tires were as environmentally friendly as practically possible. Regular tires were priced at $188.32 each. Another brand of tires, which contained approximately 87% natural compounding ingredients, was priced 31% higher.

(a) What was the cost of one of the environmentally friendly tires?

59a. _____

(b) What was the cost of four of the above tires?

59b. _____

60. In an effort to cut costs, improve vehicle handling, and improve vehicle fuel economy, many manufacturers have introduced electrically assisted steering (EAS). Average fuel economy improvements are approximately 4%. If a vehicle gets 32.6 mpg with a traditional hydro-mechanical steering assist system, what increase in fuel economy might this vehicle experience if it is equipped with an EAS system?

60. _____

61. This vehicle has come in for maintenance work. The labor rate is $72.50/hour and the sales tax rate for parts and materials is 9.5%. Complete and total the repair order in Figure 13–11.
    This vehicle is a 2006 Zephyr with a 3.0 L engine.

61. _____

Quan	Part No.	Name of Part	Sales Amount	
		Total Parts		

REPAIR ORDER

Name John Martinez

Address 3259 Circle Dr.

Phone No. 555-4583   **Date** _____

Speedometer Reading _____ **VIN No.** _____

Year, Make, and Model	License No. and State	Engine
06 Zephyr		3.0 L

Operation Number	SERVICE INSTRUCTIONS	TIME	AMOUNT
	R & R All Brake Pads		
	Reface All Rotors		
	R & R Master Cylinder		
	R & R Both Front Speed Sensors		
	R & R Both Struts		
	R & R Left Front Wheel Bearing		
	R & R Front Left Stabilizer Bar Link		
	Steering Sys. Diagnosis		
	R & R Both Rear Shock Absorbers		
	4 Wheel Alignment		

					Accessories	Amount	Total Labor	
		@					Total Parts	
1	PT. BK. Fluid	@ 7.11					Materials	
	Qts. Oil	@					Accessories	
	Lbs. Grease	@					Tires, Tubes	
		Total Materials			Total Accessories		Outside Work	

I hereby authorize the above repair work to be done along with the necessary material, and hereby grant you and/or your employees permission to operate the car, truck or vehicle herein described on streets, highways or elsewhere for the purpose of testing and/or inspection. An express mechanic's lien is hereby acknowledged on above car, truck or vehicle to secure the amount repairs thereto.

NOT RESPONSIBLE FOR LOSS OR DAMAGE TO CARS OR ARTICLES LEFT IN CARS IN CASE OF FIRE, THEFT OR ANY OTHER CAUSE BEYOND OUR CONTROL.

WORK AUTHORIZED BY ___*John Martinez*___ DATE

Total	
Tax	
TOTAL AMOUNT	

Figure 13–11

14

AUTOMOBILE HEATING, VENTILATION, AND AIR CONDITIONING

Objectives: After studying this chapter, you should be able to:

- Add and subtract positive and negative numbers when working with temperatures.
- Learn how the capacity of air-conditioning systems is measured in British thermal units (Btu) and understand how this relates to temperature change.
- Convert Btu to the metric system.
- Explain the units of heat measurement.
- Determine R-134a pressure at any given temperature.
- Determine the boiling point of engine coolant.
- Read air-conditioning pressure gauge values.
- Convert temperature values between the customary and the metric system.

The modern automobile has become a highly complex vehicle. It is designed to be dependable, economical, and safe and, at the same time, to be comfortable.

One of the systems that makes driving both safe and comfortable is the heating, ventilation, and air-conditioning system. It makes driving bearable, even enjoyable, during cold winter weather or hot, humid summer conditions. The problems in Chapter 14 deal with some design considerations and service situations that are associated with heating and air conditioning.

The capacity of air-conditioning systems is rated in a standard measurement system of British thermal units or British thermal units per hour. The abbreviation *Btu* stands for *British thermal unit*. One Btu is equal to the amount of heat that is required to change the temperature of 1 pound of water 1 degree Fahrenheit (°F). If heat is added to 5 pounds of water so that the water temperature is raised 15°F, we can determine the number of Btus required to make this change by multiplying the pounds of water by the temperature change. In this case, 5 pounds × 15°F = 75 Btu.

■ EXAMPLE 14–1

How many Btus are required to change the temperature of 4 pounds of water 27°F?

Solution

Number of Btu = (Number of pounds of water)
$$\times \text{ (temperature change)}$$
$$= 4 \text{ lb} \times 27°F$$
$$4 \times 27 = 108.$$

A total of 108 Btus are required.

The capacity of air-conditioning units can also be rated using the metric system, in calories. The symbol "cal" is used for calories. One Btu equals 252 calories. One cal is the amount of heat that is required to change the temperature of 1 gram of water 1 degree Celsius (°C). A kilocalorie (kcal) is 1000 calories and is the amount of heat needed to change 1 kilogram of water 1°C.

243

■ EXAMPLE 14–2

How many calories are required to change the temperature of 6 kg of water 4°C? How many kilocalories?

Solution

Number of calories = (number of grams of water)
$\times$ (temperature change)

6 kg = 6000 grams (g)

Number of calories = 6000 g $\times$ 4°C

6000 $\times$ 4 = 24,000

A total of 24,000 calories are required.

Number of calories = (number of kg of water)
$\times$ (temperature change)

= 6 kg $\times$ 4°C

6 $\times$ 4 = 24

A total of 24 kilocalories are required.

In the last few years, the capacity of air conditioning has been rated in the metric system in terms of joules or kilojoules. The symbol "J" is used for joules and "kJ" for kilojoules. Remember, the prefix *kilo* means that 1 kJ = 1000 J.

One calorie is equivalent to 4.184 joules.

■ EXAMPLE 14–3

How many joules are equivalent to 3865 calories?

Solution

This can be set up as a proportion.

$$\frac{1 \text{ cal}}{4.184 \text{ J}} = \frac{3865 \text{ cal}}{? \text{ J}}$$

4.184 $\times$ 3865 = 16,171.16

Step 1

16,171.16 ÷ 1

Step 2

16,171.16 ÷ 1 = 16,171.16

Step 3 So, $^{1 \text{ cal}}/_{4.184 \text{ J}} = {}^{3685 \text{ cal}}/_{16\,171.16 \text{ J}}$.

3865 cal are equivalent to 16 171.16 J.

Many larger air-conditioning systems have their capacity rated in Btus, but expressed in terms of "tons of refrigeration effect." The average automobile has an air-conditioning system rated at approximately 1 ton. This term "tons of refrigeration" is based on the amount of heat that is required to melt 1 ton of ice in 24 hours. Ice is obviously different from water and has a different heat capacity.

During the melting process, the temperature does not change as the ice changes to water. To change 1 pound of ice to water without changing the temperature requires 144 Btu. Changing ice to water without changing the temperature is called "heat to fusion."

■ EXAMPLE 14–4

How many Btus are needed to change 1 ton (2000 pounds) of ice to water without changing the temperature?

Solution

This can be set up as a proportion.

$$\frac{1 \text{ pound}}{144 \text{ Btu}} = \frac{2000 \text{ pounds}}{? \text{ Btu}}$$

144 $\times$ 2000 = 288,000

Step 1

288,000 ÷ 1

Step 2

288,000 ÷ 1 = 288,000

Step 3 So, $^{1 \text{ pound}}/_{144 \text{ Btu}} = {}^{2000 \text{ pounds}}/_{288,000 \text{ Btu}}$.

It takes 288,000 Btu to change 1 ton of ice to water without raising the temperature.

To determine the value of a ton of refrigeration effect, the Btu value must be put on a time basis. If the answer to Example 14–4 is divided by 24 hours, the answer is the amount of heat a 1-ton air-conditioning system removes during 1 hour.

■ EXAMPLE 14–5

Determine the cooling rate in Btus per hour of an air-conditioning system that is rated 1 ton.

Solution

In Example 14–4, we found that it took 288,000 Btu to change 1 ton of ice to water without raising the temperature. Use the formula

$$\frac{\text{Cooling Effect}}{\text{per Hour}} = \frac{\text{Pounds of Ice} \times \text{Heat to Fusion}}{\text{Time}}$$

We already know that the product pounds of ice × heat to fusion = 288,000 Btu.

$$\text{So, cooling effect per hour} = \frac{288,000 \text{ Btu}}{24 \text{ hr}}$$

$$288,000 \div 24 = 12,000$$

The cooling effect is 12,000 Btu per hour.

Another way of saying this is that 12,000 Btu per hour is equal to 1 ton of refrigeration effect.

In the metric system, it requires 80 calories of heat to change 1 gram of ice to water without raising the temperature. This is known as the heat of fusion.

■ EXAMPLE 14–6

How many calories are needed to change 785 kilograms of ice to water without changing the temperature?

Solution

A proportion can be used to solve this problem.

$$\frac{1 \text{ gram}}{80 \text{ cal}} = \frac{785 \text{ kg}}{? \text{ cal}}$$

It is first necessary to change 785 kilograms to grams.

$$785 \text{ kg} = 785 \text{ kg} \times 1000 \text{ g/kg} = 785,000 \text{ g}$$

The proportion now becomes

$$\frac{1 \text{ gram}}{80 \text{ cal}} = \frac{785,000 \text{ g}}{? \text{ cal}}$$

$$80 \times 785,000 = 62,800,000$$

Step 1

$$62,800,000 \div 1$$

Step 2

$$62,800,000 \div 1 = 62,800,000$$

Step 3 So, $\dfrac{1 \text{ gram}}{80 \text{ cal}} = \dfrac{785,000 \text{ g}}{62,800,000 \text{ cal}}$

It takes 62,800,000 calories (or 62,800 kcal).

The metric system expresses the power of an air conditioner in watts. 1 Btu/hr = 0.293 W.

■ EXAMPLE 14–7

An air conditioner is rated at 16,000 Btu/hr. What is the rating in watts?

Solution

Again, we can use a proportion to solve this problem.

$$\frac{1 \text{ Btu/hr}}{0.293 \text{ W}} = \frac{16,000 \text{ Btu/hr}}{? \text{ W}}$$

$$16,000 \times 0.293 = 4688$$

Step 1

$$4688 \div 1$$

Step 2

$$4688 \div 1 = 4688$$

Step 3 So, $\dfrac{1 \text{ Btu/hr}}{0.293 \text{ W}} = \dfrac{16,000 \text{ Btu/hr}}{4688 \text{ W}}$

16,000 Btu/hr is equivalent to 4688 W.

Current automotive air-conditioning systems use a refrigerant fluid called R-134a. This refrigerant carries heat from the inside of the car to the outside where it is dissipated to the outside air.

The temperature-pressure relationship of R-134a is very important to proper operation and diagnosis. This relationship is expressed in Table 14–1. This table reveals several things:

- By reading the temperature, we can see the corresponding pressure.

- By reading the pressure, we can note the corresponding temperature.

- By reading a temperature in Fahrenheit, we can find its corresponding temperature in Celsius.

- Reading the temperature in Celsius can help us determine the corresponding temperature in Fahrenheit.

- A pressure in pounds per square inch can be used to determine what that same pressure would be in kilopascals.

- We can read a pressure in kilopascals to determine its equivalent pressure in pounds per square inch.

Look at Table 14–1. You can see that when R-134a has zero pressure (for example, when an open container contains R-134a) it will boil, or vaporize, at −15°F or −26.1°C.

Table 14–1
Pressure-Temperature Relationships of R–134a

The table indicates the pressure of Refrigerant 134a at various temperatures. For example, a container of R–134a at 80°F will have a pressure of 86.4 PSI (595.7 kPa). If the temperature increases to 125°F, the pressures will increase to 183.6 PSI (1265.9 kPa). This chart can also be used to determine the temperature at which R–134a will boil at various temperatures. For example, at a pressure 78.4 PSI (540.6 kPa) R–134a boils at 75°F (23.9°C). At a pressure of 40 PSI (275.8 kPa), R–134a boils at 45°F (7.2°C).

Temperature		Pressure		Temperature		Pressure		Temperature		Pressure	
°F	°C	PSI	kPa	°F	°C	PSI	kPa	°F	°C	PSI	kPa
–15	–26.1	0	0	45	7.2	40.0	275.8	105	40.6	134.3	926.0
–10	–23.3	1.9	13.1	50	10.0	45.3	312.3	110	43.3	145.6	1003.9
–5	–20.6	4.1	28.3	55	12.8	51.1	352.3	115	46.1	157.6	1086.6
0	–17.8	6.5	44.8	60	15.6	57.3	395.1	120	48.9	170.3	1174.2
5	–15.0	9.1	62.7	65	18.3	63.9	440.6	125	51.7	183.6	1265.9
10	–12.2	12.0	82.7	70	21.1	70.9	488.9	130	54.4	197.6	1362.4
15	–9.4	15.0	103.4	75	23.9	78.4	540.6	135	57.2	212.4	1464.5
20	–6.7	18.4	126.9	80	26.7	86.4	595.7	140	60.0	227.9	1571.3
25	–3.9	22.1	152.4	85	29.4	94.9	661.2	145	62.8	244.3	1684.4
30	–1.1	26.1	180.0	90	32.2	103.9	716.4	150	65.6	261.4	1802.3
35	1.7	30.4	209.6	95	35.0	113.5	782.6	155	68.3	279.5	1927.1
40	4.4	35.0	241.3	100	37.8	123.6	852.2	160	71.1	300.0	2057.4

NAME _____ DATE _____ SCORE _____

1. How many Btus are required to change the temperature of 3 pounds of water 17°F?

1. _____

2. A container with 13 pounds of water is heated for 7 minutes. A thermometer indicates the temperature to be 54°F before heating and 173°F after the heating is finished.

   (a) What was the temperature change?

   2a. _____

   (b) How many Btus were required to make the temperature change?

   2b. _____

3. The container in Problem 2 was allowed to cool. After 2 hours and 43 minutes, the water had cooled to 85°F.

   (a) How many degrees were lost during cooling?

   3a. _____

   (b) How many Btus were lost during cooling?

   3b. _____

4. A container with 5 kg of water is heated for 12 minutes. A thermometer indicates the temperature to be 7°C before heating and 93°C after heating.

   (a) What was the temperature change?

   4a. _____

   (b) How many cal were required to make the change?

   4b. _____

   (c) How many kcal were required to make the change?

   4c. _____

5. The container in Problem 4 was allowed to cool. After 2 hours and 17 minutes, the water had cooled to 34°C.

   (a) How many degrees were lost during cooling?

   5a. _____

   (b) How many cal were lost during cooling?

   5b. _____

   (c) How many kcal were lost during cooling?

   5c. _____

6. How many Btus are needed to change 747 pounds of ice into water without changing the temperature?

6. _____

7. How many pounds of ice can be melted into water at 32°F when 20,160 Btus are applied to the ice?

7. _____

8. How much heat (in Btus) is needed to melt 3 tons of ice without changing the temperature?

8. _____

9. How many calories are needed to change 12,487 grams of ice into water without changing the temperature?

9. _____

10. How many grams of ice can be melted into water at 0°C when 24,320 calories are applied to the ice?

10. _____

11. How much heat (in calories) is needed to melt 1 metric ton (1000 kg) of ice without changing the temperature?

11. _____

12. If a system has a cooling rate of 36,000 Btus per hour, how is this expressed in tons of cooling effect?

12. _____

13. How are 8000 Btus per hour expressed in terms of tons of cooling effect?

13. _____

14. Service procedures caution that containers of refrigerant should not be subjected to temperatures higher than 125°F.

   (a) What is the pressure, in psi, of R-134a at 125°F?

14a. _____

   (b) What is the pressure, in both psi and kPa, of R-134a when the temperature in °F and °C both have the same value?

14b. _____

15. Use Table 14–1 to change 125°F to °C.

15. _____

16. (a) What is the pressure, in psi, of R-134a at 51.7°C?

16a. _____

   (b) What is the pressure, in kPa, of R-134a at 51.7°C?

16b. _____

17. If a technician connected a pressure gauge to an air-conditioning system filled with R-134a on a 90°F summer day, what pressure, in psi, would the gauge indicate?

17. _____

18. What pressure in kilopascals would a pressure gauge indicate if the temperature were 35°C?

18. _____

19. One day the early-morning temperature was 45°F and the mid-afternoon temperature was 95°F. An air-conditioning system was filled with R-134a.

   (a) How many pounds per square inch pressure were in the system when the temperature was 45°F?

19a. _____

   (b) What was the pressure when the temperature was 95°F?

19b. _____

   (c) How much did the pressure change during the day?

19c. _____

20. One morning the temperature was –15°C and the mid-afternoon temperature was 10°C. The air-conditioning system was filled with R-134a.

   (a) How many kilopascals of pressure are there in the system at –15°C?

20a. _____

   (b) What was the pressure when the temperature was 10°C?

20b. _____

   (c) How much did the pressure in the air-conditioning system change during the day?

20c. _____

21. In the morning, the temperature of an air-conditioning system filled with R-134a is –5°F.

   (a) What is the temperature in °C?

21a. _____

   (b) What is the pressure, in psi, in the system?

21b. _____

   (c) What is the pressure in kPa?

21c. _____

22. By mid-afternoon, the temperature had risen to 25°F in the system in Problem 21. The system is charged with R-134a.

   (a) What is the temperature in °C?

22a. _____

   (b) What was the pressure in the system at mid-afternoon in psi?

22b. _____

   (c) What was the pressure in kPa?

22c. _____

   (d) How many °F did the temperature change?

22d. _____

   (e) How many °C did the temperature change?

22e. _____

23. At what temperature in °F do the temperature and pressure (psi) numbers have the closest match?

23. _____

24. Determine the temperature, in degrees Fahrenheit, at which R-134a boils or vaporizes in the evaporator if the system pressure is 37 psi.

24. _____

25. At what temperature, in °C, does R-134a vaporize if the system pressure is 312.3 kPa?

25. _____

26. After the air-conditioning compressor has been resealed, the service manual asks the technician to perform a manual pump-up test. The test is performed by turning the compressor crankshaft and reading the pressure gauge after 10 complete revolutions. The pump-up specification calls for a reading of 244.75 kPa or better.

   The technician only has a gauge that reads in psi, as shown in Figure 14–1.

(a) What is the reading in psi on the gauge in Figure 14–1?

26a. _____

(b) What is the reading in kPa on the gauge in Figure 14–1?

26b. _____

(c) Does this compressor meet specifications?

26c. _____

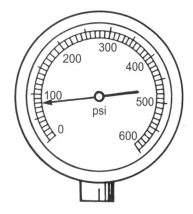

Figure 14–1

27. An engine thermostat is being tested in Figure 14–2. The thermostat is stamped 92°C. The thermometer, however, is marked in degrees F. If the thermostat water is heated, at what point could you expect the thermostat to open? (Express your answer in degrees F.)

27. _____

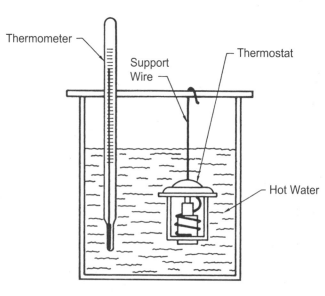

Figure 14–2

28. As the antifreeze content of the coolant is increased, the boiling point of the coolant is also increased as shown in Figure 14–3.

   (a) What is the boiling point of the coolant if the system is filled with the recommended solution of 50% water and 50% antifreeze (50:50)?

   28a. _____

   (b) 100:0

   28b. _____

   (c) 40:60

   28c. _____

   (d) 0:100

   28d. _____

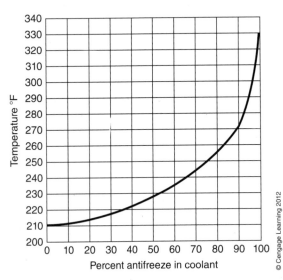

Figure 14–3

29. An automobile air-conditioning system has 5 ounces of oil in the compressor, 3 ounces in the evaporator, 1 ounce in the condenser, and 2 ounces in the accumulator. What is the total amount of oil in the system?

   29. _____

30. An automobile air-conditioner system has 283 mL of oil in the compressor, 59 mL in the evaporator, 29 mL in the condenser, and 30 mL in the filter-drier. What is the total amount of oil in the system?

   30. _____

31. New parts may come from the factory with no oil in them.

   (a) If the evaporator and accumulator are replaced, how many ounces of oil must be added to the system? (Refer to Problem 29 for capacities.)

   31a. _____

   (b) What percentage is this of the total oil that is required?

   31b. _____

32. (a) How many milliliters of oil must be added if the compressor and the condenser are replaced? (Refer to Problem 30.)

   32a. _____

   (b) What percentage is this of the total oil required?

   32b. _____

33. The amount of R-134a in an automobile air-conditioning system may vary from 2 to 4 pounds (0.9 to 1.8 kg). There are 16 ounces per pound.

   (a) How many ounces are in a system that has a capacity of 3.75 pounds?

   33a. _____

   (b) If refrigerant comes in 14 oz cans, how many cans are needed to fill this system?

   33b. _____

34. If the condenser holds ¾ pound of R-134a, the evaporator ⅝ pound, the receiver-driver ⅞ pound, the pressure-operated absolute (P.O.A.) valve 3/16 pound, and the lines 9/16 pound, what is the total weight of the R-134a in this system?

   34. _____

35. If the condenser holds 0.3 kg of R-134a, the evaporator 0.26 kg, the receiver-drier 0.36 kg, and the lines 0.23 kg, how much R-134a is in this system?

35. _____

36. A driver, trying to save gas on a hot day, drove 285 miles on 9.8 gallons of gasoline with the windows down. The service station attendant suggested that the windows should be rolled up and the air conditioner should be used to be more comfortable and save gas. The next time the tank was filled, the driver had traveled 337 miles and used 10.8 gallons. Determine if the service station attendant was correct.

    (a) What was the fuel mileage with the windows open?

36a. _____

    (b) What was the mileage with the air conditioner on and the windows closed?

36b. _____

    (c) What was the mileage difference?

36c. _____

    (d) Which method got the better mileage?

36d. _____

    (e) Was the service station attendant correct?

36e. _____

37. An automobile is having its air-conditioning system repaired. The service technician takes 1.4 hours to install a new compressor shaft seal, 0.6 hours to replace a thermostatic switch, 0.6 hours to replace a vacuum diaphragm, and 2.0 hours to evacuate and recharge the system.

    (a) What was the total time spent on this job?

37a. _____

    (b) If the labor rate was $86.00 per hour, what was the customer charged for labor on this job?

37b. _____

38. An imported-car service manual specifies an 88°C thermostat. However, the parts store only lists thermostats in degrees Fahrenheit. What rating in degrees Fahrenheit should be used for this particular car?

38. _____

39. A technician has repaired a cooling system and has accounted for his time and the parts as follows:

Operation	Customer Parts Price	Time (hrs)
R & R thermostat and check opening temp	$ 8.38	0.4
R & R heater core	68.06	1.1
R & R heater hoses	18.80	0.2
R & R radiator hoses	32.16	0.3
Back flush engine and radiator	9.81	0.6
R & R water pump	32.10	1.3
Fill with coolant	12.40	0.3

The repair shop labor rate is $75.50/hour, the technician works on a 40% commission on labor charges and 7% on parts charges, and the shop has the parts marked up 85%.

    (a) How much money did the shop take in on this job?

39a. _____

    (b) How much was the technician paid for this job?

39b. _____

    (c) What was the net income for the shop on this job?

39c. _____

40. Engineering data suggests that a direct-acting engine-powered radiator fan can use 4% of the engine output power. If an engine is rated at 335 horsepower, what horsepower gain might be expected if this fan is removed?

40. _____

41. The engine in Problem 40 was mounted in an engine dynamometer and tested. The engine produced a torque reading of 372 lb-ft at a speed of 4950 rpm. Using the following formula:

$$BHP = \frac{T \times rpm}{5252}$$

where   BHP = brake horsepower
                 T = torque
         rpm = engine revolutions per minute

(a) Determine the BHP the engine was producing.

41a. _____

(b) How many horsepower were gained by removing the fan?

41b. _____

(c) How many kilowatts is this?

41c. _____

(d) What percent horsepower gain was there?

41d. _____

42. Engine efficiency is illustrated by the graph in Figure 14–4. If we assume that a gallon of gasoline contains 117,000 Btus, how many Btus are included in each segment of the graph?

42. Cooling _____

      Exhaust _____

      Friction _____

      Air flow _____

      Work _____

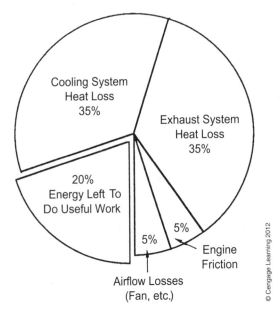

Figure 14–4

43. The effectiveness of the heater is directly related to the operating temperature of the cooling system. Figure 14–5 shows the relationship between a rise in cooling system pressure and a corresponding rise in the boiling point of the coolant. The figure shows that a cooling system pressure of 4 psi results in a 12°F rise of the boiling point. Determine the final boiling point related to the following conditions. You may also need to consult Figure 14–3.

	Pressure	Coolant Antifreeze : Water Ratio
(a)	0	0 : 100
(b)	8	0 : 100
(c)	14	0 : 100
(d)	0	50 : 50

43a. _____

43b. _____

43c. _____

43d. _____

(e)	9	50:50	43e. _____
(f)	13	50:50	43f. _____
(g)	0	70:30	43g. _____
(h)	7	70:30	43h. _____
(i)	15	70:30	43i. _____

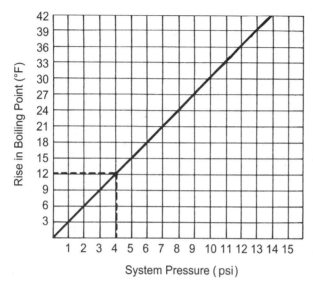

Figure 14–5

44. An air-conditioning gauge set is used to check both the low- and high-side pressures in an air-conditioning system. List the pressures indicated by the air-conditioning gauge set in Figure 14–6.

44. **Low side** _____

   High side _____

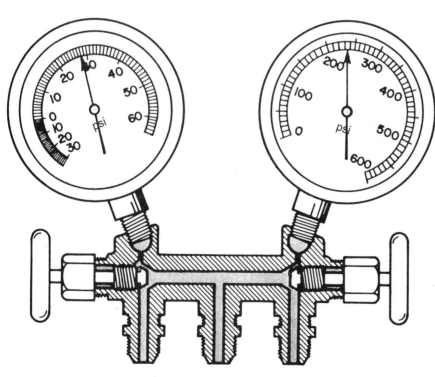

Figure 14–6

45. A 2006 Fusion had 73,000 miles on it and was brought in to a repair facility with the following complaints:

- Engine overheating

- AC did not blow cold air

- HVAC blower howled when on high speed

The technician did a visual inspection and noticed both that the serpentine belt needed replacement and that something had penetrated the grill and put a hole in the AC condenser causing a low refrigerant level. Since this hole allowed moisture to get into the AC system, the AC drier also had to be replaced. The technician also noticed that the engine cooling system was low on coolant and conducted a cooling system pressure test. This test showed that the water pump was leaking and needed to be replaced. When the HVAC blower was put on high speed, the blower motor armature began vibrating and caused the howling noise indicating it also needed to be replaced. The repair includes two cans of R-134a refrigerant at $8.79/can and 6 quarts of antifreeze at $5.32/quart. The labor rate was $87.50/hour and the tax on parts and materials was 6.5%.

Fill in the necessary information in the repair order, Figure 14–7, and total the invoice.

Quan	Part No.	Name of Part	Sales Amount	
		Total Parts		

REPAIR ORDER

Name ___Marjorie Weatherstone___

Address ___123 Grassy Hills Place___

Phone No. ___555-3141___ Date _____

Speedometer Reading _____ VIN No. _____

Year, Make, and Model	License No. and State	Engine
06 Zephyr		3.0 L

Operation Number	SERVICE INSTRUCTIONS	TIME	AMOUNT
	R & R Serpentine Belt		
	Cooling, Sys. Press. Test		
	R & R AC Condensor		
	R & R AC Drier		
	R & R HVAC Blower Motor		
	R & R Water Pump		
	Recover, Evacuate, and Reacharge		
	AC System		

Quan	Part No.		@			Accessories	Amount		
6	Qts. A/Freeze	@	5.32						
	Qts. Oil	@							
2	Lbs. R-134A	@	8.79						
	Total Materials					Total Accessories			

	Amount
Total Labor	
Total Parts	
Materials	
Accessories	
Tires, Tubes	
Outside Work	
Total	
Tax	
TOTAL AMOUNT	

WORK AUTHORIZED BY ___Marjorie Weatherstone___ DATE _____

Figure 14–7

MEASUREMENT TOOLS

Objectives: After studying this chapter, you should be able to:

- Use various measurement tools, including the ruler, digital meter, test meter, dial indicator, and micrometer.
- Use a scale to measure linear distances.
- Recognize and read digital multimeters.
- Use inside and outside micrometers to determine specification compliance.
- Use a dial indicator to measure movement and specification compliance.
- Describe the various thread sizes.

Measurement systems are used to communicate valuable information related to almost all areas of the automobile. They help to define such things as value and pricing; speed, power, and torque; weight, area, and volume; and time, temperature, distance, and angularity. Precise measurements are vital to proper automotive performance, service, and reliability.

In the automobile industry two measurement systems are widely used. These two systems are the U.S. customary system and the international metric system. There are many tools used in the automotive trades. Each tool uses at least one of these two measurement systems.

The problems in this chapter deal with these systems and with measurement tools associated with the automotive trade.

Digital Meters

Although there are many analog test instruments still being used in the automotive service industry, modern technology currently uses the more accurate digital test instruments. Figure 15–1 and Figure 15–2 illustrate two

different digital multimeters, one rather simple, the other more complicated.

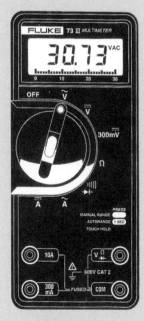

Figure 15–1

Figure 15–2

Digital test meter measurement values are easy to determine as long as you pay close attention to the measurement unit symbols also appearing on the face of the meter. See Figure 15–3 for measurement unit symbols used with the Fluke 88 automotive meter.

Measure Unit Symbols

Symbol	Meaning
AC	Alternating current or voltage
DC	Direct current or voltage
V	Volts
mV	Millivolts (1/1000 volts)
A	Ampere (amps). Current
mA	Millampere (1/1000 amps)
%	Percent (for duty cycle readings only)
Ω	Ohms. Resistance
kΩ	Kilohm (1000 ohms). Resistance
MΩ	Megaohm (1,000,000 ohms). Resistance
Hz	Hertz (1 cycle/sec.). Frequency
kHz	Kilohertz (1000 cycle/sec). Frequency
RPM 1	Revolutions/minute. Counting one cycle per spark
RPM 2	Revolutions/minute. Counting two cycles per spark
ms	Milliseconds (1/1000 sec) for pulse width measurements

Figure 15–3

When reading multimeter measurement values you must note two things on the screen; the numerical value being displayed and the measurement unit symbol.

■ EXAMPLE 15–1

Looking at Figure 15–1, you see that the measurement value is 30.73. Also note the measurement unit symbol VAC. When you look at Figure 15–3 you will not find the symbol VAC. However, the first letter, V, as listed in Figure 15–3 represents volts or voltage. The second and third letters, AC, represent alternating current. Because the complete symbol is VAC, it means that you are measuring voltage of alternating current or, as is commonly said, alternating current voltage. The value in this example is 30.73 volts.

■ EXAMPLE 15–2

In this example you will be using Figure 15–4 and Figure 15–5. Both figures look quite similar; however, if

you look closely, you can see that Figure 15–5 has an additional letter or symbol showing. Figure 15–4 shows a meter value of 18.800 Ω. By looking in the measurement unit symbol table in Figure 15–3 you will find that the Ω symbol represents ohms, a unit of electrical resistance. The reading on the meter in Figure 15–4 then is 18.800 or 18.80 ohms.

The meter in Figure 15–5 shows the same numerical value as the meter in Figure 15–4; however, this meter screen shows an additional symbol. In this figure you will again see the Ω symbol, but you also see that it is preceded by the k symbol. Referring back to the table in Figure 15–3 you can see that when both k and Ω are used together (kΩ) it changes the actual value of the meter reading by 1000, so in Figure 15–5 the meter is reading 18,800 ohms.

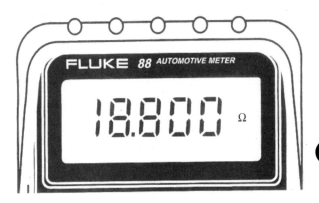

Figure 15–4

Figure 15–5

NAME _____ DATE _____ SCORE _____

In the following problems, write the correct meter readings in the blank opposite the figures.

1.

1. _____

2.

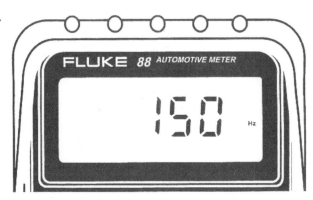

2. _____

3.

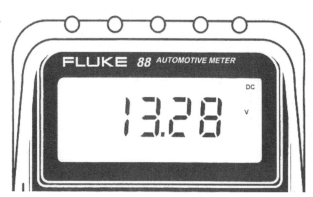

3. _____

4.

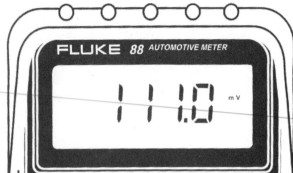

4. _____

5.

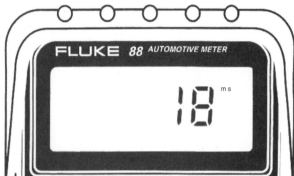

5. _____

6.

6. _____

7.

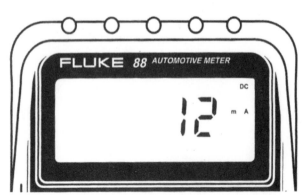

7. _____

8.

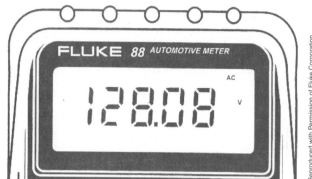

8. _____

Scale Measurement

A ruler is one of the most basic tools for measurement. In many cases, both the U.S. customary scale and the metric may be on the same ruler. The upper scale of the ruler in Figure 15–6 is marked in the U.S. customary unit of inches. The bottom scale is in the metric system unit of centimeters.

On the ruler in Figure 15–6, the upper inch scale has the first 3 inches marked in 32nds of an inch ($\frac{1}{32}$ inch); the remainder are marked in 16ths of an inch (or $\frac{1}{16}$ inch) increments. The bottom scale is marked in 1 millimeter (mm) increments.

■ EXAMPLE 15–3

For each of the marked dimensions in Figure 15–7, record the dimension in its lowest fraction of an inch.

Solution

Look at dimension A. It is less than 1″ long. Each mark represents $\frac{1}{32}$″ and A is $\frac{1}{32}$″ less than 1″. Since 1″ = $\frac{32}{32}$″, dimension A is $\frac{32}{32}$″ − $\frac{1}{32}$″ = $\frac{31}{32}$″.

Now, look at dimension B. It is between 2″ and 3″ long. You can count and see that it is $\frac{12}{32}$″ past the 2″ mark. But $\frac{12}{32}$″ = $\frac{3}{8}$″, so dimension B is $2\frac{3}{8}$″ long.

The metric scale is somewhat easier to read since it has no fractions. One millimeter is equal to 0.1 ($\frac{1}{10}$) of a centimeter (cm); consequently, each 10 mm are equal to 1 cm. Thus, 10 mm may also be read as 1 cm, 20 mm may be read as 2 cm, 30 mm may be read as 3 cm, and so on. A reading of 145 mm may also be stated as 14.5 cm.

■ EXAMPLE 15–4

Use the metric rule in Figure 15–8 to measure each of the marked dimensions in the proper form. This rule is marked in mm increments.

Solution

Look at the dimension marked A in Figure 15–8. The dimension is 19 mm long.

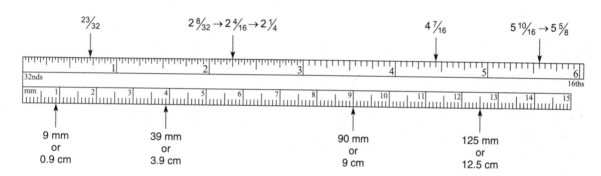

Figure 15–6

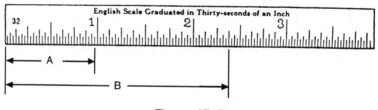

Figure 15–7

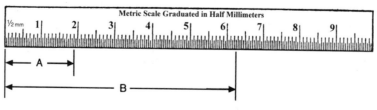

Figure 15–8

This is 1 cm + 9 mm. There are 10 mm in each cm, so 1 mm = 0.1 cm and so 9 mm = 0.9 cm. Thus,

$$19 \text{ mm} = 1 \text{ cm} + 9 \text{ mm}$$
$$= 1 \text{ cm} + 0.9 \text{ cm}$$
$$= 1.9 \text{ cm}$$

Therefore, 19 mm is the same as 1.9 cm.

Now, measure dimension B. It is 62.5 mm long. This is the same as 6 cm + 2.5 mm. Since each millimeter is 0.1 cm, 2.5 mm = 0.25 cm. So dimension B is

$$62.5 \text{ mm} = 6 \text{ cm} + 2.5 \text{ mm}$$
$$= 6 \text{ cm} + 0.25 \text{ cm}$$
$$= 6.25 \text{ cm}$$

NAME _____ DATE _____ SCORE _____

1. Use the ruler in Figure 15–9 to determine the length of each of the marked
   dimensions in its lowest inch fraction.

   1a. _____ in.

   1b. _____ in.

   1c. _____ in.

   1d. _____ in.

   1e. _____ in.

   1f. _____ in.

   1g. _____ in.

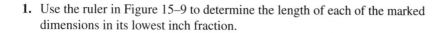
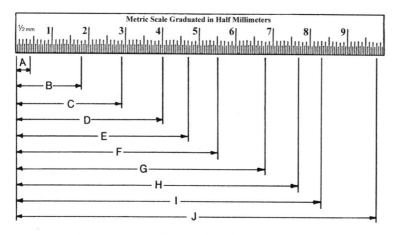

Figure 15–9

   1h. _____ in.

2. Use the ruler in Figure 15–10 to determine the length of each of the marked
   dimensions in the indicated metric unit.

   2a. _____ mm

   2b. _____ cm

   2c. _____ mm

   2d. _____ cm

   2e. _____ cm

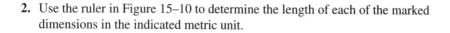
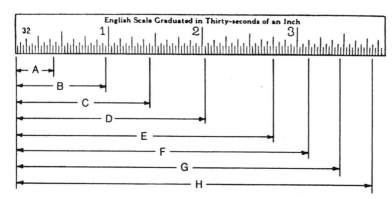

Figure 15–10

   2f. _____ mm

   2g. _____ cm

   2h. _____ mm

   2i. _____ mm

   2j. _____ cm

3. Use an inch ruler to measure the lengths of each of the objects in Figure 15–11.

3a. _____

3b. _____

3c. _____

3d. _____

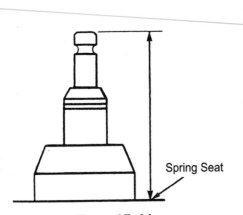

Figure 15–11a

Measure valve spring seat to valve tip.

Figure 15–11b

Measure rocker arm stud length.

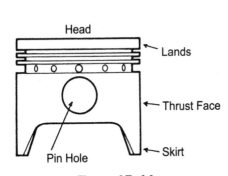

Figure 15–11c

Measure piston diameter.

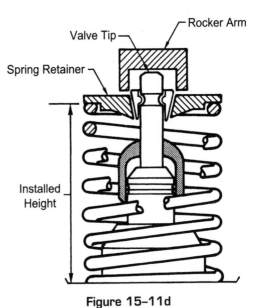

Figure 15–11d

Measure valve spring installed height.

4. Use a metric ruler to measure the lengths of each of the objects in Figure 15–12.

4a. _____

4b. _____

4c. _____

4d. _____

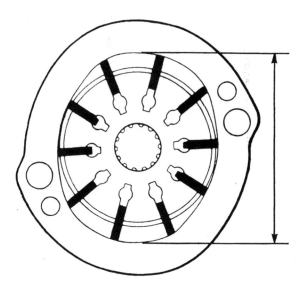

Figure 15–12a

Measure maximum blade extension diameter of power steering pump rotor.

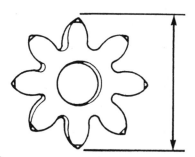

Figure 15–12b

Measure oil pump rotor tip-to-tip distance.

Figure 15–12c

Measure rocker stud length.

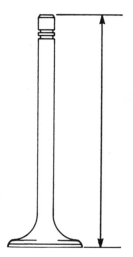

Figure 15–12d

Measure total valve length.

Dial Indicators

Dial indicators can be used to measure movement of a part in relation to another part. They are used to measure shaft end play, gear back lash, brake disc run out, and so on. Several kinds of dial indicators are shown in Figure 15–13. Figure 15–13a is a continuous reading dial indicator and Figure 15–13b is a balanced dial indicator.

A continuous reading dial indicator can be designed to measure in either a clockwise or a counterclockwise direction. Figure 15–13c is a clockwise reading dial indicator and Figure 15–13d is read counterclockwise. The balanced dial indicator may be used to measure movement in either direction. With either of these indicators you must remember how many revolutions the needle makes to determine the total measurement.

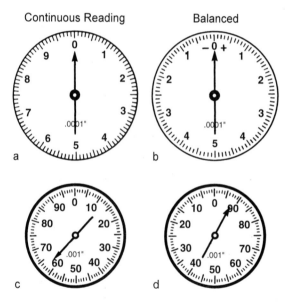

Figure 15–13

The face of the indicator is marked to show the value of each mark on the dial. Indicators in Figure 15–13a and Figure 15–13b show that each mark on the dial is equal to 0.0001″. Only every 10th mark is numbered. That means that each numbered mark (1, 2, 3, etc.), stands for 10, 20, 30, and so on. Thus, if the needle points at 2, the reading is 20 × 0.0001″ = 0.0020″. The total dial reading is equal to 0.0100″.

The dial indicators in Figure 15–13c and Figure 15–13d are marked in 0.001″ increments. Every 10th mark is numbered to represent 10, 20, 30, or 40 thousandths of an inch and so on. Thus if, as in Figure 15–13c, the needle is pointing at the third mark (starting at 60) between 60 and 70, the reading is 60 × 0.001″ = 0.060″ plus 3 × 0.001″ = 0.003″. The total reading is 0.060″ + 0.003″ = 0.063″. The total dial reading is equal to 0.0100″.

■ EXAMPLE 15–5

The needle on the indicator in Figure 15–14 has made two complete revolutions plus what is showing on the dial. What is the total measurement this dial indicator has made?

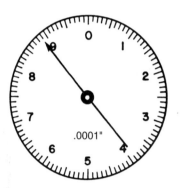

Figure 15–14

Solution

Two complete revolutions have a value of 0.010″ × 2 = 0.020″.

The dial is pointing at the ninth mark between 8 and 9. So, the dial has a reading of 89 × 0.0001″ = 0.0089″.

The total reading is 0.0200″ + 0.0089″ = 0.0289″.

■ EXAMPLE 15–6

The needle on the indicator in Figure 15–15 has made three complete revolutions in a counterclockwise direction plus what is showing on the dial. What is the total measurement this balance dial indicator has made?

Figure 15–15

Solution

Three complete revolutions have a value of 0.010″ × 3 = 0.030″.

The dial is pointing at the fifth mark between 1 and 2. So, the dial has a reading of 15 × 0.0001″ = 0.0015″.

The total reading is 0.0300″ + 0.0015″ = 0.0315″.

■ EXAMPLE 15–7

The needle on the dial indicator in Figure 15–16 has made two complete revolutions in a counterclockwise direction plus what is showing on the dial. What is the total measurement this balanced dial indicator has made?

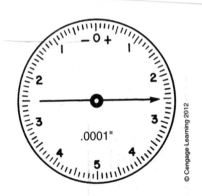

Figure 15–16

Solution

Two complete revolutions have a value of $0.010'' \times 2 = 0.020''$.

The pointer (moving counterclockwise) has moved an additional 75 marks. So, the dial has a reading of $75 \times 0.0001'' = 0.0075''$.

The total reading is

$$
\begin{array}{r}
0.020'' \\
+\ 0.0075'' \\
\hline
0.0275''
\end{array}
$$

NAME _____ DATE _____ SCORE _____

Record your readings for each of the dial indicators in Figure 15–17.

1. **(a)** Five revolutions plus the dial reading   1a. _____

   (b) The dial reading   1b. _____

   (c) Two revolutions plus the dial reading   1c. _____

2. **(a)** One clockwise revolution plus the dial reading   2a. _____

   (b) Three counterclockwise revolutions plus the dial reading   2b. _____

   (c) The dial reading (with counterclockwise revolution)   2c. _____

3. **(a)** Four revolutions plus the dial reading   3a. _____

   (b) The dial reading   3b. _____

   (c) Two revolutions plus the dial reading   3c. _____

   (d) Five revolutions plus the dial reading   3d. _____

4. **(a)** The dial reading (with clockwise revolution)   4a. _____

   (b) Four counterclockwise revolutions plus the dial reading   4b. _____

   (c) Two clockwise revolutions plus the dial reading 4c. _____

Problem 1

Problem 2

Problem 3

Problem 4

© Cengage Learning 2012

Figure 15–17

5. The illustrations in Figure 15–18 show different dial indicators that have been used to measure various engine specifications. In the spaces provided, list the correct specification as found in Appendix A, the dial indicator reading as illustrated, and whether it meets or exceeds specifications. Rotation is listed as CW for clockwise or CCW for counterclockwise. (Some specifications are in metric units and the dial indicators are in inches, you may need to convert the specifications from the metric system to customary units.)

   When dealing with runout, specifications show a maximum amount. When dealing with endplay, specifications show a range that the measurement is expected to fall within, unless otherwise stated.

Reading— CW, Dial Reading

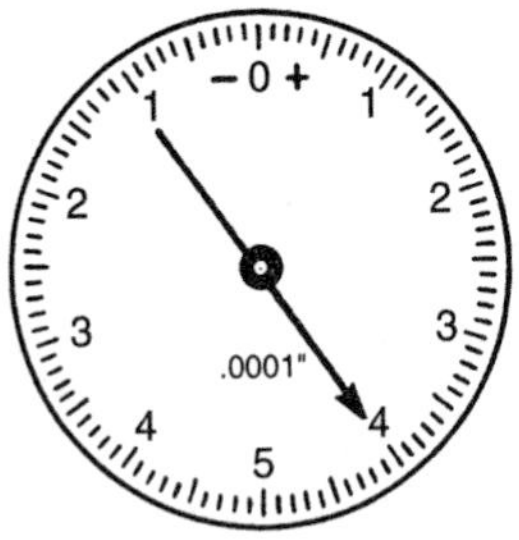

Car/Component— NISSAN: Valve Deflection

Specification— ______________________

Reading— ______________________

Pass/Fail ______________________

Reading— CW, Dial Reading

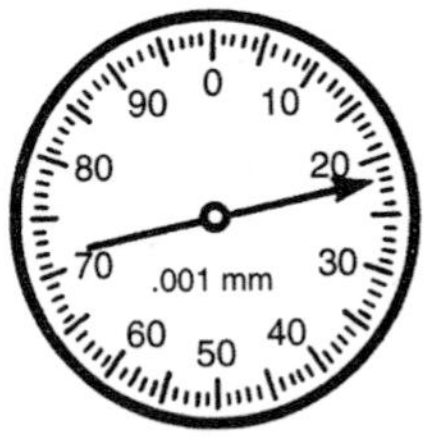

Car/Component— BUICK: Connecting Rod Side Clearance

Specification— ______________________

Reading— ______________________

Pass/Fail ______________________

Reading— CCW, Dial Reading

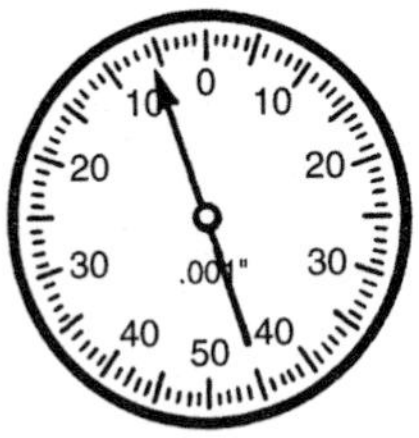

Car/Component— FORD: Camshaft Endplay

Specification— ______________________

Reading— ______________________

Pass/Fail ______________________

Reading— CCW, Dial Reading

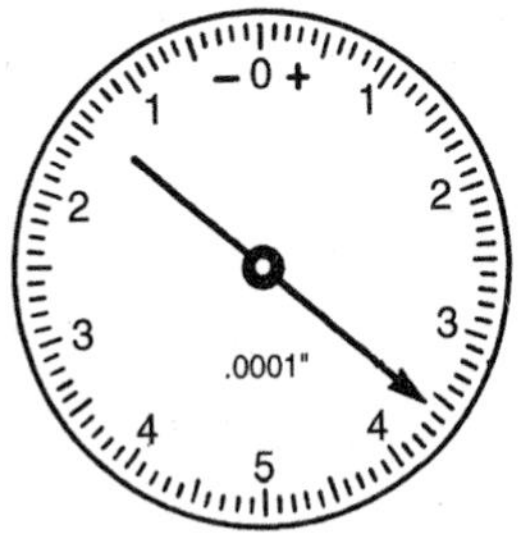

Car/Component— DODGE: Crankshaft Endplay

Specification— ______________________

Reading— ______________________

Pass/Fail ______________________

Reading— CCW 2 Revolutions + Dial Reading

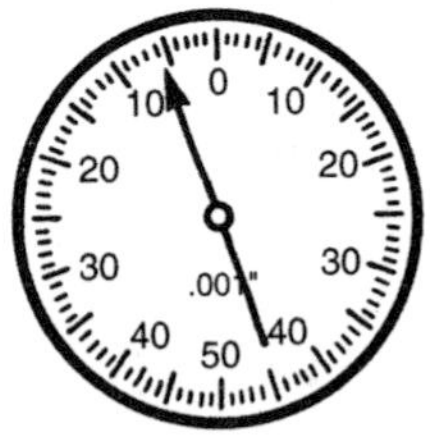

Car/Component— FORD: Lobe Lift

Specification— ______________________

Reading— ______________________

Pass/Fail ______________________

Reading— CW, Dial Reading

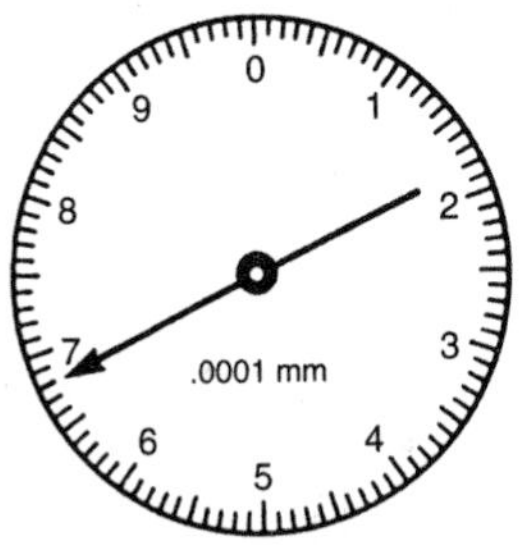

Car/Component— BUICK: Crankshaft Endplay

Specification— ______________________

Reading— ______________________

Pass/Fail ______________________

Figure 15–18a

Reading—   CW, Dial Reading   CW, Dial Reading   CW, Dial Reading

Car/Component—   FORD:
Camshaft Runout

DODGE:
Camshaft Endplay

BUICK:
Valve Seat Runout

Specification—   _____   _____   _____

Reading—   _____   _____   _____

Pass/Fail   _____   _____   _____

Reading—   CCW 2 Revolutions
+ Dial Reading   CW, Dial Reading   CCW, Dial Reading

Car/Component—   BUICK:
Camshaft Lift

NISSAN:
Camshaft Endplay

DODGE:
Front Brake Disc
Runout

Specification—   _____   _____   _____

Reading—   _____   _____   _____

Pass/Fail   _____   _____   _____

Figure 15–18b

Micrometers

An outside micrometer, as in Figure 15–19, is a tool used to measure outside diameters with great precision. It is the tool of choice when measuring camshaft and crankshaft surfaces. These measurements require great precision, usually to a thousandth of an inch (0.001″) or a hundredth of a millimeter (0.01 mm). These are very small measurements. This page is approximately 0.003″ or 0.08 mm thick.

major divisions are further marked into four smaller divisions called minor divisions. Each minor division is equal to 0.025″, as shown in Figure 15–21.

The thimble is marked in 0.001″ increments and is mounted to the sleeve with screw threads so that each complete revolution of the thimble is equal to 0.025″ (Figure 15–22) and moves exactly one minor division on the sleeve.

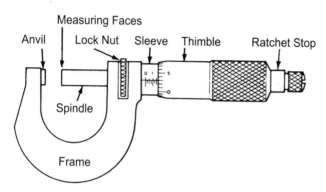

Figure 15–19

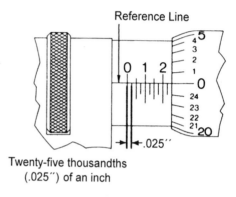

Twenty-five thousandths
(.025″) of an inch

Figure 15–21

When the thimble is turned, the spindle is moved toward the anvil, pinching the part to be measured between these two surfaces. The amount of the measurement is read from the sleeve and thimble. The sleeve of a non-metric micrometer is marked with 10 major divisions. Each major division is equal to 0.100″. All ten major divisions equal 1 inch (Figure 15–20). These

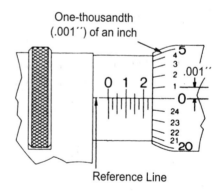

Figure 15–22

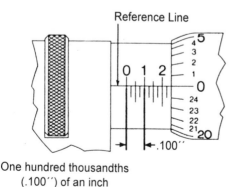

One hundred thousandths
(.100″) of an inch

Figure 15–20

To read a micrometer, you must add together the two marks on the sleeve and the reading on the thimble. That means that you add the major division reading, the minor division reading after the highest major division mark, and the number of marks on the thimble that are indexed with the reference line.

■ EXAMPLE 15–8

In Figure 15–23, you see the sleeve and thimble of a micrometer that has measured a part and is ready to be read. Determine the reading on this micrometer.

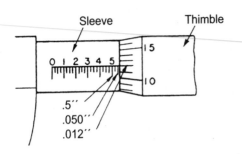

Figure 15–23

Solution

First, notice that there are five major divisions showing. Since each major division equals 0.100″, five divisions equal 0.500″. Next, notice that two minor divisions are showing *past* the 5 mark. Each minor division is equal to 0.025″, so two minor divisions equal 0.050″. Finally, the reading on the thimble shows that the 12 mark is indexed with the reference line, or 0.012″.

Adding these three figures together, you find

$$
\begin{array}{r}
0.500 \\
0.050 \\
+\ 0.012 \\
\hline
0.562
\end{array}
$$

Thus, the micrometer in Figure 15–23 reads 0.5620.

Outside micrometers come in many different sizes. The micrometer shown in Figure 15–19 is a 0″–1″ model. This means that the readings are a direct measurement. If it was a 1″–2″ micrometer, the sleeve and thimble reading would be added to 1 inch. If it was a 2″–3″ model, the reading would be added to 2 inches. This would continue for larger models.

Metric outside micrometers are read much the same as those using the U.S. customary system. The sleeve is marked off in 1 mm marks above the reference line (Figure 15–24) and 0.5 mm marks below the line (Figure 15–25). The thimble has 50 marks on it, and each mark is equal to 0.01 mm (Figure 15–26).

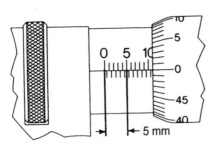

Figure 15–24

To read a metric micrometer, you must add together the upper and lower divisions or marks and the marks on the thimble.

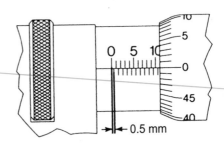

Figure 15–25

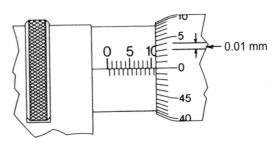

Figure 15–26

■ EXAMPLE 15–9

What is the reading on (the metric micrometer in Figure 15–27)?

Solution

There are 13 marks above the reference line. This equals 13 mm. You can also just see one lower line showing *past* the 13 mm line. This means that you must add 0.5 mm to your reading. Finally, you can see the reference line is aligned with the 28 mark on the thimble. That means that there are 0.28 mm showing on the thimble. When all three readings are added, you obtain

$$
\begin{array}{r}
13.00\ \text{mm} \\
0.50\ \text{mm} \\
+\ 0.28\ \text{mm} \\
\hline
13.78\ \text{mm}
\end{array}
$$

So, the micrometer in Figure 15–27 has a reading of 13.78 mm.

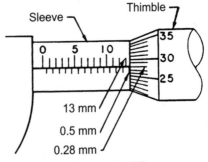

Figure 15–27

Reading an inside micrometer is similar to reading an outside model. An inside micrometer can be used to measure the wear on an engine cylinder. This type of micrometer consists of a sleeve and thimble similar to an outside micrometer. It also has extensions of different lengths and a 0.5″ spacer, as shown in Figure 15–28 and Figure 15–29.

Because of its compact design, the thimble can only be screwed 0.5″, or, in other words, the sleeve has only 0.5″ worth of increments on it. The inside micrometer in Figure 15–28 can be used to measure from 3″ to 3.5″. If the 0.5″ spacer is inserted, as in Figure 15–29, it can be used to measure from 3.5″ to 4″. To measure something slightly larger than 4″, the 0.5″ spacer and the 3″–4″ extension are removed and a 4″–5″ extension *only* must be inserted.

■ EXAMPLE 15–10

What is the reading on the micrometer in Figure 15–28?

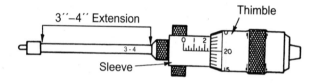

3″–4″ Extension

Thimble

Sleeve

Figure 15–28

Solution

This micrometer has a 3″–4″ extension but does not have a 0.5″ spacer, so we know the reading is between 3″ and 3.5″. There are two major divisions, so we add 0.200″. There is one minor division showing past the 2 mark, so another 0.025″ is added. Finally, you can see the reference line is aligned with the 20 mark on the thimble. That means that there are 0.020″ showing on the thimble. When all four readings are added, you obtain

3″ extension	3.000″
2 major divisions	0.200″
1 minor division	0.025″
thimble	0.020″
total	3.245″

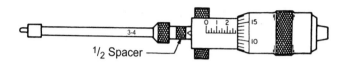

1/2 Spacer

Figure 15–29

■ EXAMPLE 15–11

What is the reading on the micrometer in Figure 15–29?

Solution

This micrometer has a 3″–4″ extension and a 0.5″ spacer, so we know the reading is between 3.5″ and 4″. There are 2 major divisions, so we add 0.200″. There are three minor divisions showing *past* the 2 mark, so another 0.075″ is added. Finally, you can see the reference line is aligned with the 12 mark on the thimble. That means that there are 0.012″ showing on the thimble. When all four readings are added, you obtain

3″ extension	3.000″
0.5″ spacer	0.500″
2 major divisions	0.200″
3 minor divisions	0.075″
thimble	0.012″
total	3.787″

NAME _____ DATE _____ SCORE _____

1–10. In Figure 15–30, 10 micrometers are shown. The size of each micrometer is indicated under it. Determine the correct reading for each and record it in the space provided.

1. _____

2. _____

3. _____

4. _____

5. _____

6. _____

7. _____

8. _____

9. _____

10. _____

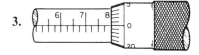
1. 0″–1″

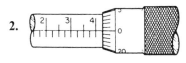
2. 0″–1″

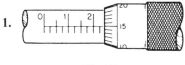
3. 1″–2″

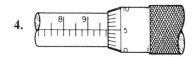
4. 1″–2″

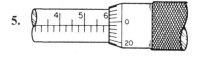
5. 2″–3″

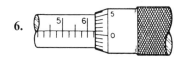
6. 2″–3″

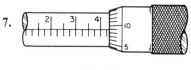
7. 3″–4″

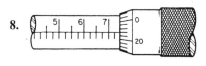
8. 3″–4″

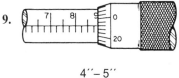
9. 4″–5″

10. 4″–5″

Figure 15–30

11–18. Eight micrometers are shown in Figure 15–31. Determine the correct reading for each and record it in the space provided.

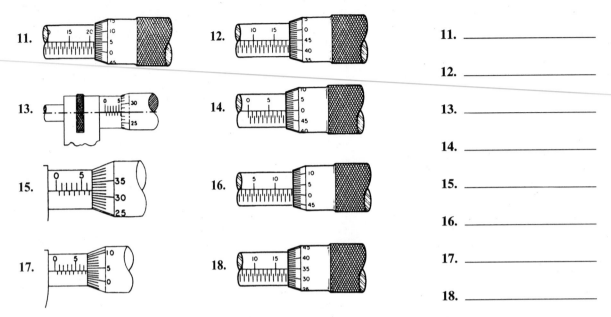

11. _____

12. _____

13. _____

14. _____

15. _____

16. _____

17. _____

18. _____

Figure 15–31

19–23. Five micrometers are shown in Figure 15–32. Determine the correct reading for each and record it in the space provided.

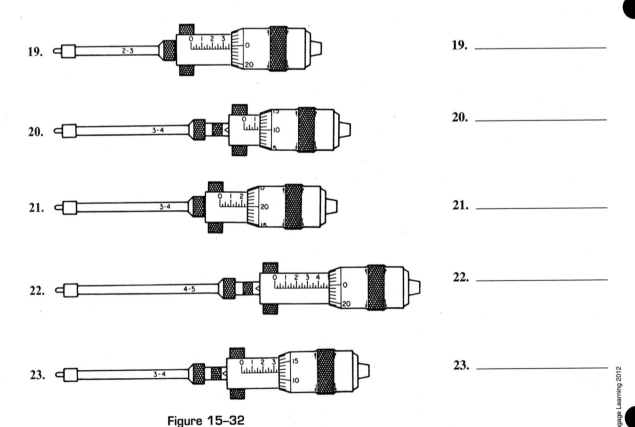

19. _____

20. _____

21. _____

22. _____

23. _____

Figure 15–32

Drill Sizes

Metalworking drills are sized several different ways. The most common designations in the U.S. customary system of measurement are either by drill number, drill letter, or fractional size. Drills may be used to drill holes for fasteners, used for holes to be tapped for threads, or used as gauging spacers for such things as carburetor adjustments. The chart in Figure 15–33 shows numbers, letters, and fractional drill sizes with decimal equivalents.

Number	Fraction	Decimal
80		0.0135
79		0.0145
—	1/64	0.0156
78		0.0160
77		0.0180
76		0.0200
75		0.0210
74		0.0225
73		0.0240
72		0.0250
71		0.0260
70		0.0280
69		0.0292
68		0.0310
—	1/32	0.0312
67		0.0320
66		0.0330
65		0.0350
64		0.0360
63		0.0370
62		0.0380
61		0.0390
60		0.0400
59		0.0410
58		0.0420
57		0.0430
56		0.0465
—	3/64	0.0469
55		0.0520
54		0.0550
53		0.0595
—	1/16	0.0625
52		0.0635
51		0.0670
50		0.0700
49		0.0730
48		0.0760
—	5/64	0.0781
47		0.0785
46		0.0810
45		0.0820
44		0.0860
43		0.0890
42		0.0935
—	3/32	0.0937
41		0.0960
40		0.0980
39		0.0995
38		0.1015
37		0.1040
36		0.1065
—	7/64	0.1094

No./Letter	Fraction	Decimal
35		0.1100
34		0.1110
33		0.1130
32		0.1160
31		0.1200
—	1/8	0.1250
30		0.1285
29		0.1360
28		0.1405
—	9/64	0.1406
27		0.1440
26		0.1470
25		0.1495
24		0.1520
23		0.1540
—	5/32	0.1562
22		0.1570
21		0.1590
20		0.1610
19		0.1660
18		0.1695
—	11/64	0.1719
17		0.1730
16		0.1770
15		0.1800
14		0.1820
13		0.1850
—	3/16	0.1875
12		0.1890
11		0.1910
10		0.1935
9		0.1960
8		0.1990
7		0.2010
—	13/64	0.2031
6		0.2040
5		0.2055
4		0.2090
3		0.2130
—	7/32	0.2187
2		0.2210
1		0.2280
A		0.2340
—	15/64	0.2344
B		0.2380
C		0.2420
D		0.2460
E	1/4	0.2500
F		0.2570
G		0.2610
—	17/64	0.2656
H		0.2660
I		0.2720
J		0.2770
K		0.2810
—	9/32	0.2812
L		0.2900
M		0.2950
—	19/64	0.2969

Letter	Fraction	Decimal
N		0.3020
—	5/16	0.3125
O		0.3160
P		0.3230
—	21/64	0.3281
Q		0.3320
R		0.3390
—	11/32	0.3437
S		0.3480
T		0.3580
—	23/64	0.3594
U		0.3680
—	3/8	0.3750
V		0.3770
W		0.3860
—	25/64	0.3906
X		0.3970
Y		0.4040
—	13/32	0.4062
Z		0.4130
	27/64	0.4219
	7/16	0.4375
	29/64	0.4531
	15/32	0.4687
	31/64	0.4844
	1/2	0.5000
	33/64	0.5156
	17/32	0.5312
	35/64	0.5469
	9/16	0.5625
	37/64	0.5781
	19/32	0.5927
	39/64	0.6094
	5/8	0.6250
	41/64	0.6406
	21/32	0.6562
	43/64	0.6719
	11/16	0.6875
	45/64	0.7031
	23/32	0.7187
	47/64	0.7344
	3/4	0.7500
	49/64	0.7656
	25/32	0.7812
	51/64	0.7969
	13/16	0.8125
	53/64	0.8281
	27/32	0.8437
	55/64	0.8594
	7/8	0.8750
	57/64	0.8906
	29/32	0.9062
	59/64	0.9219
	15/16	0.9375
	61/64	0.9531
	31/32	0.9687
	63/64	0.9844
	1	1.0000

© Cengage Learning 2012

Figure 15–33

Number and letter sizes of drills with decimal equivalents.

■ EXAMPLE 15–12

Here is a list of tap drill sizes needed for tapping threads for various fasteners. What is the number, letter, or fractional drill equivalent for each size?

Thread Size	Decimal Equivalent of Correct Drill Size	Drill Size
(a) $\frac{3}{8}''$ National Fine	0.3320	_____
(b) $\frac{11}{16}''$ National Coarse	0.5927	_____

Number	Fraction	Decimal	No./Letter	Fraction	Decimal	Letter	Fraction	Decimal
80		0.0135	35		0.1100	N		0.3020
79		0.0145	34		0.1110	–	5/16	0.3125
–	1/64	0.0156	33		0.1130	O		0.3160
78		0.0160	32		0.1160	P		0.3230
77		0.0180	31		0.1200	–	21/64	0.3281
76		0.0200	–	1/8	0.1250	Q		0.3320
75		0.0210	30		0.1285	R		0.3390
74		0.0225	29		0.1360	–	11/32	0.3437
73		0.0240	28		0.1405	S		0.3480
72		0.0250	–	9/64	0.1406	T		0.3580
71		0.0260	27		0.1440	–	23/64	0.3594
70		0.0280	26		0.1470	U		0.3680
69		0.0292	25		0.1495	–	3/8	0.3750
68		0.0310	24		0.1520	V		0.3770
	1/32	0.0312			0.1540			
					0.1562			

Figure 15–34

Number	Fraction	Decimal	No./Letter	Fraction	Decimal	Letter	Fraction	Decimal
80		0.0135	35		0.1100	N		0.3020
79		0.0145	34		0.1110	–	5/16	0.3125
–	1/64	0.0156	33		0.1130	O		0.3160
78		0.0160	32		0.1160	P		0.3230
–		0.0180	31		0.1200	–	21/64	0.3281
		0.0200	–		0.1250	Q		
56						R		
–	3/64	0.0469	13	3/16	0.1875		33/64	0.5156
55		0.0520	–		0.1890		17/32	0.5312
54		0.0550	12		0.1910		35/64	0.5469
53		0.0595	11		0.1935		9/16	0.5625
–	1/16	0.0625	10		0.1960		37/64	0.5781
52		0.0635	9		0.1990		19/32	0.5927
51		0.0670	8		0.2010		39/64	0.6094
50		0.0700	7	13/64	0.2031		5/8	0.6250
49		0.0730	–		0.2040		41/64	0.6406
48		0.0760	6		0.2055		21/32	0.6562
–	5/64	0.0781	5		0.2090		43/64	0.6719
47		0.0785	4		0.2090		11/16	
		0.0810	3		0.2130		45/64	
		0.0820						

Figure 15–35

Solution

We are given the decimal equivalent for each of these thread sizes.

(a) Looking at the table in Figure 15–33, locate 0.332 in the 3rd column, headed "Decimal." Look at the boxed region in Figure 15–34. On the right, we see the decimal equivalent of 0.3320. At the left, in the column headed "Letter," we see the letter is Q. Thus, we see that the tap drill size for ⅜″ National Fine is Q.

(b) For an ¹¹⁄₁₈″ National Coarse, we are given the decimal equivalent of a hole diameter of 0.5927″. Looking at the table in Figure 15–33, locate 0.5927 in the 3rd column, headed "Decimal." Look at the boxed region in Figure 15–35. On the right, we see the decimal equivalent of 0.5927. The left column, the one headed "Letter," is blank. In the middle column, headed "Fraction," we find the fraction ¹⁹⁄₃₂″.

Thus, we see that the tap drill size for ¹¹⁄₁₆″ National Coarse is ¹⁹⁄₃₂″.

■ EXAMPLE 15–13

A carburetor adjustment specification calls for an air valve rod adjustment dimension of 0.0250″. What is the exact or closest drill size to the listed dimension?

Solution

Once again we look at the table in Figure 15–33. This time, we locate .0250 in the first column, headed "Decimal." Look at the boxed region in Figure 15–36. On the right, we see the decimal equivalent of 0.0250. The left column, the one headed "Number," contains the numeral 72. Thus, we see that the tap drill size for this air valve rod is 72.

Number	Fraction	Decimal	No./Letter	Fraction	Decimal	Letter	Fraction	Decimal
80		0.0135	35		0.1100	N		0.3020
79		0.0145	34		0.1110	–	5/16	0.3125
–	1/64	0.0156	33		0.1130	O		0.3160
78		0.0160	32		0.1160	P		0.3230
77		0.0180	31		0.1200	–	21/64	0.3281
76		0.0200	–	1/8	0.1250	Q		0.3320
75		0.0210	30		0.1285	R		0.3390
74		0.0225	29		0.1360	–	11/32	0.3437
73		0.0240	28		0.1405	S		0.3480
72		0.0250	–	9/64	0.1406	T		0.3580
71		0.0260	27		0.1440	–	23/64	0.3594
70		0.0280	26		0.1470	U		0.3680
69		0.0292	25		0.1495	–	3/8	0.3750
68			24					
			23					

Figure 15–36

NAME _____ DATE _____ SCORE _____

1. The following is a list of tap drill sizes needed for tapping threads for various fasteners. Record the number, letter, or fractional drill equivalent for each size. (Use Figure 15–33.)

Thread Size	Decimal Equivalent of Tap Drill Size	Drill Size
(a) 10-32 National Fine	0.3320	1a. _____
(b) ⁷⁄₁₆″ National Coarse	0.3680	1b. _____
(c) ¾″ National Fine	0.6875	1c. _____
(d) 2-28 National Fine	0.1820	1d. _____
(e) ¼″ National Coarse	0.2010	1e. _____
(f) ½″ National Coarse	0.4219	1f. _____
(g) ¼″ National Fine	0.2130	1g. _____
(h) ⁵⁄₁₆″ National Coarse	0.2570	1h. _____
(i) ⅞″ National Fine	0.8125	1i. _____
(j) 1″ National Coarse	0.8750	1j. _____

2. The following is a list of carburetor adjustment specifications. Record the exact or closest drill size to the listed dimension.

Specification	Drill Size
(a) Float level, 0.250	2a. _____
(b) Vacuum piston gauge, 0.035	2b. _____
(c) Accelerator pump stroke, 0.500	2c. _____
(d) Fast idle cam, 0.095	2d. _____
(e) Choke vacuum kick, 0.128	2e. _____
(f) Choke unloader, 0.280	2f. _____
(g) Cold enrichment, 1st stage, 0.490	2g. _____
(h) Cold enrichment, 2nd stage, 0.475	2h. _____
(i) Cold enrichment, 3rd stage, 0.350	2i. _____
(j) Cold enrichment, 4th stage, 0.125	2j. _____
(k) Venturi limiter, 0.750	2k. _____
(l) Secondary lockout, 0.065	2l. _____

APPENDIX A

PAGES FROM SPECIFICATION MANUALS AND TIME/PARTS GUIDE

BUICK

ENGINE SPECIFICATIONS

GENERAL DATA:

Type	In Line
Displacement	1.8/2.0
RPO	L46/LR9
Bore	89
Stroke	74/80
Compression Ratio	9.0:1
Firing Order	1-3-4-2

CYLINDER BORE:

Diameter	88.992-89.070
Out of Round	.02 Max
Taper-Thrust Side	.02 Max

PISTON:

Clearance	.020-.046

NOTICE: All dimensions are in millimetres (mm) unless otherwise specified.

PISTON RING:

C o m p r e s s i o n	Groove Clearance	Top	.030-.068
		2nd	.030-.086
	Gap	Top	.25-.50
		2nd	.25-.50
O i l	Groove Clearance		.199
	Gap		

PISTON PIN:

Diameter	22.9937-23.0015
Clearance	.0065-.0091
Fit in Rod	.019-.052 Press

(Buick continued)

CAMSHAFT:

± Lift .05	In	6.65
	Ex	6.65
	Journal Diameter	47.44-47.49
	Journal Clearance	.026-.101

CRANKSHAFT:

Main Journal	Diameter	#1,2,3,4	63.360-63.384
		#5	63.340-63.364
	Taper		.005 Max.
	Out Of Round		.005 Max.
Main Bearing Clearance		1-4	.015-.047
		5	.036-.068
Crankshaft End Play			.05-.18
Crankpin	Diameter		50.758-50.784
	Taper		.005 Max.
	Out Of Round		.005 Max.
	Rod Bearing Clearance		.025-.079
	Rod Side Clearance		.10-.61

VALVE SYSTEM:

Lifter		Hydraulic
Rocker Arm Ratio		1.5
Valve Lash (All)		1-1/2 Turns From Zero Lash
Face Angle (All)		45° ± 0° 15' Valve
Seat Angle (All)		46° ± 0° 15' Head
Seat Runout		.05
Seat Width	In	1.25-1.50
	Ex	1.60-1.90
Stem Clearance	In	.028-.066
	Ex	.035-.078
Valve Spring		48.5
Pressure N @ mm	Closed	342 ± 18 @ 40.6
	Open	810 ± 27 @ 33.9
Installed Height		+ Shim 40.6

	SPECIFICATIONS FOR DIAGNOSIS FOR WARRANTY REPAIRS OR CUSTOMER PAID SERVICE
TOE DEG (PER WHEEL)	.125° TOE-OUT ±.125°
CAMBER	+.60° ±.50°

NISSAN

SERVICE DATA AND SPECIFICATIONS (S.D.S.) — E15, E16 & CD17

Destination	California		Non-California		M.P.G.	Canada	
Engine	E16		E16		E15	E16	
Transaxle	M/T	A/T	M/T	A/T	M/T	M/T	A/T
Carburetor model	DFC328-1	DFC328-2	DCZ328-1	DCZ328-2	DFP306-2	DCZ328-11	DCZ328-12
Choke type	Automatic		Automatic			Automatic	
Fuel level adjustment mm (in) Gap between float and carburetor body "H"	12 (0.47)						
Gap between valve stem and float seat "h"	1.3 - 1.7 (0.051 - 0.067)						
Fast idle adjustment Clearance "A" (*4 at 2nd cam step) mm (in)	0.86±0.07 (0.0339± 0.0028)	1.15±0.07 (0.0453± 0.0028)	0.86±0.07 (0.0339± 0.0028)	1.15±0.07 (0.0453± 0.0028)	0.8±0.07 (0.0315± 0.0028)	0.72±0.07 (0.0283± 0.0028)	1.00±0.07 (0.0394± 0.0028)
Fast idle speed rpm	2,600 - 3,400	2,900 - 3,700	2,400 - 3,200	2,700 - 3,500	2,400 - 3,200	1,900 - 2,700	2,400 - 3,200

(Nissan continued)

Front axle and front suspension

Wheel alignment (Unladen*) Camber	degree	−35' - 1°05'
Caster	degree	45' - 2°15'
Kingpin inclination	degree	12°10' - 13°40'
Toe-in	mm (in)	3 - 5 (0.12 - 0.20)
Side slip (Reference data)	mm/m (in/ft)	Out 3 - In 3 (Out 0.036 - In 0.036)
Standard side rod length	mm (in)	175.9 (6.93)
Front wheel turning angle Toe-out turns (Inside/Outside)	degree	20/17°30'
Full turn (Inside/Outside)	degree	
E15		40° - 44°/31° - 35°
CD17 M/T		37° - 41°/29° - 33°
CD17 A/T		33° - 37°/27° - 31°

Valve guide

Unit: mm (in)

	Standard	Service
Valve guide Outer diameter	11.023 - 11.034 (0.4340 - 0.4344)	
Valve guide Inner diameter [Finished size]	7.000 - 7.018 (0.2756 - 0.2763)	
Cylinder head valve guide hole diameter	10.975 - 10.996 (0.4321 - 0.4329)	
Interference fit of valve guide	0.027 - 0.059 (0.0011 - 0.0023)	

	Standard	Max. tolerance
Stem to guide clearance Intake	0.020 - 0.053 (0.0008 - 0.0021)	0.1 (0.04)
Exhaust	0.040 - 0.073 (0.0016 - 0.0029)	0.1 (0.04)
Valve deflection limit		0.2 (0.008)

CAMSHAFT AND CAMSHAFT BEARING

Unit: mm (in)

	Standard	Max. tolerance
Camshaft journal to bearing clearance	0.02 - 0.06 (0.0008 - 0.0024)	0.1 (0.004)
Inner diameter of camshaft bearing	30.000 - 30.021 (1.1811 - 1.1819)	–
Outer diameter of camshaft journal	29.960 - 29.980 (1.1795 - 1.1803)	–
Camshaft bend	Less than 0.02 (0.0008)	0.05 (0.0020)
Camshaft end play		0.17 (0.0067)

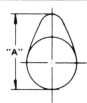

"A"

EM671

Cam height "A" Intake	44.495 - 44.500 (1.7518 - 1.7520)
Exhaust	45.495 - 45.500 (1.7911 - 1.7913)
Wear limit of cam height	0.15 (0.0059)

CYLINDER BLOCK

Unit: mm (in)

Surface flatness Wear limit	0.1 (0.004)
Cylinder bore Inner diameter Standard	80.000 - 80.050 (3.1496 - 3.1516)
Wear limit	0.2 (0.008)
Out-of-round (X-Y) Wear limit	0.2 (0.008)
Taper (A-B) Wear limit	0.2 (0.008)
Difference in inner diameter between cylinders Standard	Less than 0.05 (0.0020)
Wear limit	0.2 (0.008)
Piston to cylinder clearance Standard	0.05 - 0.07 (0.0020 - 0.0028)
Feeler gauge extracting force [with gauge thickness 0.06 mm (0.0024 in)] N (kg, lb)	7.8 - 14.7 (0.8 - 1.5, 1.8 - 3.3)

FORD

GENERAL SPECIFICATIONS

DISPLACEMENT	2.3L
NUMBER OF CYLINDERS	4
BORE AND STROKE	3.780 x 3.126
FIRING ORDER	1-3-4-2
OIL PRESSURE (HOT @ 2000 RPM)	40 - 60

CYLINDER HEAD AND VALVE TRAIN

COMBUSTION CHAMBER VOLUME (cc)	59.8 - 62.8
VALVE GUIDE BORE DIAMETER	0.3433 - 0.3443
VALVE SEATS	
Width – Intake	.060 - .080
Width – Exhaust	.070 - .090
Angle	45°
RUNOUT LIMIT (T. I. R. MAX)	0.0016
VALVE ARRANGEMENT (Front to Rear)	E-I-E-I-E-I-E-I
VALVE LASH ADJUSTER BORE DIAMETER	0.8430 - 0.9449
VALVE STEM TO GUIDE CLEARANCE	
Intake	0.0010 - 0.0027
Exhaust	0.0015 - 0.0032
Service Clearance Limit	0.0055 Max.
VALVE HEAD DIAMETER	
Intake	1.73 - 1.74
Exhaust	1.49 - 1.51
VALVE FACE RUNOUT LIMIT	0.002 Max.
VALVE FACE ANGLE LIMIT	44°
VALVE STEM DIAMETER (STANDARD)	
Intake	.3416-.3423
Exhaust	.3411-.3418
(0.015 Oversize)	
Intake	.3566-.3573
Exhaust	.3561-.3568
(0.030 Oversize)	
Intake	.3716-.3723
Exhaust	.3711-.3718
VALVE SPRINGS	
Compression Pressure (Lb. @ Spec. Length)	
Intake	71 - 79 @ 1.56
Exhaust	159 - 175 @ 1.16
Free Length (Approximate)	1.89
Assembled Height	1-17/32'' - 1-19/32
Service Limit	10° Pressure Loss @ Specified Length
Out of Square Service Limit	5/64 (0.078)
ROCKER ARM (Cam Follower)	
Ratio	1.64:1
VALVE TAPPET, LIFTER OR ADJUSTER	
Diameter (Standard)	0.8422 - 0.8427
Clearance-to-Bore	0.0007 - 0.0027
Service Limit	0.005 Max.
Hydraulic Leakdown Rate	2 - 8 Seconds
Collapsed Tappet Gap	
Allowable	0.035 - 0.055 @ Cam
Desired	0.040 - 0.050 @ Cam

CAMSHAFT

LOBE LIFT	
Intake	0.2437
Exhaust	0.2437
Allowable Lobe Lift Loss	0.005 Max.
THEORETICAL VALVE LIFT @ ZERO LASH	
Intake	0.3997
Exhaust	0.3997
ENDPLAY	0.001 - 0.007
Service Limit	0.009
JOURNAL-TO-BEARING CLEARANCE	0.001 - 0.003
Service Limit	0.006

CAMSHAFT (Continued)

JOURNAL DIAMETER	
#1	1.7713 - 1.7720
#2	1.7713 - 1.7720
#3	1.7713 - 1.7720
#4	1.7713 - 1.7720
Runout Limit	0.005 Max. T. I. R.
Out-of-Round Limit	0.005 T. I. R. Max.
Front Bearing Location	0.000 - 0.010

CYLINDER BLOCK

HEAD GASKET SURFACE FLATNESS	0.003 in any 6'' – 0.006 overall
HEAD GASKET SURFACE FINISH (RMS)	60 - 150
CYLINDER BORE	
Diameter	3.7795 - 3.7831
Surface Finish (RMS)	18 - 38
Out-of-Round Limit	0.0015
Out-of-Round Service Limit	0.005
Taper Service Limit	0.010
MAIN BEARING BORE DIAMETER	2.5902 - 2.5910
DISTRIBUTOR SHAFT BEARING BORE DIAMETER	.5155 - .5170

PISTONS AND RINGS

PISTON	
Diameter	
Coded Red	Non-Turbo 3.7780 - 3.7786
	Turbo 3.7760 - 3.7766
Coded Blue	Non-Turbo 3.7792 - 3.7798
	Turbo 3.7772 - 3.7778
0.003 Oversize	Non-Turbo 3.7804 - 3.7810
	Turbo 3.7784 - 3.7790
Piston-to-Bore-Clearance (Select Fit)	Non-Turbo 0.0014 - 0.0022
	Turbo 0.0034 - 0.0042
Pin Bore Diameter	Non-Turbo 0.9123 - 0.9126
	Turbo 0.9124 - 0.9127
Ring Groove Width	
Compression (Top)	0.080 - 0.081
Compression (Bottom)	0.080 - 0.081
Oil	0.188 - 0.189
PISTON PIN	
Length	3.010 - 3.040
Diameter	
Standard	0.9119 - .9124
0.001 Oversize	0.9130 - .9133
0.002 Oversize	0.9140 - .9143
Piston-to-Pin Clearance	0.0002 - 0.0004
Pin-to-Rod Clearance	Interference Fit
PISTON RINGS	
Ring Width	
Compression (Top)	0.077 - 0.078
Compression (Bottom)	0.077 - 0.078
Side Clearance	
Compression (Top)	0.002 - 0.004
Compression (Bottom)	0.002 - 0.004
Oil Ring	Snug Fit
Service Limit	0.006 Max.
Ring Gap	
Compression (Top)	0.010 - 0.020
Compression (Bottom)	0.010 - 0.020
Oil (Steel Rail)	0.015 - 0.055

FRONT WHEEL ALIGNMENT

Vehicle Model	Alignment Factors	Nominal	Minimum	Maximum
Mustang	Caster	+1°	+1/4°	+1-3/4°
	Camber	+1/4°	−1/2°	+1°
	Toe-inches	3/16''	1/16''	5/16''

DODGE

FRONT WHEEL ALIGNMENT	Acceptable Alignment Range	Preferred Setting
CAMBER—All Models	−0.2° to +0.8° (−1/4° to +3/4°)	+0.3° (+5/16°)
TOE—All Models		
Specified in Inches	7/32″ **OUT** to 1/8″ **IN**	1/16″ **OUT** ±1/16″
Specified in Degrees	.4° **OUT** to .2° **IN**	0.1° **OUT** ±.1°

REAR WHEEL ALIGNMENT	Acceptable Alignment Range	Preferred Setting
CAMBER		
M, Z	−1.25° to −.25° (−1-1/4° to −1/4°)	−.75° ± .5° (1/2°)
P, D, C, V	−1.0° to 0° (−1° to 0°)	−.5° ± .5° (1/2°)
Z-28	−1.1° to −.1° (−1-1/8° to −1/8°)	−.6° ± .5° (1/2°)
TOE*		
M, Z, Z-28 Specified in Inches	5/32″ **OUT** to 11/32″ **IN**	3/32″ **IN**
Specified in Degrees	0.3° **OUT** to 0.7° **IN**	0.2° **IN**
P, D, C, V Specified in Inches	3/16″ **OUT** to 3/16″ **IN**	0″ ± 1/8″
Specified in Degrees	.38° **OUT** to .38° **IN**	0° ± .25°

*TOE OUT when backed on alignment rack is TOE IN when driving.

Description	Standard Dimension	Service Limit
Compression Pressure	1.03 MPa (149 psi) at 250 RPM	
Maximum Variation Between Cylinders	0.1 MPa (15 psi)	
Valve Clearance—Hot Engine		
Intake Valve	0.15mm (.006 in.)	
Exhaust Valve	0.25mm (.010 in.)	
Jet Valve	0.15mm (.006 in.)	
Flatness of Cylinder Head Gasket Surface	Less than 0.05mm (.002 in.)	Less than 0.1mm (.004 in.)
Camshaft		
Bearing Clearance	0.05/0.09mm (.002/.004 in.)	
Height of Cam Lobe		
Intake and Exhaust	42.2mm (1.6614 in.)	41.7mm—min. (1.6414 in.—min.)
End Play	0.1/0.2mm (.004/.008 in.)	
Valves		
Thickness of Valve Head (Margin)		
Intake	1.2mm (.047 in.)	0.7mm (.028 in.)
Exhaust	2mm (.079 in.)	1mm (.039 in.)
Valve Stem to Valve Guide Clearance		
Intake	0.03/0.06mm (.0012/.0024 in.)	0.1mm (.004 in.)
Exhaust	0.05/0.09mm (.0020/.0035 in.)	0.15mm (.006 in.)
Valve Spring, Free Length	47.5mm (1.869 in.)	46.5mm (1.479 in.)
Valve Spring Load	270 N at 40.4mm (61 lbs. at 1.59 in.)	
Valve Spring Perpendicularity	1.5° max.	3° max
Valve Spring Installed Height	40.4mm (1.590 in.)	41.4mm—max. (1.629 in.—max.)
Jet Valve		
Stem O.D.	4.3000mm (.1693 in.)	
Seat Angle	45°	
Spring Free Length	29.60mm (1.1654 in.)	
Load	34.3 N at 21.5mm (5.5 lbs. at .846 in.)	

(Dodge continued)

Description	Standard Dimension	Service Limit
Piston		
O. D.	91.1mm (3.5866 in.)	
Ring Side Clearance		
No. 1 Ring	0.06/0.10mm (.0024/.0039 in.)	0.15mm (.006 in.)
No. 2 Ring	0.02/0.06mm (.0008/.0024 in.)	0.1mm (.0039 in.)
Piston Pin Press-in Pressure	7,350/17,100 N (1,614/3,859 lbs.)	
End Gap		
No. 1 Ring	0.25/0.45mm (.01/.018 in.)	1.0mm (.039 in.)
No. 2 Ring	0.25/0.45mm (.01/.018 in.)	1.0mm (.039 in.)
Side Rail—Oil Ring	0.2/0.9mm (.0078/.035 in.)	1.5mm (.059 in.)
Connecting Rod		
Bend	0.05mm in 100mm (.002 in. in 3.937 in.)	
Twist	0.1mm in 100mm (.0039 in. in 3.937 in.)	
Connecting Rod to Crankshaft Side Clearance	0.1/0.25mm (.004/.010 in.)	
Bearing Clearance	0.02/0.07mm (.0008/.0028 in.)	
Crankshaft		
Conecting Rod Journal O.D.	53mm (2.0866 in.)	
Main Bearing Journal O.D.	60mm (2.3622 in.)	
Out-of-Roundness and Taper of Bearing Surface	0.01mm (.0004 in.)	
Main Bearing Clearance	0.02/0.07mm (.0008/.0028 in.)	
End Play	0.05/0.18mm (.002/.007 in.)	
Silent Shaft		
Front Bearing Journal O.D.	23mm (.906 in.)	
Front Bearing Clearance	0.02/0.06mm (.0008/.0024 in.)	
Rear Bearing Journal O.D.	43mm (1.693 in.)	
Rear Bearing Clearance	0.05/0.09mm (.0020/.0035 in.)	
Timing Belt Tensioner		
Spring Free Length	65.7mm (2.587 in.)	
Spring Load	19.6 N at 36.9mm (4.4 lbs. at 1.453 in.)	
Oil Pump		
Relief Valve Opening Pressure	343 to 441 kPa (49.8 to 64.0 psi)	
Gear to Housing Clearance	0.11/0.15mm (.0043/.0059 in.)	
Driven Gear to Bearing Clearance	0.02/0.05mm (.0008/.0020 in.)	
Driven Gear to Bearing Clearance (Oil Pump Body) .	0.02/0.05mm (.0008/.0020 in.)	
Driven Gear to Bearing Clearance (Oil Pump Cover) .	0.04/0.07mm (.0016/.0028 in.)	
Gear End Play	0.06/0.12mm (.0024/.0047 in.)	
Relief Spring Free Length	47mm (1.850 in.)	
Relief Spring Load	42.2 N at 40mm (9.5 lbs. at 1.575 in.)	
Oil Pressure Switch Minimum Actuating Pressure ..	28 kPa (4 psi) or less	

(Dodge continued)

SPECIFICATIONS METRIC

FRONT BRAKES

Type	Single Piston—Pin Slider—Disc
Caliper Bore Diameter	54mm
Adjustment	Automatic
Piston Material	Glass Filled Phenolic
Piston Boot Type...................................	Press in EPDM Rubber
Disc Type L body	Solid
All—except L body	Vented
Diameter—Outside L	228mm—Standard
P & K/P,D...........................	240mm—Standard. . . . 260mm Heavy Duty
C,E,G,H,J & K/C	260mm—Standard
Runout—Maximum Allowable T.I.R.	.1016mm
Parallelism—Total Variation in Thickness in 360° of Rotation	.01270mm

REAR BRAKES—DRUM

Type	Leading Trailing
Adjustment	Automatic
Diameter—H,J,K,L,P	200mm—Standard
E,G	220mm—Standard
C,K	220mm—Heavy Duty and C-45
Wheel Cylinder Diameter—L,H,J,K,P,C	15.9mm and C-45
E,G	14.3mm

REAR BRAKES—DISC

Type	Single Piston—Pin Slider
Caliper Bore Diameter	33 and 36mm
Adjustment	Automatic
Piston Material	Metal
Piston Boot Type...................................	Press in EPDM Rubber
Disc Type	Solid
Diameter—outside	275mm

1 FORD 1

FUSION, MILAN (06-09), ZEPHYR (06), MKZ (07-09), MILAN HYBRID (10)

SECTION INDEX

ALPHABETICAL INDEX

1 FORD
FUSION, MILAN (06-09), ZEPHYR (06), MKZ (07-09), MILAN HYBRID (10)

FORD 1

FUSION, MILAN (06-09), ZEPHYR (06), MKZ (07-09), MILAN HYBRID (10)

1 FORD
FUSION, MILAN (06-09), ZEPHYR (06), MKZ (07-09), MILAN HYBRID (10)

LABOR 1 MAINTENANCE & LUBRICATION 1 LABOR

OPERATION INDEX

AIR FILTER, R&R 9
AIR FILTER, SERVICE 8
BELT TENSIONER, R&R 6
CHASSIS, LUBRICATE 1
COOLING SYSTEM, SERVICE 11
IDLER PULLEY, R&R 7
LUBE & FILTER, SERVICE 12
MAINTENANCE SCHEDULE, SERVICE 13
OIL FILTER, R&R 10
SERPENTINE BELT, R&R 5
TRANS FLUID, CHANGE 14
TRANS FLUID, CHANGE 15
WHEELS, BALANCE 3
WHEELS, R&R 2
WHEELS, ROTATE 4

CHASSIS

CHASSIS & WHEELS

		(Factory) Time	Motor Time

1-CHASSIS, LUBRICATE
Fusion, Milan, Zephyr, MKZ
See LUBE & FILTER, SERVICE

2-WHEELS, R&R
Fusion, Milan, Zephyr, MKZ
One .... 06-09 C 0.5
NOTES FOR: WHEELS, R&R
To R&R TPMS Attachment Kit, Add
One .... 06-09 .... 0.2

3-WHEELS, BALANCE
Fusion, Milan, Zephyr, MKZ
One .... 06-09 C 0.3
Each Adtnl. .... 06-09 C 0.2

4-WHEELS, ROTATE
Fusion, Milan, Zephyr, MKZ
4 Wheels .... 06-09 C 0.4
NOTES FOR: WHEELS, ROTATE
To Include Spare, Add .... 06-09 .... 0.1

ENGINE SERVICE

BELTS & PULLEYS

5-SERPENTINE BELT, R&R
Fusion, Milan, Zephyr
2.3L .... 06-09 C (0.3) 0.4
3.0L .... 06-09 C (0.3) 0.4
MKZ
One .... 07-09 C (0.4) 0.5
Both .... 07-09 C (0.4) 0.5

6-BELT TENSIONER, R&R
Fusion, Milan, Zephyr
2.3L .... 06-09 B (0.4) 0.5
3.0L .... 06-09 B (0.4) 0.5

		(Factory) Time	Motor Time

MKZ .... 07-09 B (0.9) 1.3

7-IDLER PULLEY, R&R
Fusion, Milan, Zephyr
2.3L .... 06-09 B (0.5) 0.6
3.0L
One .... 06-09 B (0.4) 0.5
Both .... 06-09 B (0.5) 0.6

FILTERS

8-AIR FILTER, SERVICE
Fusion, Milan, Zephyr, MKZ .... 06-09 C (0.2) 0.3

9-AIR FILTER, R&R
Fusion, Milan, Zephyr, MKZ .... 06-09 C (0.2) 0.3

10-OIL FILTER, R&R
Fusion, Milan, Zephyr
2.3L .... 06-09 B (0.6) 0.6
3.0L .... 06-09 B (0.4) 0.4
MKZ .... 07-09 B (0.4) 0.4
NOTES FOR: OIL FILTER, R&R
To Change Oil, Add .... 06-09 .... 0.1

PERIODICAL MAINTENANCE

PERIODICAL MAINTENANCE

11-COOLING SYSTEM, SERVICE
Fusion, Milan, Zephyr, MKZ .... 06-09 B 0.9
NOTES FOR: COOLING SYSTEM, SERVICE
Includes: Pressure Test System For Leaks, Check Thermostat & Heater Operation, Check All Hoses & Belts. Drain & Flush System And Add Coolant.

12-LUBE & FILTER, SERVICE
Fusion, Milan, Zephyr
2.3L .... 06-09 B 1.0
3.0L .... 06-09 B 0.8
MKZ .... 07-09 B 0.8
NOTES FOR: LUBE & FILTER, SERVICE
Includes: Inspect And Correct All Fluid Levels And Tire Pressure.
To Install Grease Fittings,
Add .... 06-09 .... 0.1

13-MAINTENANCE SCHEDULE, SERVICE
5000 MILES
Fusion, Milan, Zephyr, MKZ .... 06-08 C (0.7) 0.9
7500 MILES
Fusion, Milan, Zephyr, MKZ .... 09 C (0.8) 1.0
10,000 MILES
Fusion, Milan, Zephyr, MKZ .... 06-08 C (0.7) 0.9
15,000 MILES
Fusion, Milan, Zephyr, MKZ .... 06-08 C (0.9) 1.4
.... 09 C (1.0) 1.4
20,000 MILES
Fusion, Milan, Zephyr, MKZ .... 06-08 C (0.7) 0.9
22,500
Fusion, Milan, Zephyr, MKZ .... 09 C (0.7) 0.9
25,000 MILES
Fusion, Milan, Zephyr, MKZ .... 06-08 C (0.7) 0.9

		(Factory) Time	Motor Time

30,000 MILES
Fusion, Milan, Zephyr, MKZ .... 09 C (1.0) 1.4
35,000 MILES
Fusion, Milan, Zephyr, MKZ .... 06-08 C (0.7) 0.9
37,500 MILES
Fusion, Milan, Zephyr, MKZ .... 09 C (0.8) 1.0
40,000 MILES
Fusion, Milan, Zephyr, MKZ .... 06-08 C (0.7) 0.9
45,000 MILES
Fusion, Milan, Zephyr, MKZ .... 06-08 C (0.9) 1.4
.... 09 C (1.0) 1.4
50,000 MILES
Fusion, Milan, Zephyr, MKZ .... 06-08 C (0.7) 0.9
52,500 MILES
Fusion, Milan, Zephyr, MKZ .... 09 C (0.8) 1.0
55,000 MILES
Fusion, Milan, Zephyr, MKZ .... 06-08 C (0.7) 0.9
60,000 MILES
Fusion, Milan, Zephyr, MKZ .... 09 C (1.0) 1.4
65,000 MILES
Fusion, Milan, Zephyr, MKZ .... 06-08 C (0.7) 0.9
67,500 MILES
Fusion, Milan, Zephyr, MKZ .... 09 C (0.8) 1.0
70,000 MILES
Fusion, Milan, Zephyr, MKZ .... 06-08 C (0.7) 0.9
75,000 MILES
Fusion, Milan, Zephyr, MKZ .... 06-08 C (0.9) 1.4
.... 09 C (1.0) 1.4
80,000 MILES
Fusion, Milan, Zephyr, MKZ .... 06-08 C (0.7) 0.9
82,500 MILES
Fusion, Milan, Zephyr, MKZ .... 09 C (0.8) 1.0
85,000 MILES
Fusion, Milan, Zephyr, MKZ .... 06-08 C (0.7) 0.9
90,000 MILES
Fusion, Milan, Zephyr, MKZ
2.3L .... 09 C (1.5) 2.3
3.0L .... 09 C (2.4) 3.3
95,000 MILES
Fusion, Milan, Zephyr, MKZ .... 06-08 C (0.7) 0.9
97,500 MILES
Fusion, Milan, Zephyr, MKZ .... 09 C (0.8) 1.0
NOTES FOR: MAINTENANCE SCHEDULE, SERVICE
For Vehicles Following A Severe Maintenance Schedule, Use 1.0 For Every 3,000 Mile Schedule Interval UNLESS The Normal Maintenance Schedule Listed Above Is Already Divisible By 3,000, Then Use The Time Already Indicated.

TRANSAXLE

AUTOMATIC TRANSAXLE

14-TRANS FLUID, CHANGE
Fusion, Milan, Zephyr, MKZ .... 06-09 B 1.4
NOTES FOR: TRANS FLUID, CHANGE
Includes: R&I Oil Pan, Drain And Refill Transmission.
To Change Fluid, Add 0.1

MANUAL TRANSAXLE

15-TRANS FLUID, CHANGE
Fusion, Milan .... 06-09 B 0.5

PARTS 1 MAINTENANCE & LUBRICATION 1 PARTS

PARTS INDEX

AIR FILTER 5
FILTER 7
OIL FILTER 6
SERPENTINE BELT 2
SERPENTINE IDLER PULLEY 3
SERPENTINE TENSIONER 4
WATER PUMP BELT 1

ENGINE SERVICE

BELTS & PULLEYS

			Part No.	Price

1-WATER PUMP BELT
Fusion, Milan
3.0L 06-09 6E5Z8620CA 28.76
Zephyr 06 6E5Z8620CA 28.76
MKZ 07-09 7T4Z8620A 21.09

			Part No.	Price

2-SERPENTINE BELT
Fusion, Milan
2.3L
Manual Trans 06-09 6E5Z8620C 53.02
Auto Trans 06-09 6E5Z8620B 57.51
3.0L 06-09 6E5Z8620BB 35.96
Zephyr 06 6E5Z8620BB 35.96
MKZ 07-09 BT4Z8620A 18.38

3-SERPENTINE IDLER PULLEY
Fusion, Milan
2.3L 06-09 3M4Z8678AB 20.15
3.0L
Smooth Pulley 06-09 6E5Z8678AA 26.53
Grooved Pulley 06-09 6E5Z6C348A 31.15
Zephyr
Smooth Pulley 06 6E5Z8678AA 26.53
Grooved Pulley 06 6E5Z6C348A 31.15

4-SERPENTINE TENSIONER
Fusion, Milan
2.3L 06-09 6E5Z6A228A 93.00
3.0L 06-09 4L8Z6B209AA 106.35
Zephyr 06 4L8Z6B209AA 106.35

			Part No.	Price

MKZ 07-09 BT4Z6B209B 61.67

FILTERS

5-AIR FILTER
Fusion, Milan
2.3L 06-09 6E5Z9601EA 25.02
3.0L 06-09 6E5Z9601GA 25.04
Zephyr 06 6E5Z9601GA 25.04
MKZ 07-09 7T4Z9601A 25.04A

6-OIL FILTER
Fusion, Milan
2.3L 06-09 3S7Z6731A 10.51
3.0L 06-09 F1AZ6731BD 7.69
Zephyr 06 F1AZ6731BD 7.69
MKZ 07-09 E4FZ6731AB 7.22

TRANSAXLE

AUTOMATIC TRANSAXLE

7-FILTER
Fusion, Milan
5 Speed Auto 06-09 8E5Z7B155A 93.67

FORD 1

FUSION, MILAN (06-09), ZEPHYR (06), MKZ (07-09), MILAN HYBRID (10)

LABOR | 2 EMISSION SYSTEM 2 | LABOR

OPERATION INDEX

A.I.R. PUMP, R&R	2
CONTROL VALVE, R&R	7
DETECTION PUMP, R&R	6
EGR VALVE, R&R	3
PCV VALVE, R&R	1
SOLENOID VALVE, R&R	4
TUBE, R&R	5
VAPOR CANISTER, R&R	8

EMISSION SYSTEM

EMISSION SYSTEM

		(Factory Time)	Motor Time
1-PCV VALVE, R&R			
Fusion, Milan, Zephyr			
2.3L	06-08 B		2.1

			(Factory Time)	Motor Time
3.0L		09 B	(1.6)	2.1
		06-09 B	(0.2)	0.3
MKZ		06-08 B	(0.2)	0.3
		09 B	(0.3)	0.4

A.I.R. SYSTEM

2-A.I.R. PUMP, R&R
Fusion, Milan, Zephyr

2.3L		06-09 B		0.8

EGR SYSTEM

3-EGR VALVE, R&R
Fusion, Milan, Zephyr

2.3L		06-09 B	(0.6)	0.7
3.0L		06-09 B	(0.5)	0.6

4-SOLENOID VALVE, R&R
Fusion, Milan, Zephyr | 06-09 B | 0.3

			(Factory Time)	Motor Time
5-TUBE, R&R				
Fusion, Milan, Zephyr				
2.3L		06-09 B		2.1
3.0L		06-09 B	(0.7)	0.7

VAPOR CANISTER

6-DETECTION PUMP, R&R
Fusion, Milan, Zephyr, MKZ .. | 06-09 B | | 2.7

7-CONTROL VALVE, R&R
VENT CONTROL VALVE
Fusion, Milan, Zephyr, MKZ ... | 06-09 B | (0.6) | 0.7
VAPOR MANAGEMENT
Fusion, Milan, Zephyr

2.3L		06-09 B	(0.4)	0.5
3.0L		06-09 B	(0.9)	1.4
MKZ		07-09 B	(0.9)	1.4

8-VAPOR CANISTER, R&R
Fusion, Milan, Zephyr, MKZ .. | 06-09 B | (0.7) | 0.9

PARTS | 2 EMISSION SYSTEM 2 | PARTS

PARTS INDEX

A.I.R. PUMP	8
CONTROL VALVE	9
EGR VALVE	2
GASKET	3
PCV VALVE	1
PURGE CONTROL VALVE	6
TUBE	4
VAPOR CANISTER	5
VENT CONTROL SOLENOID	7

EMISSION SYSTEM

EMISSION SYSTEM

		Part No.	Price
1-PCV VALVE			
Fusion, Milan			
2.3L	06-09	4L5Z6A666BA	19.67
3.0L	06-09	5L3Z6A666BA	78.56
Zephyr	06	5L3Z6A666BA	78.56
MKZ	07-09	2C5Z6A666AA	56.64

EGR SYSTEM

		Part No.	Price
2-EGR VALVE			
Fusion, Milan			
2.3L	06-09	1S7Z9D475A	82.78

			Part No.	Price
3.0L		06-09	6E5Z9D475BA	182.44
Zephyr		06	6E5Z9D475BA	182.44

3-GASKET
Fusion, Milan

		Part No.	Price
2.3L	06-09	1S7Z9D476AA	10.78
3.0L	06-09	E6AZ9D476B	3.55
Zephyr	06	E6AZ9D476B	3.55

4-TUBE
Fusion, Milan

		Part No.	Price
3.0L	06	6E5Z9D477GA	26.65
	07-09	7E5Z9D477AA	27.48
Zephyr	06	6E5Z9D477GA	26.65

VAPOR CANISTER

5-VAPOR CANISTER
Fusion, Milan
2.3L
To 9/4/06 .... 06-07 | 6E5Z9D653AB | 301.73
From 9/4/06
49 State/Non Green State

		Part No.	Price
To 6/30/08	08-09	9E5Z9D653A	234.69
From 6/30/08	09	9E5Z9D653C	301.73

California/Green State

		Part No.	Price
To 6/30/08	08-09	9E5Z9D653A	234.69
From 6/30/08	09	9E5Z9D653A	234.69

3.0L

			Part No.	Price
To 9/4/06		06-07	6E5Z9D653AB	301.73
From 9/4/06		07-09	9E5Z9D653C	301.73
Zephyr		06	6E5Z9D653AB	301.73
MKZ		07-09	9E5Z9D653C	301.73

6-PURGE CONTROL VALVE
Fusion, Milan

			Part No.	Price
To 9/4/06		06-07	4U5Z9C915CA	37.53
From 9/4/06		07-09	7U5Z9C915D	47.00
Zephyr		06	4U5Z9C915CA	37.53
MKZ		07-09	7U5Z9C915D	47.00

7-VENT CONTROL SOLENOID

		Part No.	Price
Fusion, Milan	06-09	9U5Z9F945E	48.75
Zephyr	06	9U5Z9F945E	48.75
MKZ	07-09	9U5Z9F945E	48.75

A.I.R. SYSTEM

8-A.I.R. PUMP
Fusion, Milan

		Part No.	Price
2.3L	06-09	6E5Z9A486BA	307.00

9-CONTROL VALVE
Fusion, Milan

		Part No.	Price
2.3L	06-09	6E5Z9F491AA	297.00

LABOR | 3 ELECTRICAL 3 | LABOR

OPERATION INDEX

AMPLIFIER, R&R	69	LOCK SWITCH, R&R	75	WINDOW REGULATOR, R&R	78
ANTENNA, R&R	73	MAIN RELAY, R&R	16	WINDOW REGULATOR, R&R	81
BACK-UP SWITCH, R&R	36	MASS AIR FLOW SENSOR, R&R	17	WINDOW SWITCH, R&R	77
BATTERY TERMINALS, CLEAN	24	MEMORY MODULE, R&R	87	WINDOW SWITCH, R&R	82
BATTERY, CHARGE/TEST	22	MEMORY MODULE, R&R	91	WIPER ARM, R&R	50
BATTERY, R&R	23	MIRROR SWITCH, R&R	86	WIPER BLADE, R&R	51
CABLE ASSY, R&R	25	MODULE, R&R	56	WIPER MOTOR, R&R	47
CAMSHAFT POSITION SENSOR, R&R	9	MODULE, R&R	72	WIPER RELAY, R&R	52
CENTER SENSOR, R&R	59	MONITOR, DIAGNOSIS	18	WIPER SWITCH, R&R	45
CLOCKSPRING, R&R	58	MOTOR, R&R	83	WIPER TRANSMISSION, R&R	48
CLUTCH SWITCH, R&R	37	MOTOR, R&R	89	
COMPOSITE ASSY, R&R	32	MULTIFUNCTION SWITCH, R&R	41	
COMPRESSION, TEST	1	OIL PRESSURE SENDING UNIT, R&R	35	
COOLANT TEMP SENSOR, R&R	10	OXYGEN SENSOR, R&R	19	
CRANKSHAFT POSITION SENSOR, R&R	11	PARK BRAKE WARNING SWITCH, R&R	43	
DIAGNOSTIC CIRCUIT, INSPECT	12	PASSENGER DISCRIMINATING SENSOR, R&R	64	
DIAGNOSTIC MODULE, R&R	60	PASSENGER INFLATOR MODULE, R&R	62	
DRIVER INFLATOR MODULE, R&R	61	PASSENGER MODULE, R&R	63	
ECM, R&R	13	PINPOINT, TEST	20	
ECM, REPROGRAM	14	POSITION SENSOR, R&R	88	
ELEMENT, R&R	85	RADIO, R&R	70	
EXTERIOR BULBS, R&R	26	RANGE SENSOR, R&R	42	
FOG LAMP ASSY, R&R	31	RESERVOIR, R&R	46	
FUEL GAUGE SENDING UNIT, R&R	34	SENSOR, R&R	57	
HAZARD SWITCH, R&R	38	SIDE IMPACT INFLATOR MODULE, R&R	67	
HEADLAMP BULB, R&R	28	SIDE SENSOR, R&R	66	
HEADLAMP SWITCH, R&R	39	SPARK PLUGS, R&I	3	
HEADLAMP, ALIGN	33	SPARK PLUGS, R&R	4	
HORN RELAY, R&R	30	SPEAKER, R&R	55	
HORN, R&R	29	SPEAKER, R&R	71	
IGNITION COIL, R&R	2	STOPLAMP SWITCH, R&R	44	
IGNITION LOCK CYLINDER, R&R	7	SWITCH, R&R	84	
IGNITION SWITCH, R&R	8	SWITCH, R&R	90	
INFLATOR CURTAIN, R&R	65	SYSTEM, DIAGNOSIS	5	
INSTRUMENT CLUSTER, R&R	53	SYSTEM, DIAGNOSIS	68	
INSTRUMENT LIGHT RHEOSTAT, R&R	40	SYSTEM, TUNE-UP	6	
INTERIOR BULB, R&R	27	TIRE PRESSURE SENSOR, R&R	54	
KEYBOARD ACTUATOR, R&R	92	TRANSMITTER, R&R	93	
KNOCK SENSOR, R&R	15	VEHICLE SPEED SENSOR, R&R	21	
LOCK ACTUATOR, R&R	74	WASHER PUMP, R&R	49	
LOCK ACTUATOR, R&R	79	WINDOW MOTOR, R&R	76	
		WINDOW MOTOR, R&R	80	

IGNITION SYSTEM

IGNITION SYSTEM

			(Factory Time)	Motor Time
1-COMPRESSION, TEST				
Fusion, Milan, Zephyr				
2.3L		06-09 B		0.7
3.0L		06-09 B		2.3
MKZ		07-09 C		2.3

NOTES FOR: COMPRESSION, TEST
Includes: R&I Spark Plugs.

2-IGNITION COIL, R&R
Fusion, Milan, Zephyr
2.3L

One		06-09 B	(0.2)	0.3
All		06-09 B	(0.3)	0.4

3.0L
Right Side

One		06-08 B	(1.2)	1.6
		09 B	(1.0)	1.4
All		06-08 B	(1.4)	1.6
		09 B	(1.1)	1.6

Left Side

One		06-08 B	(0.3)	0.3
		09 B	(0.1)	0.3
All		06-08 B	(0.5)	0.6

(CONTINUED)

1 FORD
FUSION, MILAN (06-09), ZEPHYR (06), MKZ (07-09), MILAN HYBRID (10)

ELECTRICAL - Time Cont'd	(Factory) Time	Motor Time
	09 B (0.2)	0.3
Both Sides		
All	06-08 B (1.6)	1.8
	09 B (1.2)	1.7
MKZ		
Right Side		
One	07-09 B (0.6)	0.9
All	07-09 B (0.7)	1.0
Left Side		
One	07-09 B (0.2)	0.3
All	07-09 B (0.3)	0.6
Both Sides		
All	07-09 B (0.8)	1.8
3-SPARK PLUGS, R&I		
Fusion, Milan, Zephyr		
2.3L	06-09 C (0.6)	0.9
3.0L	06-09 C (1.6)	2.0
MKZ	07-09 C (1.2)	2.0
4-SPARK PLUGS, R&R		
Fusion, Milan, Zephyr		
2.3L	06-09 C (0.6)	0.9
3.0L	06-09 C (1.6)	2.0
MKZ	07-09 C (1.2)	2.0
5-SYSTEM, DIAGNOSIS		
Fusion, Milan, Zephyr, MKZ ..	06-09 B	0.6
NOTES FOR: SYSTEM, DIAGNOSIS		
Includes: Check Ignition Timing And Fuel Adjustments.		
6-SYSTEM, TUNE-UP		
Fusion, Milan, Zephyr, MKZ		
See Service		
Maintenance Schedule 06-09		

IGNITION LOCK

	(Factory) Time	Motor Time
7-IGNITION LOCK CYLINDER, R&R		
Fusion, Milan, Zephyr, MKZ ..	06-09 B (0.5)	0.6
8-IGNITION SWITCH, R&R		
Fusion, Milan, Zephyr, MKZ ..	06-09 B (0.5)	0.6

POWERTRAIN CONTROL

POWERTRAIN CONTROL

	(Factory) Time	Motor Time
9-CAMSHAFT POSITION SENSOR, R&R		
Fusion, Milan, Zephyr		
2.3L	06-09 B (0.2)	0.3
3.0L		
Right Side	06-09 B (0.2)	0.4
Left Side	06-09 B (0.2)	0.3
Both Sides	06-09 B (0.4)	0.5
MKZ		
Right Side	07-09 B (0.2)	0.4
Left Side	07-09 B (0.2)	0.3
Both Sides	07-09 B (0.3)	0.5
10-COOLANT TEMP SENSOR, R&R		
Fusion, Milan, Zephyr		
2.3L	06-08 B (0.2)	0.3
	09 B (0.1)	0.3
3.0L	06-09 B (0.4)	0.5
MKZ	07-09 B	0.5
11-CRANKSHAFT POSITION SENSOR, R&R		
Fusion, Milan, Zephyr		
2.3L	06-09 B (0.4)	0.8
3.0L	06-09 B (0.3)	0.5
MKZ	07-09 B	1.8
12-DIAGNOSTIC CIRCUIT, INSPECT		
Fusion, Milan, Zephyr, MKZ ..	06-09 B	0.5
NOTES FOR: DIAGNOSTIC CIRCUIT, INSPECT		
Includes: Time To Hook-Up And Disconnect Test Equipment And Perform Test. Does Not Include: Pinpoint Test Or Re-Test Upon Completion Of Repair.		
To Re-Test System, Add 06-09		0.3
13-ECM, R&R		
Fusion, Milan, Zephyr, MKZ ..	06-09 B (0.6)	0.8
14-ECM, REPROGRAM		
Fusion, Milan, Zephyr, MKZ ..	06-09 B (0.1)	0.2
15-KNOCK SENSOR, R&R		
Fusion, Milan, Zephyr		
2.3L	06-08 B	2.3
3.0L	06-09 B	2.9
MKZ	07-09 B	2.9
16-MAIN RELAY, R&R		
Fusion, Milan, Zephyr, MKZ ..	06-09 B	0.2
17-MASS AIR FLOW SENSOR, R&R		
Fusion, Milan, Zephyr		
2.3L	06-08 B (0.2)	0.3
	09 B (0.1)	0.2
3.0L	06-09 B (0.1)	0.2
MKZ	07-09 B (0.1)	0.2

	(Factory) Time	Motor Time
18-MONITOR, DIAGNOSIS		
Fusion, Milan, Zephyr, MKZ ..	06-09 B	1.4
NOTES FOR: MONITOR, DIAGNOSIS		
To Repeat Final Quick Test,		
Add 06-09		0.2
19-OXYGEN SENSOR, R&R		
FRONT		
Fusion, Milan, Zephyr		
2.3L	06-08 B (0.3)	0.4
	09 B (0.2)	0.4
3.0L		
Right Bank	06-09 B (0.4)	0.5
Left Bank	06-09 B (0.3)	0.4
Both Banks	06-09 B (0.5)	0.6
MKZ		
Right Bank	07-09 B (0.4)	0.5
Left Bank	07-09 B (0.2)	0.4
Both Bank	07-09 B (0.5)	0.6
REAR		
Fusion, Milan, Zephyr		
2.3L	06-08 B (0.3)	0.4
	09 B (0.2)	0.4
3.0L		
Right Side	06-09 B (0.4)	0.5
Left Side	06-09 B (0.3)	0.4
Both Sides	06-09 B (0.4)	0.5
MKZ		
Right Bank	07-09 B (0.3)	0.5
Left Bank	07-09 B (0.4)	0.6
Both Banks	07-09 B (0.6)	0.8
20-PINPOINT, TEST		
Fusion, Milan, Zephyr, MKZ ..	06-09 B	0.5
NOTES FOR: PINPOINT, TEST		
To Test One Component Or Circuit Of A System.		
21-VEHICLE SPEED SENSOR, R&R		
Fusion, Milan, Zephyr, MKZ ..	06-09 B	0.5

CHASSIS ELECTRICAL

BATTERY

	(Factory) Time	Motor Time
22-BATTERY, CHARGE/TEST		
Fusion, Milan, Zephyr, MKZ ..	06-09 B	0.4
NOTES FOR: BATTERY, CHARGE/TEST		
Includes: Test Battery And Charge If Necessary. Does Not Include: Battery, Renew		
23-BATTERY, R&R		
Fusion, Milan, Zephyr, MKZ ..	06-09 C	0.3
NOTES FOR: BATTERY, R&R		
Does Not Include: Any Test, Check Or Charge.		
24-BATTERY TERMINALS, CLEAN		
Fusion, Milan, Zephyr, MKZ ..	06-09 C	0.3
25-CABLE ASSY, R&R		
Fusion, Milan, Zephyr, MKZ ..	06-09 C	0.7

BULBS

	(Factory) Time	Motor Time
26-EXTERIOR BULBS, R&R		
Fusion, Milan, Zephyr, MKZ		
One	06-09 C (0.2)	0.3
Two Or More	06-09 C (0.3)	0.4
27-INTERIOR BULB, R&R		
Fusion, Milan, Zephyr, MKZ ..	06-09 C (0.2)	0.2
28-HEADLAMP BULB, R&R		
Fusion, Milan, Zephyr, MKZ		
Right Side	06-09 C (0.6)	0.8
Left Side	06-09 C (0.5)	0.6
Both Sides	06-09 C (0.8)	1.0

HORN

	(Factory) Time	Motor Time
29-HORN, R&R		
Fusion, Milan, Zephyr, MKZ		
One Or Both	06-09 B (0.3)	0.6
30-HORN RELAY, R&R		
Fusion, Milan, Zephyr, MKZ ..	06-09 B	0.2

FOG LAMPS

	(Factory) Time	Motor Time
31-FOG LAMP ASSY, R&R		
Fusion, Milan, Zephyr, MKZ		
One Side	06-09 B (0.3)	0.5
Both Sides	06-09 B (0.4)	0.8

HEADLAMPS

	(Factory) Time	Motor Time
32-COMPOSITE ASSY, R&R		
Fusion, Milan, Zephyr, MKZ		
One Side	06-09 C (0.9)	1.0
Both Sides	06-09 C (0.9)	1.3
33-HEADLAMP, ALIGN		
Fusion, Milan, Zephyr, MKZ		
Two	06-09 C	0.4

SENDERS

	(Factory) Time	Motor Time
34-FUEL GAUGE SENDING UNIT, R&R		
Fusion, Milan, Zephyr, MKZ		
FWD	06-09 B	2.7
AWD	07-09 B	3.3
NOTES FOR: FUEL GAUGE SENDING UNIT, R&R		
Does Not Include: Drain & Refill Fuel Tank.		
35-OIL PRESSURE SENDING UNIT, R&R		
Fusion, Milan, Zephyr	06-09 B (0.5)	0.6
MKZ	07-09 B (0.5)	0.6

SWITCHES

	(Factory) Time	Motor Time
36-BACK-UP SWITCH, R&R		
Fusion, Milan	06-09 B	0.5
37-CLUTCH SWITCH, R&R		
Fusion, Milan	06-09 B	0.3
38-HAZARD SWITCH, R&R		
Fusion, Milan, Zephyr, MKZ	06-09 B	0.5
39-HEADLAMP SWITCH, R&R		
Fusion, Milan, Zephyr, MKZ ..	06-09 B (0.2)	0.4
40-INSTRUMENT LIGHT RHEOSTAT, R&R		
Fusion, Milan, Zephyr, MKZ ..	06-09 B (0.2)	0.4
41-MULTIFUNCTION SWITCH, R&R		
Fusion, Milan, Zephyr, MKZ ..	06-09 B (0.7)	0.9
42-RANGE SENSOR, R&R		
Fusion, Milan, Zephyr		
2.3L	06-09 B (0.5)	0.6
3.0L	06-09 B	1.6
MKZ	07-09 B	1.6
43-PARK BRAKE WARNING SWITCH, R&R		
Fusion, Milan, Zephyr, MKZ ..	06-09 B	0.8
44-STOPLAMP SWITCH, R&R		
Fusion, Milan, Zephyr, MKZ ..	06-09 B (0.2)	0.3
45-WIPER SWITCH, R&R		
Fusion, Milan, Zephyr, MKZ ..	06-09 B (0.7)	0.9

WIPERS

	(Factory) Time	Motor Time
46-RESERVOIR, R&R		
Fusion, Milan, Zephyr, MKZ ..	06-09 B (0.7)	0.9
47-WIPER MOTOR, R&R		
Fusion, Milan, Zephyr, MKZ ..	06-09 B (0.6)	0.8
48-WIPER TRANSMISSION, R&R		
Fusion, Milan, Zephyr, MKZ ..	06-09 B	0.8
49-WASHER PUMP, R&R		
Fusion, Milan, Zephyr, MKZ ..	06-09 B	0.8
50-WIPER ARM, R&R		
Fusion, Milan, Zephyr, MKZ		
Each	06-09 C	0.2
51-WIPER BLADE, R&R		
Fusion, Milan, Zephyr, MKZ		
Each	06-09 C	0.1
52-WIPER RELAY, R&R		
Fusion, Milan, Zephyr, MKZ ..	06-09 B	0.4

INSTRUMENTS & GAUGES

INSTRUMENTS & GAUGES

	(Factory) Time	Motor Time
53-INSTRUMENT CLUSTER, R&R		
Fusion, Milan, Zephyr, MKZ ..	06-09 B (0.7)	0.9

TIRE PRESSURING MONITORING

	(Factory) Time	Motor Time
54-TIRE PRESSURE SENSOR, R&R		
Fusion, Milan, Zephyr, MKZ		
One	06-09 B (0.4)	0.6
NOTES FOR: TIRE PRESSURE SENSOR, R&R		
Includes: System Relearn.		
To R&R TPMS Attachment		
Kit, Add		
One	06-09	0.2

PARKING AID

	(Factory) Time	Motor Time
55-SPEAKER, R&R		
Fusion, Milan, Zephyr, MKZ ..	06-09 B	0.9
56-MODULE, R&R		
Fusion, Milan, Zephyr, MKZ ..	06-09 B	1.0
57-SENSOR, R&R		
Fusion, Milan, Zephyr, MKZ		
One	06-09 B	1.4
Each Add	06-09 B	0.2

(CONTINUED)

FORD 1

FUSION, MILAN (06-09), ZEPHYR (06), MKZ (07-09), MILAN HYBRID (10)

ELECTRICAL - Time Cont'd

			(Factory Time)	Motor Time

RESTRAINT SYSTEMS

..... SUPPLEMENTAL RESTRAINT SYSTEM .....

58-CLOCKSPRING, R&R
Fusion, Milan, Zephyr, MKZ .. 06-09 B (0.5) 0.7
NOTES FOR: CLOCKSPRING, R&R
WARNING: Before Repairing Any Air Restraint System, The Battery Cables And Any Back-up Power Supplies To The System Must Be Disconnected In Order To Prevent Accidental Deployment.

59-CENTER SENSOR, R&R
Fusion, Milan, Zephyr, MKZ .. 06-09 B (0.2) 0.3
NOTES FOR: CENTER SENSOR, R&R
WARNING: Before Repairing Any Air Restraint System, The Battery Cables And Any Back-up Power Supplies To The System Must Be Disconnected In Order To Prevent Accidental Deployment.

60-DIAGNOSTIC MODULE, R&R
Fusion, Milan, Zephyr, MKZ .. 06-09 B (0.5) 0.6
NOTES FOR: DIAGNOSTIC MODULE, R&R
WARNING: Before Repairing Any Air Restraint System, The Battery Cables And Any Back-up Power Supplies To The System Must Be Disconnected In Order To Prevent Accidental Deployment.

61-DRIVER INFLATOR MODULE, R&R
Fusion, Milan, Zephyr, MKZ .. 06-09 B (0.1) 0.4
NOTES FOR: DRIVER INFLATOR MODULE, R&R
WARNING: Before Repairing Any Air Restraint System, The Battery Cables And Any Back-up Power Supplies To The System Must Be Disconnected In Order To Prevent Accidental Deployment.

62-PASSENGER INFLATOR MODULE, R&R
Fusion, Milan, Zephyr .... 06-09 B (0.3) 0.4
MKZ .... 07-09 B (0.6) 0.8
NOTES FOR: PASSENGER INFLATOR MODULE, R&R
WARNING: Before Repairing Any Air Restraint System, The Battery Cables And Any Back-up Power Supplies To The System Must Be Disconnected In Order To Prevent Accidental Deployment.

63-PASSENGER MODULE, R&R
Fusion, Milan, Zephyr .... 06-09 B (0.4) 0.6
MKZ .... 07-09 B (0.5) 0.6
NOTES FOR: PASSENGER MODULE, R&R
WARNING: Before Repairing Any Air Restraint System, The Battery Cables And Any Back-up Power Supplies To The System Must Be Disconnected In Order To Prevent Accidental Deployment.

64-PASSENGER DISCRIMINATING SENSOR, R&R
Fusion, Milan, Zephyr
  One .... 06-09 B (0.7) 0.9
  Both .... 06-09 B (1.1) 1.4
MKZ
  One .... 07-09 B (0.9) 1.1
  Both .... 07-09 B (1.1) 1.6

65-INFLATOR CURTAIN, R&R
Fusion, Milan, Zephyr
  One Side .... 06-09 B (2.6) 3.1
  Both Sides .... 06-09 B (2.9) 3.5
NOTES FOR: INFLATOR CURTAIN, R&R
WARNING: Before Repairing Any Air Restraint System, The Battery Cables And Any Back-up Power Supplies To The System Must Be Disconnected In Order To Prevent Accidental Deployment.

66-SIDE SENSOR, R&R
FRONT
  Fusion, Milan, Zephyr, MKZ
    One Side .... 06-09 B (0.3) 0.4
    Both Sides .... 06-09 B (0.5) 0.6
REAR
  Fusion, Milan, Zephyr, MKZ
    One Side .... 06-09 B (0.4) 0.5

    Both Sides .... 06-09 B (0.6) 0.7
ALL
  Fusion, Milan, Zephyr, MKZ ... 06-09 B (1.0) 1.3
NOTES FOR: SIDE SENSOR, R&R
WARNING: Before Repairing Any Air Restraint System, The Battery Cables And Any Back-up Power Supplies To The System Must Be Disconnected In Order To Prevent Accidental Deployment.

67-SIDE IMPACT INFLATOR MODULE, R&R
Fusion, Milan, Zephyr, MKZ
  One Side .... 06-09 B (0.2) 0.3
  Both Sides .... 06-09 B (0.3) 0.5
NOTES FOR: SIDE IMPACT INFLATOR MODULE, R&R
WARNING: Before Repairing Any Air Restraint System, The Battery Cables And Any Back-up Power Supplies To The System Must Be Disconnected In Order To Prevent Accidental Deployment.

68-SYSTEM, DIAGNOSIS
Fusion, Milan, Zephyr, MKZ ... 06-09 B 0.5
NOTES FOR: SYSTEM, DIAGNOSIS
WARNING: Before Repairing Any Air Restraint System, The Battery Cables And Any Back-up Power Supplies To The System Must Be Disconnected In Order To Prevent Accidental Deployment.

BODY ELECTRICAL

.... ANTENNA & RADIO ....

69-AMPLIFIER, R&R
Fusion, Milan, Zephyr .... 06-09 B (0.5) 0.8
MKZ .... 07-09 B (1.1) 1.3

70-RADIO, R&R
Fusion, Milan, Zephyr .... 06-09 B (0.5) 0.8
MKZ .... 07-09 B (0.7) 1.0

71-SPEAKER, R&R
FRONT DOOR
  Fusion, Milan, Zephyr, MKZ
    One Side .... 06-09 B (0.4) 0.5
    Both Sides .... 06-09 B (0.7) 0.9
CENTER
  MKZ
    Front .... 07-09 B (0.2) 0.3
    Rear .... 07-09 B (0.9) 1.3
REAR DOOR
  Fusion, Milan, Zephyr, MKZ
    One Side .... 06-09 B (0.3) 0.4
    Both Sides .... 06-09 B (0.5) 0.6
WOOFER
  Fusion, Milan, Zephyr, MKZ
    One Side .... 06-09 B (0.9) 1.3
    Both Sides .... 06-09 B (1.0) 1.4

.... ANTI-THEFT COMPONENTS ....

72-MODULE, R&R
Fusion, Milan, Zephyr, MKZ .. 06-09 B (0.3) 0.7

.... IMMOBILIZER ....

73-ANTENNA, R&R
Fusion, Milan, Zephyr, MKZ .. 06-09 B (0.3) 0.7

.... FRONT DOOR ....

74-LOCK ACTUATOR, R&R
Fusion, Milan, Zephyr, MKZ
  One Side .... 06-09 B 1.0
  Both Sides .... 06-09 B 1.9

75-LOCK SWITCH, R&R
Fusion, Milan, Zephyr, MKZ .. 06-09 C 0.3

76-WINDOW MOTOR, R&R
Fusion, Milan, Zephyr
  One Side .... 06-09 B (0.4) 0.6
  Both Sides .... 06-09 B (0.8) 1.0
MKZ
  One Side .... 07-09 B (0.5) 0.7

  Both Sides .... 07-09 B (1.0) 1.1

77-WINDOW SWITCH, R&R
Fusion, Milan, Zephyr, MKZ
  One Side .... 06-09 B (0.2) 0.4
  All .... 06-09 B (0.5) 0.7

78-WINDOW REGULATOR, R&R
Fusion, Milan, Zephyr
  One Side .... 06-09 B (0.9) 1.2
  Both Sides .... 06-09 B (1.8) 2.2
MKZ
  One Side .... 07-09 B (1.0) 1.3
  Both Sides .... 07-09 B (1.8) 2.3

.... REAR DOOR ....

79-LOCK ACTUATOR, R&R
Fusion, Milan, Zephyr, MKZ
  One Side .... 06-09 B 1.6
  Both Sides .... 06-09 B 2.8

80-WINDOW MOTOR, R&R
Fusion, Milan, Zephyr, MKZ
  One Side .... 06-09 B (0.4) 0.6
  Both Sides .... 06-09 B (0.7) 1.0

81-WINDOW REGULATOR, R&R
Fusion, Milan, Zephyr, MKZ
  One Side .... 06-09 B (0.8) 1.0
  Both Sides .... 06-09 B (1.5) 1.9

82-WINDOW SWITCH, R&R
Fusion, Milan, Zephyr, MKZ
  One Side .... 06-09 B (0.2) 0.4
  All .... 06-09 B (0.5) 0.7

.... SUNROOF ....

83-MOTOR, R&R
Fusion, Milan, Zephyr, MKZ .. 06-09 B (2.2) 3.0

84-SWITCH, R&R
Fusion, Milan, Zephyr, MKZ .. 06-09 B 0.3

.... HEATED SEATS ....

85-ELEMENT, R&R
SEAT BACK
  Fusion, Milan, Zephyr, MKZ
    Right Side .... 06-09 B (1.5) 2.0
    Left Side .... 06-09 B (1.4) 2.0
    Both Sides .... 06-09 B (2.5) 3.2

.... MIRRORS ....

86-MIRROR SWITCH, R&R
Fusion, Milan, Zephyr, MKZ .. 06-09 B (0.2) 0.3

87-MEMORY MODULE, R&R
MKZ .... 07-09 B (0.3) 0.4

.... POWER SEATS ....

88-POSITION SENSOR, R&R
Fusion, Milan, Zephyr, MKZ .. 06-09 B 0.8

89-MOTOR, R&R
Fusion, Milan, Zephyr, MKZ .. 06-09 B 1.3

90-SWITCH, R&R
Fusion, Milan, Zephyr .... 06-09 B (0.4) 0.5
MKZ
  One .... 07-09 B (0.7) 1.0
  Both .... 07-09 B (1.0) 1.3

91-MEMORY MODULE, R&R
MKZ .... 07-09 B (0.6) 0.8

·· KEYLESS ENTRY COMPONENTS ··

92-KEYBOARD ACTUATOR, R&R
MKZ .... 07-09 B (0.4) 0.6

93-TRANSMITTER, R&R
Fusion, Milan, Zephyr, MKZ .. 06-09 B (0.2) 0.3

(CONTINUED)

1 FORD
FUSION, MILAN (06-09), ZEPHYR (06), MKZ (07-09), MILAN HYBRID (10)

PARTS — 3 ELECTRICAL 3 — PARTS

IGNITION SYSTEM

IGNITION SYSTEM

		Part No.	Price

1-IGNITION COIL

Fusion, Milan			
2.3L	06-09	6E5Z12029AA	96.20
3.0L	06-09	6E5Z12029BA	96.20
Zephyr	06	6E5Z12029BA	96.20
MKZ	07-09	7T4Z12029E	102.11

IGNITION LOCK

2-IGNITION LOCK CYLINDER

Fusion, Milan	06-09	5S4Z11582BB	152.84
Zephyr	06	5S4Z11582BB	152.84
MKZ	07-09	5S4Z11582BB	152.84

3-IGNITION SWITCH

Fusion, Milan	06-08	98AZ11572A	86.20
Zephyr	06	98AZ11572A	86.20
MKZ	07-09	98AZ11572A	86.20

POWERTRAIN CONTROL

		Part No.	Price

POWERTRAIN CONTROL

4-COOLANT TEMP SENSOR

Fusion, Milan			
2.3L	06-09	F8CZ12A648AA	35.24
3.0L	06-09	3L8Z12A648AA	62.69
Zephyr	06	3L8Z12A648AA	62.69

5-CAMSHAFT POSITION SENSOR

Fusion, Milan			
2.3L	06-09	6M8Z12K073AA	42.64
3.0L			
Right	06-09	3M4Z6B288AB	46.56
Left	06-09	3M4Z6B288BB	46.56
Zephyr			
Right	06	3M4Z6B288AB	46.56
Left	06	3M4Z6B288BB	46.56
MKZ	07-09	AT4Z6B288A	92.09

6-CRANKSHAFT POSITION SENSOR

Fusion, Milan			
2.3L	06-09	5M6Z6C315A	68.56
3.0L	06-09	5L8Z6C315AA	45.11
Zephyr	06	5L8Z6C315AA	45.11
MKZ	07-09	AA5Z6C315A	37.18

7-ECM

Fusion, Milan			
2.3L			
MSA4, MSA5	06-07	6E5Z12A650ZF	714.45
RGP3, APP0, APP1	06-07	6U7Z12A650BEC	714.98
VJA3, ZVD0, ZVD1, ZVD2	06-07	6U7Z12A650BFC	538.98
HXT0, HXT1	06-07	6U7Z12A650BGC	676.91
ZSC1, ZSC2, ZSC3	07-08	7E5Z12A650SD	500.11
GNS1, GNS2	07-08	7E5Z12A650AGC	512.06
TJP2, TJP3	07-08	7E5Z12A650PD	500.98
CPR2	07	7U7Z12A650CTA	483.08
ZFC1, DXY0	08	8E5Z12A650XA	676.91
GVU1. UUU0	08	8E5Z12A650UA	466.45
HHF1	08	8E5Z12A650HB	399.98
VPF1	08	8E5Z12A650VB	502.46
49 State/Non Green State			
Manual Trans	09	9E5Z12A650GB	475.54
Auto Trans	09	9E5Z12A650HB	466.91
California/Green State	09	9E5Z12A650FB	475.54
3.0L			
Manual Trans			
Federal	06-07	6U7Z12A650BJC	592.52
California	06-07	6U7Z12A650BHC	592.52
Auto Trans			
Federal	06-07	6U7Z12A650BHC	592.52
California	06-07	6U7Z12A650BHC	592.52
All			
SSE0	08-09	8E5Z12A650SA	551.29
RNR0	08-09	8E5Z12A650RA	551.17
RXY1	08-09	8E5Z12A650LB	442.02
EJY1	08-09	8E5Z12A650JB	442.02
MTS1	08-09	8E5Z12A650MB	399.98
KPV1	08-09	8E5Z12A650KB	512.06
Zephyr			
Federal	06	6U7Z12A650BHC	592.52
California	06	6U7Z12A650BHC	592.52
MKZ			
Federal			
FWD	07	7H6Z12A650NB	461.52
AWD	07	7H6Z12A650LB	512.06
Califirnia			
FWD	07	7H6Z12A650PB	461.52
AWD	07	7H6Z12A650MB	512.06
All			
GJG1	08-09	8H6Z12A650GB	489.22
FHK1	08-09	8H6Z12A650FB	417.00
JRE1	08-09	8H6Z12A650EB	442.02
ADC1	08-09	8H6Z12A650DB	442.02

8-KNOCK SENSOR

Fusion, Milan			
2.3L	06-09	1S7Z12A699BB	52.00
3.0L	06-09	3M8Z12A699AA	56.98
Zephyr	06	3M8Z12A699AA	56.98
MKZ	07-09	7T4Z12A699AB	59.98

9-MASS AIR FLOW SENSOR

Fusion, Milan	06-09	3L3Z12B579BA	244.91
Zephyr	06	3L3Z12B579BA	244.91
MKZ	07-09	3L3Z12B579BA	244.91

10-MAP SENSOR

Fusion, Milan			
2.3L	06-09	4S4Z9F479AA	90.11
3.0L	06-09	6E5Z9F479BA	66.60
Zephyr	06	6E5Z9F479BA	66.60

11-OXYGEN SENSOR

Fusion, Milan			
2.3L			
Upper	06-09	5L8Z9F472BA	133.24
Lower	06-09	5W6Z9G444BA	105.80
3.0L			
Upper	06-07	5L8Z9F472BA	133.24
From 9/4/06			
FWD			
Front	07-09	5L8Z9F472AA	101.87
Rear	07-09	5L8Z9F472BA	133.24

		Part No.	Price
AWD	07-09	5L8Z9F472AA	101.87

Lower			
To 9/4/06	06-07	6E5Z9G444AA	103.84
From 9/4/06			
FWD			
Front	07-09	6E5Z9G444AA	103.84
Rear	07-09	5L8Z9G444F	98.69
AWD			
Front	07-09	6E5Z9G444AA	103.84
Rear	07-09	5W6Z9G444BA	105.80
Zephyr			
Upper	06	5L8Z9F472BA	133.24
Lower	06	6E5Z9G444AA	103.84
MKZ			
Upper	07-09	7T4Z9F472A	97.96
Lower	07-09	5F9Z9G444AB	97.96

CHASSIS ELECTRICAL

BATTERY

12-POSITIVE CABLE

Fusion, Milan			
2.3L			
To 9/4/06	06-07	6E5Z14300AA	113.10
From 9/4/06	07-09	7E5Z14300AA	104.08
3.0L			
To 4/7/06	06-07	6E5Z14300CP	93.12
From 4/7/06 To 9/4/06	06-07	6E5Z14300CR	101.84
From 9/4/06	07-09	7E5Z14300BA	104.98
Zephyr			
To 4/7/06	06	6E5Z14300CP	93.12
From 4/7/06	06	6E5Z14300CR	101.84
MKZ	07-09	7H6Z14300AA	201.48

NOTES FOR: POSITIVE CABLE
Includes Negative Cable.

BULBS

13-EXTERIOR BULBS

Fusion			
Front Turn Signal	06-09	6E5Z13466BA	5.98
Rear Combination			
4157K	06-09	6E5Z13466AC	2.54
194	06-09	C2AZ13466C	0.80
Back-Up Lamp	06-09	E6DZ13466B	3.50
License Plate Lamp	06-09	F5RZ13466C	1.70
High Mount Stop Lamp	06-09	E6DZ13466B	3.50
Milan			
Front Turn Signal	06-09	6E5Z13466BA	5.98
Rear Combination			
3157NA	06-09	YR3Z13466AD	5.30
Back-Up Lamp	06-09	E6DZ13466B	3.50
License Plate Lamp	06-09	F5RZ13466C	1.70
High Mount Stop Lamp	06-09	E6DZ13466B	3.50
Zephyr			
Front Turn Signal	06	6E5Z13466BA	5.98
Rear Combination			
3157NA	06	YR3Z13466AD	5.30
194	06	C2AZ13466C	0.80
Back-Up Lamp	06	E6DZ13466B	3.50
License Plate Lamp	06	F5RZ13466C	1.70
High Mount Stop Lamp	06	E6DZ13466B	3.50
MKZ			
Front Turn Signal	07-09	6E5Z13466BA	5.98
Rear Combination			
3157NA	07-09	YR3Z13466AD	5.30
194	06-09	C2AZ13466C	0.80
Back-Up Lamp	06-09	E6DZ13466B	3.50
License Plate Lamp	07-09	F5RZ13466C	1.70
High Mount Stop Lamp	07-09	E6DZ13466B	3.50

14-FOG LAMP BULB

Fusion, Milan	06-08	2C5Z13N021AA	25.98
Zephyr	06	2C5Z13N021AA	25.98
MKZ	07-08	2C5Z13N021AA	25.98

15-HEADLAMP BULB

Fusion, Milan			
Low Beam	06-08	2C5Z13N021AA	25.98
High Beam	06-08	3W1Z13N021BA	26.98
Zephyr			
Low Beam			
w/o HID	06	2C5Z13N021AA	25.98
w/HID	06	2U5Z13N021AA	339.18
High Beam	06	3W1Z13N021BA	26.98
MKZ			
Low Beam			
w/o HID	07-09	2C5Z13N021AA	25.98
w/HID	07-09	2U5Z13N021AA	339.18
High Beam	07-09	3W1Z13N021BA	26.98

HORN

16-HORN

Fusion, Milan			
To 12/12/05	06	7E5Z13832A	52.20
From 12/12/05	06-08	7E5Z13832A	52.20
Zephyr			
To 12/12/05	06	7E5Z13832A	52.20
From 12/12/05	06	7E5Z13832A	52.20
MKZ	07-08	7E5Z13832A	52.20

(CONTINUED)

FORD 1

FUSION, MILAN (06-09), ZEPHYR (06), MKZ (07-09), MILAN HYBRID (10)

ELECTRICAL - Parts Cont'd

		Part No.	Price
·········· HEADLAMPS ··········			
17-COMPOSITE ASSY			
Fusion			
Right	06-09	6E5Z13008AD	184.89
Left	06-09	6E5Z13008BD	172.01
Milan			
Right	06-09	6N7Z13008AC	357.92
Left	06-09	6N7Z13008BC	363.40
Zephyr			
w/o High Intensity			
Right	06	6H6Z13008AC	375.90
Left	06	6H6Z13008DC	378.32
w/High Intensity			
Right	06	6H6Z13008BC	953.40
Left	06	6H6Z13008CC	953.40
MKZ			
w/o High Intensity			
Right	07-09	6H6Z13008AC	375.90
Left	07-09	6H6Z13008DC	378.32
w/High Intensity			
Right	07-09	6H6Z13008BC	953.40
Left	07-09	6H6Z13008CC	953.40
18-SENTINEL SENSOR			
Fusion, Milan	06-09	5F9Z13A018BA	38.98
Zephyr	06	6H6Z13A018AA	29.98
MKZ	07-09	6H6Z13A018AA	29.98
NOTES FOR: SENTINEL SENSOR			
Includes Sunload Sensor.			
19-CONTROL MODULE			
HID CONTROL			
MODULE			
Zephyr	06	6H6Z13C170A	263.75
MKZ	07-09	6H6Z13C170A	263.75
·········· SENDERS ··········			
20-FUEL GAUGE SENDING UNIT			
Fusion, Milan			
2.3L	06-07	7E5Z9A299S	101.76
	08-09	8E5Z9A299S	78.38
3.0L	06	7E5Z9A299S	101.76
FWD	07	7E5Z9A299S	101.76
	08-09	8E5Z9A299S	78.38
AWD			
Primary	07-09	7E5Z9A299T	109.87
Secondary	07-09	7E5Z9275C	96.00
Zephyr	06	7E5Z9A299S	101.76
MKZ			
FWD	07	7E5Z9A299S	101.76
	08-09	8E5Z9A299S	78.38
AWD			
Primary	07-09	7E5Z9A299T	109.87
Secondary	07-09	7E5Z9275C	96.00
21-OIL PRESSURE SENDING UNIT			
Fusion, Milan			
2.3L	06-09	6U5Z9278G	22.51
3.0L	06-09	6U5Z9278J	23.11
Zephyr	06	6U5Z9278J	23.11
MKZ	07-09	6U5Z9278D	39.16
·········· SWITCHES ··········			
22-BACK-UP SWITCH			
Fusion, Milan	06-09	6E5Z7G072AA	73.07
23-CLUTCH SWITCH			
Fusion, Milan	06-09	6G9Z11A152A	50.91
24-HEADLAMP SWITCH			
Fusion, Milan			
w/o Fog Lamps	06-09	8E5Z11654AA	86.20
w/Fog Lamps			
w/o Auto Lamps	06-09	8E5Z11654AA	58.78
w/Auto Lamps	06-09	8E5Z11654CA	54.84
Zephyr	06	8H6Z11654CA	86.20
MKZ	07-09	8H6Z11654CA	86.20
25-HAZARD SWITCH			
Fusion, Milan			
w/o Traction Control	06-09	6E5Z13D730AA	58.58
w/Traction Control	06-08	6E5Z13D730BA	72.44
	09	9E5Z13D730AA	48.71
Zephyr	06	7A1Z13D730BA	82.73
MKZ	07-08	7A1Z13D730BA	82.73
	09	9H6Z13D730AA	72.09
26-INSTRUMENT LIGHT RHEOSTAT			
Fusion, Milan	06-09	6E5Z11691AA	23.04
Zephyr	06	6H6Z11691AA	23.89
MKZ	07-09	6H6Z11691AA	23.89
27-RANGE SENSOR			
Fusion, Milan			
2.3L	06-09	6E5Z7F293A	100.53
28-STOPLAMP SWITCH			
Fusion, Milan	06-09	7E5Z13480A	24.67
Zephyr	06	7E5Z13480A	24.67
MKZ	07-09	7E5Z13480A	24.67
29-TURN SIGNAL SWITCH			
Fusion, Milan			
To 12/5/05	06	8E5Z13K359AA	83.93
From 12/5/05	07-09	8E5Z13K359AA	83.93
Zephyr			
To 12/5/05	06	8E5Z13K359AA	83.93
From 12/5/05	06	8E5Z13K359AA	83.93
MKZ	07-09	8E5Z13K359AA	83.93

		Part No.	Price
·········· WIPERS ··········			
30-WIPER ARM			
Fusion, Milan			
Right			
To 6/23/07	06-07	7E5Z17526A	22.06
From 6/23/07	07-09	7E5Z17526A	22.06
Left	06-09	6E5Z17527AA	36.98
Zephyr			
Right	06	7E5Z17526A	22.06
Left	06	6E5Z17527AA	36.98
MKZ			
Right			
To 6/23/07	07	7E5Z17526A	22.06
From 6/23/07	07-09	7E5Z17526A	22.06
Left	06-09	6E5Z17527AA	36.98
31-WIPER BLADE			
Fusion, Milan			
Left	06-09	6E5Z17528AA	16.91
Right	06-09	6E5Z17528AB	16.91
Zephyr			
Left	06	6E5Z17528AA	16.91
Right	06	6E5Z17528AB	16.91
MKZ			
Right	07-09	6E5Z17528AB	16.91
Left	07-09	6E5Z17528AA	16.91
32-WASHER PUMP			
Fusion, Milan	06-09	6E5Z17618A	56.00
Zephyr	06	6E5Z17618A	56.00
MKZ	07-09	6E5Z17618A	56.00
NOTES FOR: WASHER PUMP			
Includes Washer Reservoir.			
33-WIPER MOTOR			
Fusion, Milan	06-09	6E5Z17508AA	116.31
Zephyr	06	6E5Z17508AA	116.31
MKZ	07-09	6E5Z17508AA	116.31
34-WIPER TRANSMISSION			
Fusion, Milan	06-09	6E5Z17566A	82.32
Zephyr	06	6E5Z17566A	82.32
MKZ	07-09	6E5Z17566A	82.32

INSTRUMENTS & GAUGES

		Part No.	Price
········ INSTRUMENTS & GAUGES ········			
35-INSTRUMENT CLUSTER			
Fusion	06	6E5Z10849AA	363.08
	07	7E5Z10849B	387.69
w/o Message Center			
To 7/5/08	08-09	8E5Z10849C	375.38
From 7/5/08	09	9E5Z10849A	375.38
w/Message Center			
To 7/22/08	08-09	8E5Z10849E	395.38
From 7/5/08	09	9E5Z10849AB	396.92
Milan	06	6E5Z10849AA	363.08
	07	7E5Z10849B	387.69
To 7/22/08	08-09	8E5Z10849G	403.08
From 7/22/08	09	9E5Z10849F	403.08
Zephyr	06	6E5Z10849FA	393.85
MKZ	07	7H6Z10849AA	390.77
To 7/23/08	08-09	8H6Z10849AA	398.46
From 7/23/08	09	9H6Z10849A	398.46
·· DRIVER INFORMATION CENTER ··			
36-SWITCH			
Fusion, Milan	06-09	6E5Z10D889AA	31.75
Zephyr	06	6H6Z10D889AA	48.62
MKZ	07-09	6H6Z10D889AA	48.62
·· TIRE PRESSURING MONITORING ··			
37-TIRE PRESSURE SENSOR			
Fusion, Milan, MKZ	07-09	6F2Z1A189A	108.92
·········· NAVIGATION SYSTEM ··········			
38-ANTENNA			
Fusion, Milan			
To 9/4/06	06-07	6H6Z10E893AA	87.75
From 9/4/06 to 5/23/08	07-08	7E5Z10E893A	118.42
From 5/23/08	08-09	8E5Z10E893A	115.50
Zephyr	06	6H6Z10E893AA	87.75
MKZ			
To 12/4/06	07	6H6Z10E893AA	87.75
From 12/4/06 To 5/23/08	07-08	7E5Z10E893A	118.42
From 5/23/08	08-09	8E5Z10E893A	115.50
·········· PARKING AID ··········			
39-SENSOR			
Fusion, Milan, MKZ	08-09	5G1Z15K859AAA	218.32
40-CONTROL MODULE			
Fusion, Milan, MKZ	08-09	8E5Z15K866A	140.00

RESTRAINT SYSTEMS

		Part No.	Price
····· SUPPLEMENTAL RESTRAINT ·····			
SYSTEM			
41-CLOCKSPRING			
Fusion, Milan	06-08	8E5Z14A664A	49.23
w/o Traction Control	09	8E5Z14A664A	49.23

		Part No.	Price
w/Traction Control	09	9E5Z14A664A	64.19
Zephyr	06	8E5Z14A664A	49.23
MKZ	07-08	8E5Z14A664A	49.23
w/o Traction Control	09	8E5Z14A664A	49.23
w/Traction Control	09	9E5Z14A664A	64.19
42-DIAGNOSTIC MODULE			
Fusion, Milan			
w/o Side Air Bags	06-08	7E5Z14B321A	199.36
w/Side Air Bags	06-09	7E5Z14B321D	199.36
Zephyr			
w/o Side Air Bags	06	7E5Z14B321A	199.36
w/Side Air Bags	06	7E5Z14B321D	199.36
MKZ			
w/o Side Air Bags	07-08	7E5Z14B321A	199.36
w/Side Air Bags	07-08	7E5Z14B321D	199.36
43-DRIVER INFLATOR MODULE			
Fusion			
Camel Interior	06-09	6E5Z54043B13AA	683.56
Med. Light Stone			
Interior	06-09	6E5Z54043B13AB	683.56
Charcoal Black Interior	06-09	6E5Z54043B13AB	683.56
Milan			
Camel Interior	06-09	6N7Z54043B13AA	683.56
Med. Light Stone			
Interior	06-09	6N7Z54043B13AB	683.56
Charcoal Black Interior	06-09	6N7Z54043B13AB	683.56
Zephyr			
Sand Interior	06	6H6Z54043B13AA	683.56
Light Stone Interior	06	6H6Z54043B13AB	683.56
Charcoal Black Interior	06	6H6Z54043B13AB	683.56
MKZ			
Sand Interior	07-09	6H6Z54043B13AA	683.56
Light Stone Interior	07-09	6H6Z54043B13AB	683.56
Charcoal Black Interior	07-09	6H6Z54043B13AB	683.56
44-FRONT SENSOR			
Fusion, Milan	06-07	6E5Z14B004A	51.97
	08-09	9E5Z14B004A	51.20
Zephyr	06	6E5Z14B004A	51.97
MKZ	07	6E5Z14B004A	51.97
	08-09	9E5Z14B004A	51.20
45-PASSENGER INFLATOR MODULE			
Fusion, Milan			
Camel Interior	06-09	6E5Z54044A74AA	697.23
Med. Light Stone			
Interior	06-09	6E5Z54044A74AB	697.23
Charcoal Black Interior	06-09	6E5Z54044A74AB	697.23
Zephyr			
Light Stone Interior	06	6H6Z54044A74AA	697.23
Charcoal Black Interior	06	6H6Z54044A74AB	697.23
Sand Interior	06	6H6Z54044A74AC	697.23
MKZ			
Light Stone Interior	07-09	6H6Z54044A74AA	697.23
Charcoal Black Interior	07-09	6H6Z54044A74AB	697.23
Sand Interior	07-09	6H6Z54044A74AC	697.23
46-SIDE IMPACT INFLATOR MODULE			
Fusion			
Right			
Med. Light Stone			
Interior			
To 12/3/07	06-08	6E5Z54611D10AD	260.29
From 12/3/07	08-09	8E5Z54611D10AA	216.90
Camel Interior			
To 12/3/07	06-08	6E5Z54611D10AE	216.90
Charcoal Black			
To 12/5/05	06	6E5Z54611D10AC	245.07
From 12/5/05 To 12/3/07	06-08	6E5Z54611D10AF	216.90
From 12/3/07	08-09	8E5Z54611D10AC	216.90
Left			
Med. Light Stone			
Interior			
To 12/3/07	06-08	6E5Z54611D11AD	216.90
From 12/3/07	08-09	8E5Z54611D11AA	216.90
Camel Interior			
To 12/5/05	06	6E5Z54611D11AB	260.29
From 12/5/05 To 12/3/07	06-08	6E5Z54611D11AE	216.90
From 12/3/07	08-09	8E5Z54611D11AB	216.90
Charcoal Black			
Interior			
To 12/5/05	06	6E5Z54611D11AC	232.30
From 12/5/05 To 12/3/07	06-08	6E5Z54611D11AF	216.90
From 12/3/07	08-09	8E5Z54611D11AC	216.90
Milan			
Right			
Med. Light Stone			
Interior			
Cloth Interior			
To 12/3/07	06-08	6E5Z54611D10AD	260.29
From 12/3/07	08-09	8E5Z54611D10AA	216.90
Leather Interior			
To 12/5/05	06	6E5Z54611D10AC	245.07
From 12/5/05 To 12/3/07	06-08	6E5Z54611D10AF	216.90
From 12/3/07	08-09	8E5Z54611D10AA	216.90
Camel Interior			
To 12/3/07	06-08	6E5Z54611D10AE	216.90
From 12/3/07	08-09	8E5Z54611D10AB	216.90
Charcoal Black			
Interior			
To 12/5/05	06	6E5Z54611D10AC	245.07
From 12/5/05 To 12/3/07	06-08	6E5Z54611D10AF	216.90

(CONTINUED)

1 FORD
FUSION, MILAN (06-09), ZEPHYR (06), MKZ (07-09), MILAN HYBRID (10)

ELECTRICAL - Parts Cont'd

		Part No.	Price
From 12/3/07	08-09	8E5Z54611D10AC	216.90

Left
Med. Light Stone Interior
Cloth Interior

		Part No.	Price
To 12/3/07	06-08	6E5Z54611D11AD	216.90
From 12/3/07	08-09	8E5Z54611D11AA	216.90

Leather Interior

To 12/5/05	06	6E5Z54611D11AC	232.30
From 12/5/05 To 12/3/07	06-08	6E5Z54611D11AF	216.90
From 12/3/07	08-09	8E5Z54611D11AA	216.90

Camel Interior

To 12/5/05	06	6E5Z54611D11AB	260.29
From 12/5/05 To 12/3/07	06-08	6E5Z54611D11AE	216.90
From 12/3/07	08-09	8E5Z54611D11AB	216.90

Charcoal Black Interior

To 12/5/05	06	6E5Z54611D11AC	232.30
From 12/5/05 To 12/3/07	06-08	6E5Z54611D11AF	216.90
From 12/3/07	08-09	8E5Z54611D11AC	216.90

Zephyr
Right

Light Stone Interior	06	6H6Z54611D10AC	260.29
Sand Interior	06	6H6Z54611D10AD	216.90

Charcoal Black Interior

To 12/5/05	06	6E5Z54611D10AC	245.07
From 12/5/05	06	6E5Z54611D10AF	216.90

Left

Light Stone Interior	06	6H6Z54611D11AC	216.90
Sand Interior	06	6H6Z54611D11AD	260.29

Charcoal Black Interior

To 12/5/05	06	6E5Z54611D11AC	232.30
From 12/5/05	06	6E5Z54611D11AF	216.90

MKZ
Right
Light Stone Interior

To 12/3/07	07-08	6H6Z54611D10AC	260.29
From 12/3/07	08-09	8H6Z54611D10AA	216.90

Sand Interior

To 12/3/07	07-08	6H6Z54611D10AD	216.90
From 12/3/07	08-09	8H6Z54611D10AB	216.90

Charcoal Black Interior

To 12/3/07	07	6E5Z54611D10AF	216.90
From 12/3/07	08-09	8E5Z54611D10AC	216.90

Left
Light Stone Interior

To 12/3/07	07-08	6H6Z54611D11AC	216.90
From 12/3/07	08-09	8H6Z54611D11AA	260.29

Sand Interior

To 12/3/07	07-08	6H6Z54611D11AD	260.29
From 12/3/07	08-09	8H6Z54611D11AB	216.90

Charcoal Black Interior

To 12/3/07	07-08	6E5Z54611D11AF	216.90
From 12/3/07	08-09	8E5Z54611D11AC	216.90

47-SIDE SENSOR

Fusion, Milan	06-07	6E5Z14B345AA	29.61
	08-09	9E5Z14B345A	30.80
Zephyr	06	6E5Z14B345AA	29.61
MKZ	07	6E5Z14B345AA	29.61
	08-09	9E5Z14B345A	30.80

48-INFLATOR CURTAIN
Fusion, Milan

Right	06-08	8E5Z54042D94A	389.10
	09	AE5Z54042D94B	389.10
Left	06-08	8E5Z54042D95A	389.10
	09	AE5Z54042D95B	389.10

Zephyr

Right	06	8E5Z54042D94A	389.10
Left	06	8E5Z54042D95A	389.10

MKZ

Right	07-08	8E5Z54042D94A	389.10
	09	AE5Z54042D94B	389.10
Left	07-08	8E5Z54042D95A	389.10
	09	AE5Z54042D95B	389.10

BODY ELECTRICAL

GAUGES
49-CLOCK

		Part No.	Price
Fusion, Milan	06-09	6N7Z15000AA	62.79
Zephyr	06	2C5Z15000AA	62.79
MKZ	07-09	2C5Z15000AA	62.79

FRONT DOOR
50-WINDOW MOTOR
Fusion, Milan

Right	06-09	6E5Z5423394AA	45.23
Left	06-09	6E5Z5423395AA	42.98

Zephyr

Right	06	6H6Z5423394AA	107.15
Left	06	6H6Z5423395AA	110.85

MKZ

Right	07-09	6H6Z5423394AA	107.15
Left	07-09	6H6Z5423395AA	110.85

51-WINDOW SWITCH
Fusion, Milan

Right	06-09	6L2Z14529AAA	21.00
Left	06-09	6L2Z14529BAA	63.73

Zephyr

Right	06	6H6Z14529BAA	29.93
Left	06	6H6Z14529AAB	65.00

MKZ

Right	07-09	6H6Z14529BAA	29.93
Left	07-09	6H6Z14529AAB	65.00

52-LOCK SWITCH

Fusion, Milan	06-09	7E5Z14028AA	20.25
Zephyr	06	6H6Z14028AAA	38.33
MKZ	07-09	6H6Z14028AAA	38.33

REAR DOOR
53-WINDOW MOTOR
Fusion, Milan

Right	06-09	6E5Z5423394BA	45.23
Left	06-09	6E5Z5423395BA	42.98

Zephyr

Right	06	6E5Z5423394BA	45.23
Left	06	6E5Z5423395BA	42.98

MKZ

Right	07-09	6E5Z5423394BA	45.23
Left	07-09	6E5Z5423395BA	42.98

54-WINDOW SWITCH

Fusion, Milan	06-09	6L2Z14529AAA	21.00
Zephyr	06	6H6Z14529AAA	24.37
MKZ	07-09	6H6Z14529AAA	24.37

ANTI-THEFT COMPONENTS
55-CONTROL MODULE
Fusion, Milan

To 9/4/06	06-07	6E5Z15604AB	254.36
From 9/4/06 To 1/30/07	07	7E5Z15604A	254.36
From 1/30/07	07	7E5Z15604B	254.36
w/o Keyless Entry	08-09	8E5Z15604A	254.36
w/Keyless Entry	08-09	8E5Z15604B	254.36
Zephyr	06	6E5Z15604CB	254.36

MKZ

To 9/4/06	07	6E5Z15604CB	254.36
From 9/4/06 To 1/30/07	07	7H6Z15604A	254.36
From 1/30/07	07-09	8E5Z15604C	254.36

56-IGNITION IMMOBILIZER MODULE

Fusion, Milan	06-07	6E5Z15607AA	40.20
Zephyr	06	6E5Z15607AA	40.20
MKZ	07	6E5Z15607AA	40.20

57-HOOD SWITCH

Fusion, Milan	06-09	1X4Z14018BA	28.00
Zephyr	06	1X4Z14018BA	28.00
MKZ	07-09	1X4Z14018BA	28.00

SUNROOF
58-MOTOR

Fusion, Milan	06-09	7W1Z15790A	221.82
Zephyr	06	7W1Z15790A	221.82
MKZ	07-09	7W1Z15790A	221.82

59-SWITCH

Fusion, Milan	06-09	6E5Z15B691A	49.87

		Part No.	Price
Zephyr	06	6H6Z15B691A	48.40
MKZ	07-08	6H6Z15B691A	48.40

HEATED SEATS
60-ELEMENT
Fusion, Milan
Upper
Right

To 9/4/06	06-07	6E5Z14D696BA	78.98
From 9/4/06	07-08	7E5Z14D696A	58.60
	09	9E5Z14D696A	48.73
Left	06-08	6E5Z14D696BA	78.98
	09	9E5Z14D696B	40.40
Lower	06-09	6E5Z14D696AA	78.98

Milan
Upper
Right

To 9/4/06	06-07	6N7Z14D696BA	82.98
From 9/4/06	07-08	7N7Z14D696A	40.00
	09	9N7Z14D696B	47.33
Left	06-08	6N7Z14D696BA	82.98
	09	9N7Z14D696C	43.93

Zephyr
w/o Cooled Seats

Upper	06	6H6Z14D696BA	111.28
Lower	06	6H6Z14D696AA	111.28
w/Cooled Seats	06	NOT LISTED	

MKZ
w/o Cooled Seats

Upper	07-08	6H6Z14D696BA	111.28
	09	NOT LISTED	
Lower	07-08	6H6Z14D696AA	111.28
	09	NOT LISTED	
w/Cooled Seats	06-09	NOT LISTED	

61-CONTROL MODULE

Fusion, Milan	06-07	7E5Z14C724A	76.18

Zephyr

w/o Cooled Seats	06	7E5Z14C724A	76.18
w/Cooled Seats	06	6H6Z14C724A	183.17

MKZ

w/o Cooled Seats	07-09	7E5Z14C724A	76.18
w/Cooled Seats	07-09	6H6Z14C724A	183.17
	08-09	9U5Z14C724B	145.17

MIRRORS
62-MIRROR SWITCH

Fusion, Milan	06-09	7L3Z17B676AA	28.43
Zephyr	06	7H6Z17B676AA	28.00
MKZ	07-09	7H6Z17B676AA	28.00

63-MOTOR
Fusion, Milan

w/o Memory	06-09	6U5Z17D696A	40.12
w/Memory	06-09	6U5Z17D696B	52.90

Zephyr

w/o Memory	06	6U5Z17D696A	40.12
w/Memory	06	6U5Z17D696B	52.90

MKZ

w/o Memory	07-09	6U5Z17D696A	40.12
w/Memory	07-09	6U5Z17D696B	52.90

POWER SEATS
64-ADJUSTER SWITCH

Fusion, Milan	06-09	9L3Z14A701A	140.95

Zephyr

Right	06	9L3Z14A701FA	61.47
Left	06	9L3Z14A701FB	330.07

MKZ

Right	07-09	9L3Z14A701FA	61.47
Left	07-09	9L3Z14A701FB	330.07

65-ADJUSTER

Fusion, Milan	06-09	AE5Z5461711A	637.85

Zephyr

Right	06	8H6Z5461710A	759.72
Left	06	6H6Z5461711AA	514.18

MKZ

Right	07-08	8H6Z5461710A	759.72
	09	8H6Z5461710A	759.72
Left			
To 9/4/06	07	6H6Z5461711AA	514.18
From 9/4/06 To 12/3/07	07-08	8H6Z5461711A	628.27
From 12/3/07	08	8H6Z5461711A	628.27
	09	9H6Z5461711A	650.23

TRUNK
66-LOCK SWITCH

Fusion, Milan	06-09	6E5Z54432A38AA	21.33
Zephyr	06	6H6Z54432A38AA	34.57
MKZ	07-09	6H6Z54432A38AA	34.57

LABOR 4 ALTERNATOR – 2.3L 4 LABOR

OPERATION INDEX

ALTERNATOR, R&R	1
CHARGING CIRCUIT, INSPECT	2

ALTERNATOR
ALTERNATOR

		(Factory Time)	Motor Time
1-ALTERNATOR, R&R			
Fusion, Milan	06-09 B	(0.9)	1.3
2-CHARGING CIRCUIT, INSPECT			
Fusion, Milan	06-09 B		0.5

NOTES FOR: CHARGING CIRCUIT, INSPECT
Includes: Check Alternator Output, Regulator & Belt Tension.

(CONTINUED)

FORD 1

FUSION, MILAN (06-09), ZEPHYR (06), MKZ (07-09), MILAN HYBRID (10)

PARTS   4   **ALTERNATOR – 2.3L**   4   **PARTS**

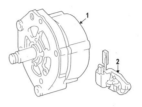

FPP010

ALTERNATOR	1-ALTERNATOR	Part No.	Price
·········· ALTERNATOR ··········	Fusion, Milan 06-09	6E5Z10346AA	339.00
	2-VOLTAGE REGULATOR		
	Fusion, Milan		
	Part Of Alternator. 06-09		

LABOR   4   **ALTERNATOR – 3.0L, 3.5L**   4   **LABOR**

OPERATION INDEX				(Factory Time)	Motor Time		(Factory Time)	Motor Time
ALTERNATOR, R&R 1	**1-ALTERNATOR, R&R**					**NOTES FOR: CHARGING CIRCUIT, INSPECT**		
CHARGING CIRCUIT, INSPECT 2	Fusion, Milan, Zephyr	06-09 B	(1.0)	1.3	*Includes: Check Alternator Output, Regulator & Belt Tension.*		
ALTERNATOR	MKZ	07-09 B	(1.8)	2.3			
·········· ALTERNATOR ··········	**2-CHARGING CIRCUIT, INSPECT**						
	Fusion, Milan, Zephyr, MKZ ..	06-09 B		0.5			

PARTS   4   **ALTERNATOR – 3.0L, 3.5L**   4   **PARTS**

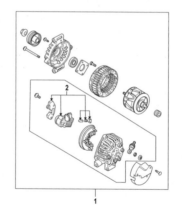

1

FPP020

ALTERNATOR		Part No.	Price		Part No.	Price
·········· ALTERNATOR ··········	Zephyr 06	6E5Z10346BA	337.37	MKZ		
	MKZ 07-09	8G1Z10346A	341.55	Part Of Alternator. 07-09		
1-ALTERNATOR	Part No. / Price	**2-VOLTAGE REGULATOR**				
Fusion, Milan 06-09	6E5Z10346BA 337.37	Fusion, Milan				
		Part Of Alternator. 06-09				
		Zephyr				
		Part Of Alternator. 06-09				

LABOR   5   **STARTER – 2.3L**   5   **LABOR**

OPERATION INDEX	STARTER		(Factory Time)	Motor Time
CIRCUIT, INSPECT 1	·········· STARTER ··········	**2-STARTER, R&R**		
STARTER DRAW, TEST 3		Fusion, Milan .... 06-09 B	(0.5)	0.6
STARTER, R&R 2		**3-STARTER DRAW, TEST**		
	1-CIRCUIT, INSPECT (Factory Time) Motor Time	Fusion, Milan .... 06-09 B		0.3
	Fusion, Milan .... 06-09 B 0.5	*(CONTINUED)*		

1 FORD
FUSION, MILAN (06-09), ZEPHYR (06), MKZ (07-09), MILAN HYBRID (10)

PARTS   5   **STARTER – 2.3L**   5   **PARTS**

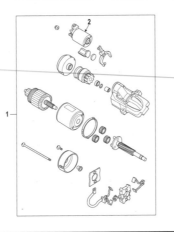

FPP030

STARTER		Part No.	Price
·········· STARTER ··········	**1-STARTER**		
	Fusion, Milan 06-09	6E5Z11002AA	223.03
	2-SOLENOID		
	Fusion, Milan		
	Part Of Starter. 06-09		

LABOR   5   **STARTER – 3.0L, 3.5L**   5   **LABOR**

OPERATION INDEX	
CIRCUIT, INSPECT	1
STARTER DRAW, TEST	3
STARTER, R&R	2

STARTER		(Factory Time)	Motor Time
·········· STARTER ··········			
1-CIRCUIT, INSPECT			
Fusion, Milan, Zephyr, MKZ .. 06-09 B			0.5

		(Factory Time)	Motor Time
2-STARTER, R&R			
Fusion, Milan, Zephyr, MKZ .. 06-09 B		(0.6)	0.7
3-STARTER DRAW, TEST			
Fusion, Milan, Zephyr, MKZ .. 06-09 B			0.3

PARTS   5   **STARTER – 3.0L, 3.5L**   5   **PARTS**

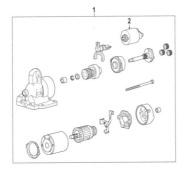

FPP040

STARTER			Part No.	Price		Part No.	Price
·········· STARTER ··········	Zephyr 06		6E5Z11002BA	223.03	MKZ		
	MKZ 07-09		7H6Z11002A	187.35	Part Of Starter. 07-09		
1-STARTER	Part No.	Price	**2-SOLENOID**				
Fusion, Milan 06-09	6E5Z11002BA	223.03	Fusion, Milan				
			Part Of Starter. 06-09				
			Zephyr				
			Part Of Starter. 06				

LABOR   6   **FUEL SYSTEM**   6   **LABOR**

OPERATION INDEX	
ACTUATOR SWITCH, R&R	16
DAMPER, R&R	9
FUEL PUMP RELAY, R&R	2
FUEL PUMP, R&R	1
FUEL PUMP, TEST	3
FUEL RAIL, R&R	10
FUEL TANK, R&R	5
GASKET, R&R	13
INERTIA SWITCH, R&R	4
INJECTOR O-RING, R&R	12
INJECTOR, R&R	11
MANIFOLD GASKET, R&R	15
PEDAL TRAVEL SENSOR, R&R	7
PRESSURE SENSOR, R&R	8

SYSTEM, DIAGNOSIS	17
SYSTEM, SERVICE	6
THROTTLE BODY, R&R	14

FUEL SUPPLY		(Factory Time)	Motor Time
·········· FUEL SUPPLY ··········			
1-FUEL PUMP, R&R			
Fusion, Milan, Zephyr, MKZ			
FWD .... 06-09 B			2.7
AWD .... 07-09 B			3.3

		(Factory Time)	Motor Time
NOTES FOR: FUEL PUMP, R&R			
Does Not Include: Drain & Refill Fuel Tank.			
2-FUEL PUMP RELAY, R&R			
Fusion, Milan, Zephyr, MKZ .. 06-09 B		(0.1)	0.2
3-FUEL PUMP, TEST			
Fusion, Milan, Zephyr, MKZ .. 06-09 B			0.5
4-INERTIA SWITCH, R&R			
Fusion, Milan, Zephyr, MKZ .. 06-09 B		(0.3)	0.4
5-FUEL TANK, R&R			
Fusion, Milan, Zephyr, MKZ			
FWD .... 06-09 B		(2.0)	2.7
(CONTINUED)			

FORD 1

FUSION, MILAN (06-09), ZEPHYR (06), MKZ (07-09), MILAN HYBRID (10)

FUEL SYSTEM - Time Cont'd

		(Factory Time)	Motor Time
AWD	07-09 B	(2.5)	3.3

NOTES FOR: FUEL TANK, R&R
Does Not Include: Drain & Refill Fuel Tank.

6-SYSTEM, SERVICE
Fusion, Milan, Zephyr, MKZ

FWD	06-09 B		3.2
AWD	07-09 B		3.8

NOTES FOR: SYSTEM, SERVICE
Includes: R&I Fuel Tank, Blow Out Lines & Service Fuel Filter.

7-PEDAL TRAVEL SENSOR, R&R
Fusion, Milan, Zephyr, MKZ

A.P.P Sensor	06-09 B	(0.4)	0.5

FUEL INDUCTION

FUEL INJECTION

8-PRESSURE SENSOR, R&R
Fusion, Milan, Zephyr

2.3L	06-09 B		0.3
3.0L	06-09 B		0.3
MKZ	07-09 B		0.3

9-DAMPER, R&R
Fusion, Milan, Zephyr

2.3L	06-09 B	(0.5)	0.6

		(Factory Time)	Motor Time
3.0L	06-09 B	(1.4)	1.8
MKZ	07-09 B	(1.4)	2.1

10-FUEL RAIL, R&R
Fusion, Milan, Zephyr

2.3L	06-09 B	(0.7)	1.0
3.0L	06-09 B	(1.5)	2.1
MKZ	07-09 B	(1.1)	1.6

11-INJECTOR, R&R
Fusion, Milan, Zephyr

2.3L	06-09 B	(0.6)	0.9
3.0L	06-09 B	(1.4)	2.1
MKZ	07-09 B	(1.0)	1.5

12-INJECTOR O-RING, R&R
Fusion, Milan, Zephyr

2.3L	06-09 B		0.9
3.0L	06-09 B		2.1
MKZ	07-09 B		1.5

THROTTLE BODY

13-GASKET, R&R
Fusion, Milan, Zephyr

2.3L	06-09 B	(0.6)	0.8
3.0L	06-09 B	(0.7)	0.9
MKZ	07-09 B	(0.5)	0.8

		(Factory Time)	Motor Time
14-THROTTLE BODY, R&R			
Fusion, Milan, Zephyr			
2.3L	06-09 B	(0.6)	0.8
3.0L	06-09 B	(0.7)	0.9
MKZ	07-09 B	(0.5)	0.8

INTAKE

15-MANIFOLD GASKET, R&R
Fusion, Milan, Zephyr

2.3L	06-09 B	(1.6)	2.1
3.0L			
Upper	06-09 B	(1.1)	1.7
Upper & Lower	06-09 B	(1.5)	2.1
MKZ			
Upper	07-09 B	(0.7)	1.2
Upper & Lower	07-09 B	(1.6)	2.2

NOTES FOR: MANIFOLD GASKET, R&R
To R&R Manifold, Add

Upper	*06-09*	*(0.1)*	*0.1*
Lower	*06-09*	*(0.2)*	*0.2*

CRUISE CONTROL

CRUISE CONTROL

16-ACTUATOR SWITCH, R&R

Fusion, Milan, Zephyr	06-09 B	(0.2)	0.3
MKZ	06-09 B	(0.4)	0.5

17-SYSTEM, DIAGNOSIS

Fusion, Milan, Zephyr	06-09 B		1.0

PARTS — 6 FUEL SYSTEM 6 — PARTS

PARTS INDEX

FUEL SUPPLY

FUEL SUPPLY

			Part No.	Price
1-FUEL PUMP				
Fusion, Milan				
2.3L		06	7E5Z9H307S	221.77
		07	7E5Z9H307S	221.77
		08-09	8E5Z9H307S	296.78
3.0L		06	6E5Z9H307G	280.77
FWD		07	7E5Z9H307T	221.77
		08-09	8E5Z9H307T	303.92
AWD		07-09	7E5Z9H307U	358.67
Zephyr		06	6E5Z9H307G	280.77
MKZ				
FWD		07	7E5Z9H307T	221.77
		08-09	8E5Z9H307T	303.92
AWD		07-09	7E5Z9H307U	358.67
2-FUEL TANK				
Fusion, Milan		06	6E5Z9002AA	1128.72
FWD				
To 9/4/06		07	6E5Z9002AA	1128.72
From 9/4/06		07-09	8E5Z9002A	1103.27
AWD		07-09	AE5Z9002C	1406.23
Zephyr		06	6E5Z9002AA	1128.72

			Part No.	Price
MKZ				
FWD		07-09	8E5Z9002A	1103.27
AWD		07-09	AE5Z9002C	1406.23

FUEL INDUCTION

FUEL INJECTION

			Part No.	Price
3-FUEL RAIL				
Fusion, Milan				
2.3L				
Manual Trans		06-09	5L8Z9D280AA	108.90
Auto Trans		06-09	6E5Z9D280BA	76.98
3.0L		06-09	6E5Z9F792BA	212.30
Zephyr		06	6E5Z9F792BA	212.30
MKZ		07-09	7T4Z9F792G	404.68
4-INJECTOR				
Fusion, Milan				
2.3L				
Manual Trans		06-09	3S4Z9F593AA	102.02
Auto Trans		06-09	7L5Z9F593AA	59.78
3.0L		06-09	6E5Z9F593AA	56.26
Zephyr		06	6E5Z9F593AA	56.26
MKZ		07-09	7T4Z9F593B	59.78

THROTTLE BODY

			Part No.	Price
5-THROTTLE BODY				
Fusion, Milan				
2.3L		06-09	6E5Z9E926BA	234.34
3.0L		06-09	6E5Z9E926AA	211.26
Zephyr		06	6E5Z9E926AA	211.26
MKZ		07-09	7T4Z9E926FA	232.98
6-GASKET				
Fusion, Milan				
2.3L		06-09	6M8Z9E936AA	34.80
3.0L		06-09	YF1Z9E936AA	9.86
Zephyr		06	YF1Z9E936AA	9.86
MKZ		07-09	7T4Z9E936A	15.70

INTAKE

			Part No.	Price
7-INTAKE MANIFOLD				
Fusion, Milan				
2.3L		06-09	3S4Z9424AM	308.32
3.0L				
Upper		06-09	6E5Z9424BA	209.72
Lower		06-09	3F1Z9424AC	103.32
Zephyr				
Upper		06	6E5Z9424BA	209.72
Lower		06	3F1Z9424AC	103.32
MKZ				
Upper				
To 9/1/07		07-08	7T4Z9424D	181.48
From 9/1/07		08-09	7T4Z9424E	168.65
Lower		07-09	7T4Z9424C	168.32
8-MANIFOLD GASKET				
Fusion, Milan				
2.3L		06-09	1S7Z9439AA	9.61
3.0L				
Upper		06-09	3F1Z9H486AA	8.45
Lower		06-09	3F1Z9439AA	8.45
Zephyr				
Upper		06	3F1Z9H486AA	8.45
Lower		06	3F1Z9439AA	8.45
MKZ				
Upper		07-09	7T4Z9H486DA	7.90
Lower				
Large Gasket		07-09	7T4Z9439A	3.95
Small Gasket		07-09	7T4Z9439B	5.92

CRUISE CONTROL

CRUISE CONTROL

			Part No.	Price
9-SWITCH				
Fusion, Milan				
w/o ATC		06-09	7E5Z9C888CA	82.22
w/ATC		06-09	7E5Z9C888DA	78.11
Zephyr		06	7H6Z9C888DA	104.89
MKZ		07-09	7H6Z9C888DA	104.89

LABOR — 7 EXHAUST SYSTEM 7 — LABOR

OPERATION INDEX

EXHAUST MANIFOLD

EXHAUST MANIFOLD

		(Factory Time)	Motor Time
1-EXHAUST MANIFOLD, R&R			
Fusion, Milan, Zephyr			
2.3L	06-09 B	(1.5)	2.0
3.0L			
Right Bank	06-09 B	(1.7)	2.3
Left Bank	06-09 B	(1.5)	2.0
Both Banks	06-09 B	(2.4)	3.5

		(Factory Time)	Motor Time
AWD			
Right Bank	06-09 B	(3.0)	6.2
Left Bank	06-09 B	(1.5)	2.0
Both Banks	06-09 B	(3.8)	7.3
MKZ			
FWD			
Right Bank	07-09 B	(1.7)	2.3
Left Bank	07-09 B	(1.5)	2.0
Both Banks	07-09 B	(2.4)	3.5
AWD			
Right Bank	07-09 B	(3.0)	6.2
Left Bank	07-09 B	(1.5)	2.0
Both Banks	07-09 B	(3.8)	7.3
2-HEAT SHIELD, R&R			
Fusion, Milan			
2.3L			
Upper	06-09 C		0.4
Lower	06-09 C		1.3
3-MANIFOLD GASKET, R&R			
Fusion, Milan, Zephyr			
2.3L	06-09 B	(1.5)	2.0

		(Factory Time)	Motor Time
3.0L			
Right Bank	06-09 B	(1.7)	2.3
Left Bank	06-09 B	(1.5)	2.0
Both Banks	06-09 B	(2.4)	3.5
AWD			
Right Bank	06-09 B	(3.0)	6.2
Left Bank	06-09 B	(1.5)	2.0
Both Banks	06-09 B	(3.8)	7.3
MKZ			
FWD			
Right Bank	07-09 B	(1.7)	2.3
Left Bank	07-09 B	(1.5)	2.0
Both Banks	07-09 B	(2.4)	3.5
AWD			
Right Bank	07-09 B	(3.0)	6.2
Left Bank	07-09 B	(1.5)	2.0
Both Banks	07-09 B	(3.8)	7.3

(CONTINUED)

1 FORD
FUSION, MILAN (06-09), ZEPHYR (06), MKZ (07-09), MILAN HYBRID (10)

EXHAUST SYSTEM - Time Cont'd

		(Factory Time)	Motor Time
Both Banks	06-09 B		4.7
Rear	06-09 B		0.6
MKZ			
FWD			
Front			
Right Bank	07-09 B		1.7
Left Bank	07-09 B		1.4
Both Banks	07-09 B		2.3
Rear	07-09 B		0.6
AWD			
Front			
Right Bank	07-09 B		4.0
Left Bank	07-09 B		2.0
Both Banks	07-09 B		4.6

EXHAUST SYSTEM

EXHAUST SYSTEM

4-CATALYTIC CONVERTER, R&R

		(Factory Time)	Motor Time
Fusion, Milan, Zephyr			
2.3L	06-09 B	(0.5)	0.6
3.0L			
Front			
Right Bank	06-09 B		3.0
Left Bank	06-09 B		2.8

		(Factory Time)	Motor Time
Rear	07-09 B		0.6
5-COMPLETE EXHAUST SYSTEM, R&R			
Fusion, Milan, Zephyr			
2.3L	06-09 C	(0.8)	0.6
3.0L	06-09 C	(0.9)	1.3
MKZ	07-09 C	(1.9)	2.5
6-FRONT PIPE, R&R			
Fusion, Milan, Zephyr	06-09 C		1.3
7-INTERMEDIATE PIPE, R&R			
Fusion, Milan, Zephyr	06-09 C	(0.7)	0.8
MKZ	07-09 C	(0.6)	0.8
8-MUFFLER, R&R			
Fusion, Milan, Zephyr, MKZ	06-09 B	(0.5)	0.6

PARTS · 7 EXHAUST SYSTEM 7 · PARTS

PARTS INDEX

EXHAUST MANIFOLD

EXHAUST MANIFOLD

1-EXHAUST MANIFOLD

		Part No.	Price
Fusion, Milan			
2.3L			
49 State/Non Green			
State			
To 9/4/06	06-07	6E5Z5G232CA	874.68
From 9/4/06	07-09	8E5Z5G232A	900.15
California/Green State			
Manual Trans			
To 9/4/06	06-07	6E5Z5G232CA	874.68
From 9/4/06	07-09	8E5Z5G232A	900.15
Auto Trans			
To 9/4/06	06-07	6E5Z5G232DA	1844.02
From 9/4/06	07-09	7E5Z5G232DA	1015.47
3.0L			
Front	06	6E5Z5G232AA	691.87
FWD			
To 12/3/07	07-08	7E5Z5G232AA	988.77
From 12/3/07			
49 State/			
Non Green	08-09	8E5Z5G232B	669.67
California/Green			
State	08-09	8E5Z5G232C	693.87
AWD			
To 12/3/07	07-08	7E5Z5G232AA	988.77
From 12/3/07			
49 State/			
Non Green	08-09	8E5Z5G232B	669.67
California/Green			
State	08-09	8E5Z5G232C	693.87
Rear			
To 12/5/06	06	6E5Z5G232AF	664.57
From 12/5/06	06	6E5Z5G232AF	664.57
FWD			
To 12/3/07	07	7E5Z5G232AB	900.98
From 12/3/07			
49 State/			
Non Green	08-09	8E5Z5G232D	680.62
California/Green			
State	08-09	8E5Z5G232E	715.40
AWD	07-09	7E5Z9430BA	127.03
Zephyr			
Front	06	6E5Z5G232AA	691.87
Rear			
To 12/5/06	06	6E5Z5G232AF	664.57
From 12/5/06	06	6E5Z5G232AF	664.57
MKZ			
Right	07-09	7T4Z9430C	179.65
Left	07-09	7T4Z9431C	143.32

EXHAUST SYSTEM

2-MANIFOLD GASKET

		Part No.	Price
Fusion, Milan			
2.3L			
49 State/Non Green			
State	06-09	1L5Z9448AB	11.83
California/Green State	06-09	3S4Z9448AA	14.60
3.0L	06-09	XW4Z9448AD	18.60
Zephyr	06	XW4Z9448AD	18.60
MKZ	07-09	7T4Z9448F	10.75

EXHAUST SYSTEM

3-CATALYTIC CONVERTER

		Part No.	Price
Fusion, Milan			
2.3L			
49 State/Non Green			
State			
To 9/4/06	06-07	6E5Z5E212CA	666.67
From 9/4/06 To			
12/4/06	07	7E5Z5E212CC	583.33
From 12/4/06			
Manual Trans	07-08	7E5Z5A289A	213.37
Auto Trans	07-08	8E5Z5A289A	298.36
All	09	9E5Z5A289A	285.19
California/Green State			
Manual Trans			
To 9/4/06	06-07	6E5Z5E212CA	666.67
From 9/4/06 To			
12/4/06	07	7E5Z5E212CC	583.33
From 12/4/06	07-08	7E5Z5A289A	213.37
	09	9E5Z5A289A	285.19
Auto Trans			
To 9/4/06	06-07	6E5Z5E212DA	1000.00
From 9/4/06	07-09	7E5Z5E212DD	833.33
3.0L	06	6E5Z5E212AF	416.67
FWD			
To 9/4/06	07	6E5Z5E212AF	416.67
From 9/4/06 To			
12/3/07	07-08	7E5Z5E212AA	583.33
From 12/3/07			
49 State/Non			
Green	08-09	8E5Z5E212C	458.33
California/Green			
State	08-09	8E5Z5E212D	416.67
AWD			
To 9/4/06	07	6E5Z5E212AF	416.67
From 9/4/06 To			
12/3/07			
Front	07-08	7E5Z5E212CA	666.67
Rear	07-08	7E5Z5E212GA	583.33
From 12/3/07			
Front			
49 State/Non			
Green	08-09	8E5Z5E212A	666.67
California/Green			
State	08-09	8E5Z5E212B	750.00
Rear			
49 State/Non			
Green	08-09	8E5Z5E212E	416.67
California/Green			
State	08-09	8E5Z5E212F	416.67
Zephyr	06	6E5Z5E212AF	416.67
MKZ			
Front	07	7T4Z5E212BC	750.00

		Part No.	Price
	08-09	7T4Z5E212A	583.33
Rear	07	7T4Z5E213BC	750.00
	08-09	7T4Z5E212A	583.33

4-FRONT PIPE

		Part No.	Price
Fusion, Milan			
2.3L			
49 State/Non Green			
State			
To 9/4/06	06-07	6E5Z5G203AA	653.72
From 9/4/06	07-09	7E5Z5G203AA	689.98
California/Green State			
To 9/4/06	06-07	6E5Z5G203BA	705.92
From 9/4/06	07-09	7E5Z5G203BA	670.00
3.0L	06	6E5Z5G274AA	430.60
FWD			
To 9/4/06	07	6E5Z5G274AA	430.60
From 9/4/06 To			
12/3/07	07-08	7E5Z5G274EA	426.72
From 12/3/07			
49 State/Non			
Green	08-09	8E5Z5G274A	431.13
California/Green			
State	08-09	8E5Z5G274B	430.91
AWD			
To 9/4/06	07	6E5Z5G274AA	430.60
From 9/4/06 To			
12/3/07	07-08	7E5Z5G274A	465.09
From 12/3/07			
49 State/Non			
Green	08-09	8E5Z5G274C	438.16
California/Green			
State	08-09	8E5Z5G274D	407.16
Zephyr	06	6E5Z5G274AA	430.60
MKZ	07-09	7H6Z5G274A	256.84

5-MUFFLER

		Part No.	Price
Fusion, Milan			
2.3L	06-08	6E5Z5230AA	469.97
Standard	09	6E5Z5230AA	469.97
Optional	09	9E5Z5230A	499.97
3.0L			
w/o Chrome Pipe			
Extension	06	6E5Z5230BA	679.98
w/Chrome Pipe			
Extension	06	9E5Z5230F	819.98
FWD			
w/o Chrome Pipe			
Ext.	07-09	6E5Z5230BA	679.98
w/Chrome Pipe Ext.	07-09	9E5Z5230F	819.98
AWD			
w/o Chrome Pipe			
Ext.			
To 9/4/06	07	6E5Z5230BA	679.98
From 9/4/06	07-09	7E5Z5230CA	964.60
w/Chrome Pipe Ext.			
To 9/4/06	07	9E5Z5230F	819.98
From 9/4/06	07-09	9E5Z5230D	968.75
Zephyr	06	9E5Z5230F	819.98
MKZ			
FWD	07-09	9H6Z5230B	846.46
AWD	07-09	9H6Z5230A	934.58

6-RESONATOR

		Part No.	Price
MKZ			
FWD	07-09	7H6Z5A289BA	219.99
AWD	07-09	7H6Z5A289AA	219.99

LABOR 8 COOLING SYSTEM – 2.3L 8 LABOR

OPERATION INDEX

COOLING SYSTEM

RADIATOR

1-LOWER HOSE, R&R

		(Factory Time)	Motor Time
Fusion, Milan	06-09 C	(0.6)	0.8

2-RADIATOR, R&R

		(Factory Time)	Motor Time
Fusion, Milan			
Manual Trans	06-09 B	(1.4)	1.8
Auto Trans	06-09 B	(1.6)	2.2

3-UPPER HOSE, R&R

		(Factory Time)	Motor Time
Fusion, Milan	06-09 C	(0.5)	0.7

COOLING SYSTEM

4-COOLING SYSTEM PRESSURE, TEST

			Motor Time
Fusion, Milan	06-09 C		0.3

5-BLOCK HEATER, R&R

			Motor Time
Fusion, Milan	06-09 B		0.5

NOTES FOR: BLOCK HEATER, R&R
Does Not Include: Time To Gain Access.

(CONTINUED)

FORD 1

FUSION, MILAN (06-09), ZEPHYR (06), MKZ (07-09), MILAN HYBRID (10)

COOLING SYSTEM – 2.3L - Time Cont'd

	(Factory Time)	Motor Time
6-EXPANSION PLUG, R&R		
Fusion, Milan		
One .... 06-09 C		0.5
Each Additional .... 06-09 C		0.4
NOTES FOR: EXPANSION PLUG, R&R		
Does Not Include: Time To Gain Access.		
7-EXPANSION TANK, R&R		
Fusion, Milan .... 06-09 C	(0.4)	0.5

	(Factory Time)	Motor Time
8-THERMOSTAT, R&R		
Fusion, Milan .... 06-09 B	(0.8)	1.4
9-THERMOSTAT GASKET, R&R		
Fusion, Milan .... 06-09 C	(0.8)	1.4
·········· **WATER PUMP** ··········		
10-WATER PUMP, R&R		
Fusion, Milan .... 06-09 B	(0.9)	1.4

	(Factory Time)	Motor Time
·········· **COOLING FAN** ··········		
11-CONTROL MODULE, R&R		
Fusion, Milan .... 06-09 B	(0.9)	1.4
12-FAN ASSY, R&R		
Fusion, Milan .... 06-09 B	(0.6)	0.9
13-SWITCH, R&R		
Fusion, Milan .... 06-09 B		0.3

PARTS 8 COOLING SYSTEM – 2.3L 8 PARTS

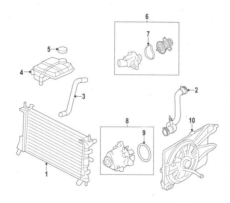

FPP045

COOLING SYSTEM

·········· RADIATOR ··········	Part No.	Price
1-RADIATOR		
Fusion, Milan		
Manual Trans 06-09	AE5Z8005G	327.37
Auto Trans 06-09	6E5Z8005C	339.40
2-UPPER HOSE		
Fusion, Milan 06-09	AE5Z8260B	23.71
3-LOWER HOSE		
Fusion, Milan 06-09	8L8Z8C362A	28.98

·········· COOLING SYSTEM ··········	Part No.	Price
4-RESERVOIR		
Fusion, Milan 06-09	6E5Z8A080AA	43.62
5-RESERVOIR CAP		
Fusion, Milan		
To 8/18/08 06-09	9C3Z8101B	10.88
From 8/18/08 09	9C3Z8101B	10.88
6-THERMOSTAT HOUSING		
Fusion, Milan 06-09	3M4Z8575B	62.14
NOTES FOR: THERMOSTAT HOUSING		
Includes Thermostat And Gasket.		
7-GASKET		
Fusion, Milan		
To 12/4/06 06-07	1S7Z8255BC	8.28

	Part No.	Price
From 12/4/06 07-09	3M4Z8255A	22.00
·········· WATER PUMP ··········		
8-WATER PUMP		
Fusion, Milan 06-09	4S4Z8501AA	89.98
NOTES FOR: WATER PUMP		
Includes Gasket.		
9-GASKET		
Fusion, Milan 06-09	1S7Z8507AE	9.20
·········· COOLING FAN ··········		
FAN RELAY		
Fusion, Milan 06-09	6E5Z8B658A	303.36
10-FAN & MOTOR		
Fusion, Milan 06-09	7E5Z8C607A	766.95
NOTES FOR: FAN & MOTOR		
Includes Fan, Motor And Shroud.		

LABOR 8 COOLING SYSTEM – 3.0L 8 LABOR

COOLING SYSTEM

OPERATION INDEX

BLOCK HEATER, R&R	5
CONTROL MODULE, R&R	11
COOLING SYSTEM PRESSURE, TEST	4
EXPANSION PLUG, R&R	6
EXPANSION TANK, R&R	7
FAN ASSY, R&R	12
LOWER HOSE, R&R	1
RADIATOR, R&R	2
SWITCH, R&R	13
THERMOSTAT GASKET, R&R	9
THERMOSTAT, R&R	8
UPPER HOSE, R&R	3
WATER PUMP, R&R	10

COOLING SYSTEM

·········· RADIATOR ··········

	(Factory Time)	Motor Time
1-LOWER HOSE, R&R		
Fusion, Milan, Zephyr .... 06-09 C	(0.7)	0.9
2-RADIATOR, R&R		
Fusion, Milan, Zephyr .... 06-09 B	(1.6)	2.2
3-UPPER HOSE, R&R		
Fusion, Milan, Zephyr .... 06-09 C	(0.6)	0.8
·········· COOLING SYSTEM ··········		
4-COOLING SYSTEM PRESSURE, TEST		
Fusion, Milan, Zephyr .... 06-09 C		0.3
5-BLOCK HEATER, R&R		
Fusion, Milan, Zephyr .... 06-09 B		0.5
NOTES FOR: BLOCK HEATER, R&R		
Does Not Include: Time To Gain Access.		
6-EXPANSION PLUG, R&R		
Fusion, Milan, Zephyr		
One .... 06-09 C		0.5
Each Additional .... 06-09 C		0.4

	(Factory Time)	Motor Time
NOTES FOR: EXPANSION PLUG, R&R		
Does Not Include: Time To Gain Access.		
7-EXPANSION TANK, R&R		
Fusion, Milan, Zephyr .... 06-09 C	(0.4)	0.5
8-THERMOSTAT, R&R		
Fusion, Milan, Zephyr .... 06-09 B	(0.5)	0.8
9-THERMOSTAT GASKET, R&R		
Fusion, Milan, Zephyr .... 06-09 C	(0.5)	0.8
·········· WATER PUMP ··········		
10-WATER PUMP, R&R		
Fusion, Milan, Zephyr .... 06-09 B	(1.0)	1.4
·········· COOLING FAN ··········		
11-CONTROL MODULE, R&R		
Fusion, Milan, Zephyr .... 06-09 B	(0.9)	1.4
12-FAN ASSY, R&R		
Fusion, Milan, Zephyr .... 06-09 B	(0.6)	0.9
13-SWITCH, R&R		
Fusion, Milan, Zephyr .... 06-09 B		0.3

PARTS 8 COOLING SYSTEM – 3.0L 8 PARTS

COOLING SYSTEM

·········· RADIATOR ··········	Part No.	Price
1-RADIATOR		
Fusion, Milan 06-09	6E5Z8005C	339.40
Zephyr 06	6E5Z8005C	339.40
2-UPPER HOSE		
Fusion, Milan 06-09	9E5Z8260F	24.93
Zephyr 06	9E5Z8260F	24.93
3-LOWER HOSE		
Fusion, Milan 06-09	6E5Z8286BA	138.27

	Part No.	Price
Zephyr 06	6E5Z8286BA	138.27
·········· COOLING SYSTEM ··········		
4-RESERVOIR CAP		
Fusion, Milan		
To 8/18/08 06-09	9C3Z8101B	10.88
From 8/18/08 09	9C3Z8101B	10.88
Zephyr 06	9C3Z8101B	10.88
5-RESERVOIR		
Fusion, Milan 06-09	6E5Z8A080AA	43.62
Zephyr 06	6E5Z8A080AA	43.62
6-WATER INLET		
Fusion, Milan 06-09	YL8Z8K528AE	22.02
Zephyr 06	YL8Z8K528AE	22.02

	Part No.	Price
7-THERMOSTAT		
Fusion, Milan 06-09	1X4Z8575EB	25.88
Zephyr 06	1X4Z8575EB	25.88
8-THERMOSTAT HOUSING		
Fusion, Milan 06-09	2M2Z8592CD	82.94
Zephyr 06	2M2Z8592CD	82.94
9-GASKET		
Fusion, Milan 06-09	F1VY8255A	8.62
Zephyr 06	F1VY8255A	8.62
·········· WATER PUMP ··········		
10-WATER PUMP		
Fusion, Milan 06-09	5M8Z8501B	175.98
	(CONTINUED)	

1 FORD
FUSION, MILAN (06-09), ZEPHYR (06), MKZ (07-09), MILAN HYBRID (10)

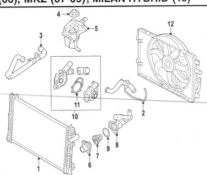

FPP047

COOLING SYSTEM – 3.0L - Parts Cont'd

		Part No.	Price
Zephyr	06	5M8Z8501B	175.98

NOTES FOR: WATER PUMP
Includes Water Inlet Hose And Water Pump Gasket.

11-GASKET

		Part No.	Price
Fusion, Milan	06-09	5F9Z8507AB	12.40

		Part No.	Price
Zephyr	06	5F9Z8507AB	12.40

·············· **COOLING FAN** ··············

FAN RELAY

		Part No.	Price
Fusion, Milan	06-09	6E5Z8B658A	303.36
Zephyr	06	6E5Z8B658A	303.36

12-FAN & MOTOR

		Part No.	Price
Fusion, Milan	06-09	7E5Z8C607A	766.95
Zephyr	06	7E5Z8C607A	766.95

NOTES FOR: FAN & MOTOR
Includes Fan, Motor And Shroud.

LABOR   8   COOLING SYSTEM – 3.5L   8   LABOR

OPERATION INDEX

BLOCK HEATER, R&R	5
CONTROL MODULE, R&R	11
COOLING SYSTEM PRESSURE, TEST	4
EXPANSION PLUG, R&R	6
EXPANSION TANK, R&R	7
FAN ASSY, R&R	12
LOWER HOSE, R&R	1
RADIATOR, R&R	2
SWITCH, R&R	13
THERMOSTAT GASKET, R&R	9
THERMOSTAT, R&R	8
UPPER HOSE, R&R	3
WATER PUMP, R&R	10

COOLING SYSTEM

·············· **RADIATOR** ··············

			(Factory Time)	Motor Time
1-LOWER HOSE, R&R				
MKZ	07-09 C		(0.7)	0.9
2-RADIATOR, R&R				
MKZ	07-09 B		(1.7)	2.2
3-UPPER HOSE, R&R				
MKZ	07-09 C		(0.6)	0.8
············ **COOLING SYSTEM** ············				
4-COOLING SYSTEM PRESSURE, TEST				
MKZ	07-09 C			0.3
5-BLOCK HEATER, R&R				
MKZ	07-09 B			0.5
NOTES FOR: BLOCK HEATER, R&R				
Does Not Include: Time To Gain Access.				
6-EXPANSION PLUG, R&R				
MKZ				
One	07-09 C			0.5
Each Additional	07-09 C			0.4

			(Factory Time)	Motor Time
NOTES FOR: EXPANSION PLUG, R&R				
Does Not Include: Time To Gain Access.				
7-EXPANSION TANK, R&R				
MKZ	07-09 C		(0.4)	0.5
8-THERMOSTAT, R&R				
MKZ	07-09 B		(0.6)	0.9
9-THERMOSTAT GASKET, R&R				
MKZ	07-09 C			0.9
············ **WATER PUMP** ············				
10-WATER PUMP, R&R				
MKZ	07-09 B		(9.8)	14.0
············ **COOLING FAN** ············				
11-CONTROL MODULE, R&R				
MKZ	07-09 B		(0.9)	1.4
12-FAN ASSY, R&R				
MKZ	07-09 B		(0.7)	0.9
13-SWITCH, R&R				
MKZ	07-09 B			0.3

PARTS   8   COOLING SYSTEM – 3.5L   8   PARTS

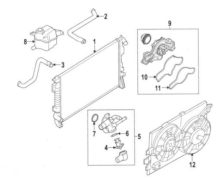

FPP055

COOLING SYSTEM

·············· **RADIATOR** ··············

		Part No.	Price
1-RADIATOR			
MKZ	07-09	AH6Z8005A	322.14
2-UPPER HOSE			
MKZ	07-09	7H6Z8260C	65.76
3-LOWER HOSE			
MKZ	07-09	AH6Z8286A	29.93

·············· **COOLING SYSTEM** ··············

RESERVOIR CAP

		Part No.	Price
MKZ	07-09	9C3Z8101B	10.88

		Part No.	Price
4-THERMOSTAT			
MKZ	07-09	1X4Z8575FA	22.42
5-THERMOSTAT HOUSING			
MKZ	07-09	8M8Z8A586D	196.50
6-THERMOSTAT GASKET			
MKZ	07-09	F1VY8255A	8.62
7-HOUSING GASKET			
MKZ	07-09	7T4Z8590A	10.26
8-RESERVOIR			
MKZ	07	7H6Z8A080A	52.08
1 Port	08-09	7H6Z8A080A	52.08
2 Port	08-09	AH6Z8A080A	48.68

·············· **WATER PUMP** ··············

		Part No.	Price
9-WATER PUMP			
MKZ			
To 12/4/06	07	7T4Z8501A	0.00
From 12/4/06	07-09	AA5Z8501A	149.74

NOTES FOR: WATER PUMP
Includes Gaskets.

		Part No.	Price
10-OUTER GASKET			
MKZ	07-09	7T4Z8507B	7.44
11-INNER GASKET			
MKZ	07-09	7T4Z8507A	7.28

·············· **COOLING FAN** ··············

FAN RELAY

		Part No.	Price
MKZ	07-09	6E5Z8B658A	303.36

(CONTINUED)

FORD 1

FUSION, MILAN (06-09), ZEPHYR (06), MKZ (07-09), MILAN HYBRID (10)

COOLING SYSTEM – 3.5L - Parts Cont'd	Part No.	Price	12-FAN & MOTOR	Part No.	Price	NOTES FOR: FAN & MOTOR	Part No.	Price
			MKZ 07-09	7H6Z8C607B	651.15	Includes: Fan, Motor And Shroud.		

LABOR   9 HVAC 9   LABOR

OPERATION INDEX

AMBIENT TEMP SENSOR, R&R	10
BLOWER MOTOR, R&R	4
CABIN AIR FILTER, R&R	3
CLUTCH, R&R	26
COMPRESSOR, R&R	27
CONDENSER, R&R	29
CONTROLLER, R&R	5
DASH CONTROL UNIT, R&R	1
DISCHARGE HOSE, R&R	16
DOOR ACTUATOR, R&R	2
DRIER, R&R	17
EVAPORATOR CORE, R&R	18
EXPANSION VALVE, R&R	19
HEATER CORE, R&R	8
HEATER HOSE, R&R	9
IN-CAR SENSOR, R&R	11
LIQUID LINE, R&R	20
MANIFOLD, R&R	28
O-RING, R&R	21
PRESSURE SWITCH, R&R	25
RELAY, R&R	15
RELAY, R&R	6
RESISTOR, R&R	13
RESISTOR, R&R	7
SUCTION HOSE, R&R	22
SUNLOAD SENSOR, R&R	12
SYSTEM, SERVICE	23
TEMP SENSOR, R&R	24
WATER SENSOR, R&R	14

CONTROLS

CONTROLS

		(Factory Time)	Motor Time
1-DASH CONTROL UNIT, R&R			
Fusion, Milan, Zephyr, MKZ ..	06-09 B	(0.2)	0.6
2-DOOR ACTUATOR, R&R			
DEFOST			
Fusion,Milan,Zephyr,MKZ ...	06-09 B	(2.2)	3.0
RECIRCULATION/FRESH AIR			
Fusion,Milan,Zephyr,MKZ ...	06-09 B	(0.3)	0.5
TEMPERATURE			
Fusion, Milan, Zephyr	06-09 B	(0.2)	0.4
MKZ			
Right Side	07-09 B	(0.8)	1.1
Left Side	07-09 B	(0.2)	0.3
Both Sides	07-09 B	(0.9)	1.4

BLOWER MOTOR

		(Factory Time)	Motor Time
3-CABIN AIR FILTER, R&R			
Milan, Zephyr, MKZ	06-09 B		0.3
4-BLOWER MOTOR, R&R			
Fusion, Milan, Zephyr, MKZ ..	06-09 B	(0.3)	0.4
5-CONTROLLER, R&R			
Fusion, Milan, Zephyr, MKZ ..	06-09 B	(0.2)	0.3
6-RELAY, R&R			
Fusion, Milan, Zephyr, MKZ ..	06-09 B		0.3
7-RESISTOR, R&R			
Fusion, Milan, Zephyr, MKZ ..	06-09 B	(0.2)	0.3

HEATER

HEATER

		(Factory Time)	Motor Time
8-HEATER CORE, R&R			
Fusion, Milan, Zephyr			
2.3L	06-09 B	(3.0)	4.6
3.0L	06-09 B	(3.6)	4.6
MKZ	07-09 B		4.6

		(Factory Time)	Motor Time
9-HEATER HOSE, R&R			
Fusion, Milan, Zephyr			
2.3L			
One Side	06-09 C	(0.8)	1.1
Both Sides	06-09 C	(1.0)	1.4
3.0L			
One Side	06-09 C	(1.0)	1.4
Both Sides	06-09 C	(1.2)	1.7
MKZ			
One Side	07-09 C		1.4
Both Sides	07-09 C		1.7

AIR CONDITIONER

...... AUTOMATIC TEMPERATURE CONTROLS ......

		(Factory Time)	Motor Time
10-AMBIENT TEMP SENSOR, R&R			
Fusion, Milan, Zephyr, MKZ	06-09 B	(0.5)	0.6
11-IN-CAR SENSOR, R&R			
Fusion, Milan, Zephyr, MKZ ..	06-09 B		0.4
12-SUNLOAD SENSOR, R&R			
Fusion, Milan, Zephyr, MKZ ..	06-09 B		0.3
13-RESISTOR, R&R			
Fusion, Milan, Zephyr, MKZ ..	06-09 B		0.4
14-WATER SENSOR, R&R			
Fusion, Milan, Zephyr, MKZ ..	06-09 B		3.3

.... AIR CONDITIONER ....

		(Factory Time)	Motor Time
15-RELAY, R&R			
Fusion, Milan, Zephyr, MKZ ..	06-09 B		0.3
16-DISCHARGE HOSE, R&R			
Fusion, Milan, Zephyr			
2.3L	06-09 B		0.7
3.0L	06-09 B	(0.3)	0.4
MKZ	07-09 B		0.4

NOTES FOR: DISCHARGE HOSE, R&R
Does Not Include: Refrigerant Recovery Or Evacuate & Recharge AC System.

		(Factory Time)	Motor Time
17-DRIER, R&R			
Fusion, Milan, Zephyr, MKZ ..	06-09 B		0.8

NOTES FOR: DRIER, R&R
Does Not Include: Refrigerant Recovery Or Evacuate & Recharge AC System.

		(Factory Time)	Motor Time
18-EVAPORATOR CORE, R&R			
Fusion, Milan, Zephyr, MKZ ..	06-09 B		4.6

NOTES FOR: EVAPORATOR CORE, R&R
Does Not Include: Refrigerant Recovery Or Evacuate & Recharge AC System.

		(Factory Time)	Motor Time
19-EXPANSION VALVE, R&R			
Fusion, Milan, Zephyr, MKZ	06-09 B	(0.4)	0.5

NOTES FOR: EXPANSION VALVE, R&R
Does Not Include: Refrigerant Recovery Or Evacuate & Recharge AC System.

		(Factory Time)	Motor Time
20-LIQUID LINE, R&R			
Fusion, Milan, Zephyr, MKZ ..	06-09 B		0.6

NOTES FOR: LIQUID LINE, R&R
Does Not Include: Refrigerant Recovery Or Evacuate & Recharge

		(Factory Time)	Motor Time
21-O-RING, R&R			
Fusion, Milan, Zephyr, MKZ			
One Or Two	06-09 B		0.4
All	06-09 B		2.1

NOTES FOR: O-RING, R&R
Does Not Include: Refrigerant Recovery Or Evacuate & Recharge AC System.

		(Factory Time)	Motor Time
22-SUCTION HOSE, R&R			
Fusion, Milan, Zephyr, MKZ ..	06-09 B		0.6

NOTES FOR: SUCTION HOSE, R&R
Does Not Include: Refrigerant Recovery Or Evacuate & Recharge AC System.

		(Factory Time)	Motor Time
23-SYSTEM, SERVICE			
PARTIAL REFRIGERANT CHARGE			
Fusion, Milan, Zephyr, MKZ .	06-09 B		0.6
FLUSH (COMPLETE)			
Fusion, Milan, Zephyr, MKZ .	06-09 B		0.3
EVACUATE & RECHARGE SYSTEM			
Fusion, Milan, Zephyr, MKZ .	06-09 B		1.4
REFRIGERANT RECOVERY			
Fusion, Milan, Zephyr, MKZ .	06-09 B		0.4

NOTES FOR: SYSTEM, SERVICE
To Be Used In Conjunction With Component Replacement Which Could Contaminate System. Does Not Include: Evacuate & Recharge System. With Any Operation Requiring A Refrigerant Line Disconnect, Add AC Service, I.E. Evacuate, Recharge And Test For Leaks. Add For Refrigerant Cost. Many Vehicles Are Now Using R134 Refrigerant In The AC System. Extra Care Must Be Observed When Servicing This Type Of System. R12 Refrigerant MUST NOT Be Used. See Manufacturers Service Manual For Specific Repair Procedures.

.... SWITCHES & SENSORS ....

		(Factory Time)	Motor Time
24-TEMP SENSOR, R&R			
EVAPORATOR DISCHARGE AIR			
Fusion, Milan, Zephyr, MKZ .	06-09 B		4.9
25-PRESSURE SWITCH, R&R			
Fusion, Milan, Zephyr, MKZ ..	06-09 B	(0.2)	0.3

NOTES FOR: PRESSURE SWITCH, R&R
Does Not Include: Refrigerant Recovery Or Evacuate & Recharge AC System.

.... COMPRESSOR ....

		(Factory Time)	Motor Time
26-CLUTCH, R&R			
Fusion, Milan, Zephyr			
2.3L	06-09 B		2.2
3.0L	06-09 B		1.9
MKZ	07-09 B		2.3

NOTES FOR: CLUTCH, R&R
Does Not Include: Refrigerant Recovery Or Evacuate & Recharge AC System.

		(Factory Time)	Motor Time
27-COMPRESSOR, R&R			
Fusion, Milan, Zephyr			
2.3L	06-09 B	(1.0)	1.7
3.0L	06-09 B	(0.9)	1.4
MKZ	07-09 B	(1.3)	1.8

NOTES FOR: COMPRESSOR, R&R
Does Not Include: Refrigerant Recovery Or Evacuate & Recharge AC System.

		(Factory Time)	Motor Time
28-MANIFOLD, R&R			
Fusion, Milan, Zephyr, MKZ ..	06-09 B	(0.5)	0.6

NOTES FOR: MANIFOLD, R&R
Does Not Include: Refrigerant Recovery Or Evacuate & Recharge AC System.

.... CONDENSER ....

		(Factory Time)	Motor Time
29-CONDENSER, R&R			
Fusion, Milan, Zephyr			
2.3L	06-09 B	(1.5)	2.1
3.0L	06-09 B	(1.4)	1.8
MKZ	07-09 B	(1.5)	2.1

NOTES FOR: CONDENSER, R&R
Does Not Include: Refrigerant Recovery Or Evacuate & Recharge AC System.

PARTS   9 HVAC 9   PARTS

PARTS INDEX

AMBIENT TEMP SENSOR	6	DASH CONTROL UNIT	1	RESISTOR	4
BLOWER MOTOR	3	DISCHARGE HOSE	10	SUCTION HOSE	11
CABIN AIR FILTER	2	DRIER	12	SUNLOAD SENSOR	8
CLUTCH	17	EVAPORATOR CORE	13		
CLUTCH COIL	18	EXPANSION VALVE	14		
COMPRESSOR	20	HEATER CORE	9	**CONTROLS**	
CONDENSER	21	IN-CAR SENSOR	7		
CONTROL MODULE	5	PRESSURE RELIEF VALVE	15	**CONTROLS**	
		PRESSURE SWITCH	16		
		PULLEY	19	(CONTINUED)	

1 FORD
FUSION, MILAN (06-09), ZEPHYR (06), MKZ (07-09), MILAN HYBRID (10)

HVAC - Parts Cont'd

		Part No.	Price
1-DASH CONTROL UNIT			
Fusion, Milan			
w/o ATC	06-09	7E5Z19980B	218.00
w/ATC			
w/o Heated Seats	06-07	8E5Z19980A	366.00
	08-09	8E5Z19980A	366.00
w/Heated Seats	06-07	8E5Z19980B	316.32
	08-09	8E5Z19980B	316.32
Zephyr			
w/o Heated Seats			
To 6/26/06	06	6H6Z19980AD	454.97
From 6/26/06	06	6H6Z19980AD	454.97
w/Heated Seats			
w/o Cooled Seat	06	6H6Z19980CD	395.98
w/Cooled Seats	06	6H6Z19980BD	466.00
MKZ			
w/o Cooled Seat			
To 9/4/06	07	6H6Z19980CD	395.98
From 9/4/06	07	7H6Z19980B	545.98
	08	8H6Z19980BA	564.98
w/Cooled Seats			
To 9/4/06	07	6H6Z19980BD	466.00
From 9/4/06	07	7H6Z19980C	538.98
	08-09	8H6Z19980CA	556.98

········· BLOWER MOTOR ·········

		Part No.	Price
2-CABIN AIR FILTER			
Zephyr	06	6H6Z18D395AA	23.00
MKZ	07-09	6H6Z18D395AA	23.00
3-BLOWER MOTOR			
Fusion, Milan	06-09	8E5Z19805A	26.07
Zephyr	06	8E5Z19805A	26.07
MKZ	07-09	8E5Z19805A	26.07
4-RESISTOR			
Fusion, Milan	06-09	6E5Z18591AA	9.00
5-CONTROL MODULE			
Fusion, Milan	06-09	8E5Z19E624A	84.13
Zephyr	06	8E5Z19E624A	84.13
MKZ	07-09	8E5Z19E624A	84.13

······ AUTOMATIC TEMPERATURE CONTROLS ······

		Part No.	Price
6-AMBIENT TEMP SENSOR			
Fusion, Milan	06-09	6E5Z19E642AA	43.87
Zephyr	06	6E5Z19E642AA	43.87
MKZ	07-09	6E5Z19E642AA	43.87
7-IN-CAR SENSOR			
Fusion, Milan			
At Plenum Chamber	06-09	6E5Z19C734BA	17.92
In Instrument Panel	06-09	6E5Z19C734AA	12.95

		Part No.	Price
Zephyr			
At Plenum Chamber	06	6E5Z19C734BA	17.92
In Instrument Panel	06	6E5Z19C734AA	12.95
MKZ			
At Plenum Chamber	07-09	6E5Z19C734BA	17.92
In Instrument Panel	07-09	6E5Z19C734AA	12.95
8-SUNLOAD SENSOR			
Fusion, Milan	06-09	5F9Z13A018BA	38.98
Zephyr	06	6H6Z13A018AA	29.98
MKZ	07-09	6H6Z13A018AA	29.98

NOTES FOR: SUNLOAD SENSOR
Includes Sentinel Sensor.

HEATER

········· HEATER ·········

		Part No.	Price
9-HEATER CORE			
Fusion, Milan	06	6E5Z18476AA	149.00
	07-09	6E5Z18476BA	168.50
Zephyr	06	6E5Z18476AA	149.00
MKZ	07-09	6E5Z18476BA	168.50

AIR CONDITIONER

········· AIR CONDITIONER ·········

		Part No.	Price
10-DISCHARGE HOSE			
Fusion, Milan			
2.3L	06-09	6E5Z19D734AA	154.82
3.0L	06-09	6E5Z19D734BA	154.02
Zephyr	06	6E5Z19D734BA	154.02
MKZ			
To 12/4/06	07	7H6Z19D734A	140.10
From 12/4/06	07-09	7H6Z19D734B	149.26
11-SUCTION HOSE			
Fusion, Milan			
Front Hose	06-09	6E5Z19835AA	58.40
Rear Hose	06-09	6E5Z19835AA	108.56
Zephyr			
Front Hose	06	6E5Z19835AA	58.40
Rear Hose	06	6E5Z19835AA	108.56
MKZ			
Front Hose	07-09	7H6Z19835A	61.30
Rear Hose	07-09	7H6Z19835B	140.16
12-DRIER			
Fusion, Milan	06-09	6E5Z19C836A	223.74
Zephyr	06	6E5Z19C836A	223.74
MKZ	07-09	6E5Z19C836A	223.74
13-EVAPORATOR CORE			
Fusion, Milan	06-09	6E5Z19860AA	214.98
Zephyr	06	6H6Z19860BA	224.98

		Part No.	Price
MKZ	07-09	6H6Z19860BA	224.98
14-EXPANSION VALVE			
Fusion, Milan	06-09	6E5Z19849AA	120.98
Zephyr	06	6E5Z19849AA	120.98
MKZ	07-09	6E5Z19849AA	120.98
15-PRESSURE RELIEF VALVE			
Fusion, Milan	06-09	6E5Z19D644AA	14.78
Zephyr	06	6E5Z19D644AA	14.78
MKZ	07-09	6E5Z19D644AA	14.78

········· SWITCHES & SENSORS ·········

		Part No.	Price
16-PRESSURE SWITCH			
Fusion, Milan	06-09	6E5Z19D594AA	59.33
Zephyr	06	6E5Z19D594AA	59.33
MKZ	07-09	6E5Z19D594AA	59.33

········· COMPRESSOR ·········

		Part No.	Price
17-CLUTCH			
Fusion, Milan			
2.3L			
To 12/10/06	06	6E5Z19D786A	42.68
From 12/10/06 To 4/27/06	06	6E5Z19D786B	54.44
From 4/27/06	06-09	6E5Z19D786A	42.68
3.0L	06-09	6E5Z19D786B	54.44
Zephyr	06	6E5Z19D786A	42.68
MKZ	07-09	6E5Z19D786A	42.68
18-CLUTCH COIL			
Fusion, Milan			
To 9/4/06	06-07	6E5Z19D798AA	50.12
From 9/4/06	07-09	7E5Z19D798AA	75.46
Zephyr	06	6E5Z19D798AA	50.12
MKZ			
To 9/4/06	07	6E5Z19D798AA	50.12
From 9/4/06	07-09	7E5Z19D798AA	75.46
19-PULLEY			
Fusion, Milan			
2.3L			
Manual Trans	06-09	6E5Z19D784B	82.86
Auto Trans	06-09	6E5Z19D784A	76.56
3.0L	06-09	6E5Z19D784A	76.56
Zephyr	06	6E5Z19D784A	76.56
MKZ	07-09	7H6Z19D784A	89.86
20-COMPRESSOR			
Fusion, Milan	06-09	8E5Z19703A	403.18
Zephyr	06	8E5Z19703A	403.18
MKZ	07-09	8H6Z19703A	340.58

········· CONDENSER ·········

		Part No.	Price
21-CONDENSER			
Fusion, Milan	06-09	6N7Z19712A	487.22
Zephyr	06	6N7Z19712A	487.22
MKZ	07-09	6N7Z19712A	487.22

LABOR   10   ENGINE – 2.3L   10   LABOR

OPERATION INDEX

ENGINE

········· ENGINE ·········

		(Factory Time)	Motor Time
1-ENGINE, R&I			
Fusion, Milan			
Manual Trans	06-09 B	(6.4)	8.8
Auto Trans	06-09 B	(6.3)	9.8

NOTES FOR: ENGINE, R&I
Does Not Include: Transfer Of Any Part Of Engine Or Replacement Of Optional Equipment.

		(Factory Time)	Motor Time
2-ENGINE, R&R			
Fusion, Milan			
Manual Trans	06-09 B		11.8
Auto Trans	06-09 B		12.8

NOTES FOR: ENGINE, R&R
Includes: Transfer All Fuel & Electrical Units. Does Not Include Replacement Of Optional Equipment.

		(Factory Time)	Motor Time
3-ENGINE, OVERHAUL			
Fusion, Milan			
Manual Trans	06-09 A		23.3
Auto Trans	06-09 A		24.3

NOTES FOR: ENGINE, OVERHAUL
Includes: Disassemble & Clean Engine, Ridge Ream & Hone Cylinders, Fit Pistons, Rings, Pins, Main & Rod Bearings, R&I Engine, Grind Valves & Tune-Up.

		(Factory Time)	Motor Time
4-LONG BLOCK, R&R			
Fusion, Milan			
Manual Trans	06-09 B	(9.4)	13.8
Auto Trans	06-09 B	(9.7)	14.8

NOTES FOR: LONG BLOCK, R&R
Includes: R&I Engine And Transfer All Necessary Components Not Supplied With Long Block.

		(Factory Time)	Motor Time
5-OIL LEAK, DIAGNOSIS			
	06-09 B		1.0

		(Factory Time)	Motor Time
6-SHORT BLOCK, R&R			
Fusion, Milan			
Manual Trans	06-09 A	(12.9)	15.8
Auto Trans	06-09 A	(13.2)	16.8

NOTES FOR: SHORT BLOCK, R&R
Includes: R&I Engine And Replacement Of All Necessary Components.

MOUNTS

········· MOUNTS ·········

		(Factory Time)	Motor Time
7-MOUNT, R&R			
FRONT			
Fusion, Milan	06-09 B	(0.3)	0.4
UPPER			
Fusion, Milan	06-09 B	(0.8)	1.1
8-TRANS MOUNT, R&R			
Fusion, Milan			
Manual Trans	06-09 B	(1.2)	1.8
Auto Trans	06-09 B	(1.3)	1.8

CYLINDER HEAD & VALVES

········· CYLINDER HEAD & VALVES ·········

		(Factory Time)	Motor Time
9-CYLINDER HEAD, R&R			
Fusion, Milan	06-09 A		15.6

NOTES FOR: CYLINDER HEAD, R&R
Includes: R&I Cylinder Head, Grind All Valves & Make All Necessary Adjustments.

		(Factory Time)	Motor Time
10-HEAD GASKET, R&R			
Fusion, Milan	06-09 B	(8.6)	11.8
11-VALVE COVER GASKET, R&R			
Fusion, Milan	06-09 B	(0.7)	0.9
12-VALVE LIFTERS, R&R			
Fusion, Milan			
One	06-09 B	(2.4)	2.8
All	06-09 B	(2.9)	3.4

(CONTINUED)

FORD 1

FUSION, MILAN (06-09), ZEPHYR (06), MKZ (07-09), MILAN HYBRID (10)

ENGINE – 2.3L - Time Cont'd

	(Factory Time)	Motor Time
13-VALVE SEALS, R&R		
Fusion, Milan		
All Cyls	06-09 B	7.7
14-VALVE SPRINGS, R&R		
Fusion, Milan		
All Cyls	06-09 B	7.7
15-VALVES, GRIND		
Fusion, Milan		
All Cyls	06-09 A	15.6
NOTES FOR: VALVES, GRIND		
Includes: R&I Cylinder Head Grind All Valves, Seats And Make All Necessary Adjustments.		
To Ream Valve Guides, Add		
Each	06-09	0.2
To R&R Valve Guides, Add		
Each	06-09	0.3
To R&R Camshaft, Add		
One	06-09	0.2
Each Additional	06-09	0.1

CAMSHAFT & TIMING

VARIABLE VALVE TIMING

	(Factory Time)	Motor Time
16-SOLENOID, R&R		
Fusion, Milan	06-09 B (0.6)	0.7

CAMSHAFT & TIMING

	(Factory Time)	Motor Time
17-CAMSHAFT, R&R		
Fusion, Milan		
One	06-09 B (2.6)	3.5
Both	06-09 B (2.9)	4.0
18-FRONT COVER GASKET, R&R		
Fusion, Milan	06-09 B (3.6)	5.2
19-TENSIONER, R&R		
Fusion, Milan		
See TIMING CHAIN, R&R		
20-TIMING CHAIN, R&R		
Fusion, Milan	06-09 B (3.9)	5.7
NOTES FOR: TIMING CHAIN, R&R		
Includes: R&I Front Cover.		
To R&R Tensioner, Add	06-09	0.1

PISTONS, RINGS & BEARINGS

PISTONS, RINGS & BEARINGS

	(Factory Time)	Motor Time
21-CONNECTING ROD BEARING, R&R		
Fusion, Milan		
Manual Trans	06-09 A	2.9
Auto Trans	06-09 A	3.2
NOTES FOR: CONNECTING ROD BEARING, R&R		
Includes: R&I Oil Pan & Plastigage Bearings And Make All Necessary Adjustments.		
22-PISTON RINGS, R&R		
Fusion, Milan		
Manual Trans		
One Cylinder	06-09 A	16.2
Each Additional	06-09 A	0.3
Auto Trans		
One Cylinder	06-09 A	17.2
Each Additional	06-09 A	0.3
NOTES FOR: PISTON RINGS, R&R		
Includes: Remove Cylinder Top Ridge, Deglaze Cylinder Walls And Clean Carbon.		
To R&R Connecting Rod Bearings, Add		
Each	06-09	0.1
To R&R Piston, Pin Or Connecting Rod, Add		
One	06-09	0.2
Each Additional	06-09	0.1
All	06-09	0.5

CRANKSHAFT & BEARINGS

CRANKSHAFT & BEARINGS

	(Factory Time)	Motor Time
23-CRANKSHAFT, R&R		
Fusion, Milan		
Manual Trans	06-09 A	15.0
Auto Trans	06-09 A	16.0
NOTES FOR: CRANKSHAFT, R&R		
Includes: R&I Engine & Oil Pan, Renew All Bearings And Seals And Make All Necessary Adjustments.		
24-FRONT CRANK SEAL, R&R		
Fusion, Milan	06-09 B (1.7)	2.3
25-MAIN & ROD BEARINGS, R&R		
Fusion, Milan		
Manual Trans	06-09 A	15.0

	(Factory Time)	Motor Time
Auto Trans	06-09 A	16.0
NOTES FOR: MAIN & ROD BEARINGS, R&R		
Includes: R&I Engine, Oil Pan And Plastigage Bearings.		
26-MAIN BEARINGS, R&R		
Fusion, Milan		
Manual Trans	06-09 A	14.6
Auto Trans	06-09 A	15.6
NOTES FOR: MAIN BEARINGS, R&R		
Includes: R&I Engine, Oil Pan And Plastigage Bearings.		
27-PULLEY, R&R		
Fusion, Milan	06-09 B (1.6)	2.1
28-REAR MAIN SEAL, R&R		
Fusion, Milan		
Manual Trans	06-09 B	0.9
Auto Trans	06-09 B	0.5
NOTES FOR: REAR MAIN SEAL, R&R		
After Trans Is Removed.		

LUBRICATION

OIL PAN

	(Factory Time)	Motor Time
29-OIL PAN, R&R		
Fusion, Milan		
Manual Trans	06-09 B (1.5)	2.1
Auto Trans	06-09 B (1.7)	2.4
30-OIL PAN GASKET, R&R		
Fusion, Milan		
Manual Trans	06-09 B (1.5)	2.1
Auto Trans	06-09 B (1.7)	2.4

OIL PUMP

	(Factory Time)	Motor Time
31-OIL PUMP, R&R		
Fusion, Milan	06-09 B (4.6)	6.8

OIL COOLER

	(Factory Time)	Motor Time
32-O-RING, R&R		
Fusion, Milan	06-09 B	1.4
33-OIL COOLER, R&R		
Fusion, Milan	06-09 B	1.4

PARTS | 10 ENGINE – 2.3L 10 | **PARTS**

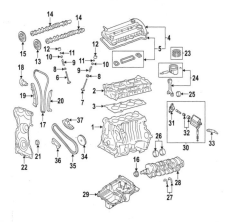

FPP060

ENGINE

ENGINE

	Part No.	Price
ENGINE		
Fusion, Milan 06-09	7E5Z6006AARM	4277.33
LOWER GASKET KIT		
Fusion, Milan 06-09	3S4Z6E078CA	32.17
1-SHORT BLOCK		
Fusion, Milan 06-09	6M8Z6009AA	2536.22

MOUNTS

MOUNTS

	Part No.	Price
MOUNT		
Fusion, Milan 06-09	6E5Z6038CK	96.81

	Part No.	Price
STRUT		
Fusion, Milan 06-09	8E5Z6068D	63.48
TRANS MOUNT		
Fusion, Milan 06-09	9E5Z6038A	87.13

CYLINDER HEAD & VALVES

CYLINDER HEAD & VALVES

	Part No.	Price
VALVE GRIND GASKET KIT		
Fusion, Milan 06-09	6M8Z6079AA	60.91
2-CYLINDER HEAD		
Fusion, Milan		
w/o Valves 06-09	6M8Z6049AA	473.90
w/Valves 06-09	6M8Z6049BA	653.38
3-HEAD GASKET		
Fusion, Milan 06-09	6M8Z6051AA	69.38

	Part No.	Price
4-VALVE COVER		
Fusion, Milan 06-09	6M8Z6582B	134.98
5-VALVE COVER GASKET		
Fusion, Milan 06-09	1S7G6K260AA	31.15
6-EXHAUST VALVE		
Fusion, Milan 06-09	1S7Z6505AA	14.38
7-INTAKE VALVE		
Fusion, Milan 06-09	9S4Z6507A	6.35
8-VALVE SEALS		
Fusion, Milan		
Intake 06-09	3S4Z6571AA	4.23
Exhaust 06-09	1S7Z6571EA	1.87
9-VALVE SPRINGS		
Fusion, Milan 06-09	3M4Z6513BA	4.10
(CONTINUED)		

1 FORD
FUSION, MILAN (06-09), ZEPHYR (06), MKZ (07-09), MILAN HYBRID (10)

ENGINE – 2.3L - Parts Cont'd

		Part No.	Price
10-VALVE SPRING RETAINERS			
Fusion, Milan 06-09		1S7Z6514AA	6.43
11-VALVE KEEPER			
Fusion, Milan 06-09		1S7Z6518AA	4.41
12-VALVE LIFTERS			
Fusion, Milan 06-09		8G9Z6500BAA	13.83

NOTES FOR: VALVE LIFTERS
Order By Size.

CAMSHAFT & TIMING

········ VARIABLE VALVE TIMING ·········

CONTROL VALVE SOLENOID			
Fusion, Milan 06-09		6M8Z6M280AA	102.65
13-ACTUATOR			
Fusion, Milan			
To 11/1/05 06		6M8Z6C525B	146.65
From 11/1/05 06-09		6M8Z6C525A	146.65

········· CAMSHAFT & TIMING ··········

14-CAMSHAFT			
Fusion, Milan			
Intake 06-09		5M8Z6250AA	60.51
Exhaust 06-09		3M4Z6250AAA	42.86
15-CAMSHAFT GEAR			
Fusion, Milan			
To 12/3/06 06-07		3L8Z6256AA	14.41
From 12/3/06 07-08		1S7Z6256AA	22.41
16-CRANKSHAFT GEAR			
Fusion, Milan			
To 12/4/06 06-07		3L8Z6306AA	28.44
From 12/4/06 07-08		1S7Z6306DA	29.46
17-TIMING CHAIN			
Fusion, Milan			
To 12/3/06 06-07		3L8Z6268AA	66.24
From 12/3/06 07-09		1L5Z6268AA	61.86
18-TENSIONER			
Fusion, Milan 06-09		1S7Z6K254AA	66.65

		Part No.	Price
19-TENSIONER GUIDE			
Fusion, Milan			
To 12/3/06 06-07		3L8Z6K255AA	27.48
From 12/3/06 07-09		6M8Z6K255A	25.67
20-CHAIN GUIDE			
Fusion, Milan			
To 1/8/07 06-07		3L8Z6K297AA	27.48
From 1/8/07 07-09		6M8Z6K297BA	23.32
21-FRONT COVER SEAL			
Fusion, Milan 06-09		1S7Z6700AA	12.54
22-FRONT COVER			
Fusion, Milan 06-09		6M8Z6019B	130.82

··· PISTONS, RINGS & BEARINGS ····

23-PISTON RINGS		
Fusion, Milan		
Part Of Short Block. .. 06-09		
24-PISTON		
Fusion, Milan		
Part Of Short Block. .. 06-09		
25-BEARINGS		
Fusion, Milan		
Part Of Short Block. .. 06-09		

CRANKSHAFT & BEARINGS

····· CRANKSHAFT & BEARINGS ·······

PULLEY			
Fusion, Milan 06-09		6U7Z6312A	60.12
REAR MAIN SEAL RETAINER			
Fusion, Milan 06-09		1S7Z6K301BA	73.09
26-BEARINGS			
Fusion, Milan			
Part Of Short Block. .. 06-09			
25-THRUST BEARING			
Fusion, Milan			
Part Of Short Block. .. 06-09			

		Part No.	Price
26-CRANKSHAFT			
Fusion, Milan			
Part Of Short Block. .. 06-09			
27-THRUST WASHER			
Part Of Short Block. .. 06-09			
28-CRANKSHAFT			
Fusion, Milan			
Part Of Short Block. .. 06-09			

LUBRICATION

········· OIL PAN ··········

OIL PAN BAFFLE			
Right 06-09		3M4Z6687AA	10.98
Left 06-09		3M4Z6687BA	10.98
29-OIL PAN			
Fusion, Milan 06-09		6M8Z6675AB	163.15

NOTES FOR: OIL PAN
Includes Oil Pan Baffle.

········· OIL PUMP ·········

30-OIL PUMP			
Fusion, Milan			
To 4/4/08 07-08		3M4Z6600BH	76.57
From 4/4/08 08-09		8E5Z6600A	99.05
31-COVER			
Fusion, Milan			
Part Of Oil Pump. .. 06-09			
32-GEAR SET			
Fusion, Milan			
Part Of Oil Pump. .. 06-09			
33-OIL PICK-UP			
Fusion, Milan 06-09		6M8Z6622AA	18.90
34-OIL PUMP GEAR			
Fusion, Milan			
To 12/3/06 06-07		3L8Z6652AA	13.52
From 12/3/06 07-09		8E5Z6652A	12.22
35-CHAIN			
Fusion, Milan 06-09		6M8Z6A895BA	34.00
36-CHAIN GUIDE			
Fusion, Milan 06-09		3M4Z6M256AA	4.41
37-ADJUSTER			
Fusion, Milan 06-09		8E5Z6K254A	10.22

LABOR 10 ENGINE – 3.0L 10 LABOR

OPERATION INDEX

ENGINE

·············· ENGINE ··············

			Factory Time	Motor Time
1-ENGINE, R&I				
Fusion, Milan, Zephyr				
FWD	06-09 B		(7.9)	10.8
AWD	07-09 B		(9.8)	13.1

			Factory Time	Motor Time
NOTES FOR: ENGINE, R&I				
Does Not Include: Transfer Of Any Part Of Engine Or Replacement Of Optional Equipment.				
2-ENGINE, R&R				
Fusion, Milan, Zephyr				
FWD	06-09 B			13.3
AWD	07-09 B			15.6
NOTES FOR: ENGINE, R&R				
Includes: Transfer All Fuel & Electrical Units. Does Not Include Replacement Of Optional Equipment.				
3-ENGINE, OVERHAUL				
Fusion, Milan, Zephyr				
FWD	06-09 A			30.3
AWD	07-09 A			32.6
NOTES FOR: ENGINE, OVERHAUL				
Includes: Disassemble & Clean Engine, Ridge Ream & Hone Cylinders, Fit Pistons, Rings, Pins, Main & Rod Bearings, R&I Engine, Grind Valves & Tune-Up.				
4-LONG BLOCK, R&R				
Fusion, Milan, Zephyr				
FWD	06-09 B	(11.1)		17.3
AWD	07-09 B			19.6
NOTES FOR: LONG BLOCK, R&R				
Includes: R&I Engine And Transfer All Necessary Components Not Supplied With Long Block.				
5-OIL LEAK, DIAGNOSIS				
Fusion, Milan, Zephyr	06-09 B			1.0
6-SHORT BLOCK, R&R				
Fusion, Milan, Zephyr				
FWD	06-09 B	(14.9)		20.3
AWD	07-09 B			22.6
NOTES FOR: SHORT BLOCK, R&R				
Includes: R&I Engine And Replacement Of All Necessary Components.				

MOUNTS

········· MOUNTS ·········

			Factory Time	Motor Time
7-MOUNT, R&R				
FRONT				
Fusion, Milan, Zephyr	06-09 B	(0.8)		1.1
UPPER				
Fusion, Milan, Zephyr	06-09 B	(2.1)		3.0
8-TRANS MOUNT, R&R				
Fusion, Milan, Zephyr	06-09 B	(1.5)		2.1

			Factory Time	Motor Time

CYLINDER HEAD & VALVES

······ CYLINDER HEAD & VALVES ·······

			Factory Time	Motor Time
9-CYLINDER HEAD, R&R				
Fusion, Milan, Zephyr				
Right Bank	06-09 A			15.6
Left Bank	06-09 A			16.2
Both Banks	06-09 A			20.5
NOTES FOR: CYLINDER HEAD, R&R				
Includes: R&I Cylinder Head, Grind All Valves & Make All Necessary Adjustments.				
10-HEAD GASKET, R&R				
Fusion, Milan, Zephyr				
Right Bank	06-09 B	(10.0)		13.7
Left Bank	06-09 B	(10.5)		14.3
Both Banks	06-09 B	(12.9)		17.1
11-ROCKER ARMS, R&R				
Fusion, Milan, Zephyr				
Right Bank	06-09 B			4.8
Left Bank	06-09 B			1.9
Both Banks	06-09 B			5.6
12-VALVE COVER GASKET, R&R				
Fusion, Milan, Zephyr				
Right Bank	06-09 B	(2.7)		4.0
Left Bank	06-09 B	(0.8)		1.1
Both Banks	06-09 B	(3.3)		4.6
13-VALVE LIFTERS, R&R				
Fusion, Milan, Zephyr				
Right Bank	06-09 B	(3.9)		5.6
Left Bank	06-09 B	(2.0)		2.7
Both Banks	06-09 B	(5.2)		6.9
14-VALVE SEALS, R&R				
Fusion, Milan, Zephyr				
All Cyls				
Right Bank	06-09 B			8.4
Left Bank	06-09 B			5.5
Both Banks	06-09 B			12.2
15-VALVE SPRINGS, R&R				
Fusion, Milan, Zephyr				
All Cyls				
Right Bank	06-09 B			8.4
Left Bank	06-09 B			5.5

(CONTINUED)

FORD 1

FUSION, MILAN (06-09), ZEPHYR (06), MKZ (07-09), MILAN HYBRID (10)

ENGINE – 3.0L - Time Cont'd

		(Factory Time)	Motor Time
Both Banks	06-09 B		12.2

16-VALVES, GRIND
Fusion, Milan, Zephyr

		(Factory Time)	Motor Time
Right Bank	06-09 A		15.6
Left Bank	06-09 A		16.2
Both Banks	06-09 A		20.5

NOTES FOR: VALVES, GRIND
Includes: R&I Cylinder Head, Timing Belt, Grind All Valves & Seats And Make All Necessary Adjustments.
To Ream Valve Guides, Add

Each	06-09	0.2

To R&R Valve Guides, Add

Each	06-09	0.3

To R&R Camshaft, Add

One	06-09	0.2
Each Additional	06-09	0.1

CAMSHAFT & TIMING

······· VARIABLE VALVE TIMING ·······

17-SOLENOID, R&R
Fusion, Milan, Zephyr

		(Factory Time)	Motor Time
Right Bank	06-09 B	(2.6)	3.5
Left Bank	06-09 B	(0.8)	1.1
Both Banks	06-09 B	(3.2)	4.6

·········· CAMSHAFT & TIMING ··········

18-CAMSHAFT, R&R
Fusion, Milan, Zephyr

		(Factory Time)	Motor Time
Right Bank	06-09 B		11.8
Left Bank	06-09 B		12.8
Both Banks	06-09 B		13.7

NOTES FOR: CAMSHAFT, R&R
Includes: R&I Valve Cover.

19-FRONT COVER GASKET, R&R

		(Factory Time)	Motor Time
Fusion, Milan, Zephyr	06-09 B	(6.7)	9.8

20-CAMSHAFT SEAL, R&R

		(Factory Time)	Motor Time
Fusion, Milan, Zephyr	06-09 B		1.6

21-TENSIONER, R&R
Fusion, Milan, Zephyr
See TIMING CHAIN, R&R

22-TIMING CHAIN, R&R

		(Factory Time)	Motor Time
Fusion, Milan, Zephyr	06-09 B		10.7

NOTES FOR: TIMING CHAIN, R&R
Includes: R&I Front Cover.

To R&R Tensioners, Add	06-09	0.1

PISTONS, RINGS & BEARINGS

····· PISTONS, RINGS & BEARINGS ·····

23-CONNECTING ROD BEARING, R&R

		(Factory Time)	Motor Time
Fusion, Milan, Zephyr	06-09 A		4.7

NOTES FOR: CONNECTING ROD BEARING, R&R
Includes: R&I Oil Pan & Plastigage Bearings And Make All Necessary Adjustments.

24-PISTON RINGS, R&R
Fusion, Milan, Zephyr
FWD

		(Factory Time)	Motor Time
One Cylinder	06-09 A		20.7
Each Additional	06-09 A		0.3

AWD

		(Factory Time)	Motor Time
One Cylinder	07-09 A		23.0
Each Additional	07-09 A		0.3

NOTES FOR: PISTON RINGS, R&R
Includes: Remove Cylinder Top Ridge, Deglaze Cylinder Walls And Clean Carbon.
To R&R Connecting Rod Bearings, Add

Each	06-09	0.1

To R&R Piston, Pin Or Connecting Rod, Add

One	06-09	0.2
Each Additional	06-09	0.1
All	06-09	0.7

CRANKSHAFT & BEARINGS

······ CRANKSHAFT & BEARINGS ······

25-CRANKSHAFT, R&R
Fusion, Milan, Zephyr
FWD

		(Factory Time)	Motor Time
	06-09 A		19.1

AWD

		(Factory Time)	Motor Time
	07-09 A		21.4

NOTES FOR: CRANKSHAFT, R&R
Includes: R&I Engine & Oil Pan, Renew All Bearings And Seals And Make All Necessary Adjustments.

26-FRONT CRANK SEAL, R&R

		(Factory Time)	Motor Time
Fusion, Milan, Zephyr	06-09 B	(0.7)	1.1

27-MAIN & ROD BEARINGS, R&R
Fusion, Milan, Zephyr

		(Factory Time)	Motor Time
FWD	06-09 A		19.1
AWD	07-09 A		21.4

NOTES FOR: MAIN & ROD BEARINGS, R&R
Includes: R&I Engine, Oil Pan And Plastigage Bearings.

28-MAIN BEARINGS, R&R
Fusion, Milan, Zephyr

		(Factory Time)	Motor Time
FWD	06-09 A		18.5
AWD	07-09 A		20.8

NOTES FOR: MAIN BEARINGS, R&R
Includes: R&I Engine, Oil Pan And Plastigage Bearings.

29-PULLEY, R&R

		(Factory Time)	Motor Time
Fusion, Milan, Zephyr	06-09 B	(0.6)	0.9

30-REAR MAIN SEAL, R&R

		(Factory Time)	Motor Time
Fusion, Milan, Zephyr	06-09 B		0.5

NOTES FOR: REAR MAIN SEAL, R&R
After Trans Is Removed.

LUBRICATION

········· OIL PAN ·········

31-OIL PAN, R&R

		(Factory Time)	Motor Time
Fusion, Milan, Zephyr	06-09 B	(2.5)	3.5

32-OIL PAN GASKET, R&R

		(Factory Time)	Motor Time
Fusion, Milan, Zephyr	06-09 B	(2.5)	3.5

········· OIL PUMP ·········

33-OIL PUMP, R&R

		(Factory Time)	Motor Time
Fusion, Milan, Zephyr	06-09 B	(9.6)	13.2

········· OIL COOLER ·········

34-O-RING, R&R

		(Factory Time)	Motor Time
Fusion, Milan, Zephyr	06-09 B		1.4

35-OIL COOLER, R&R

		(Factory Time)	Motor Time
Fusion, Milan, Zephyr	06-09 B		1.4

PARTS   **10 ENGINE – 3.0L 10**   **PARTS**

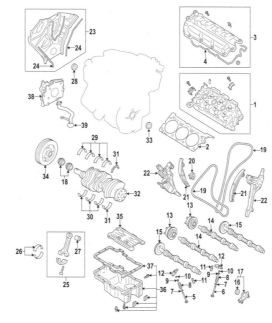

FPP070

ENGINE

········· ENGINE ·········

ENGINE	Part No.	Price
Fusion, Milan 06-09	6E5Z6006AARM	3792.00
Zephyr 06	6E5Z6006AARM	3792.00

SHORT BLOCK	Part No.	Price
Fusion, Milan 06-09	6E5Z6009AA	2463.40
Zephyr 06	6E5Z6009AA	2463.40
(CONTINUED)		

1 FORD
FUSION, MILAN (06-09), ZEPHYR (06), MKZ (07-09), MILAN HYBRID (10)

ENGINE – 3.0L – Parts Cont'd

		Part No.	Price
MOUNTS			
.... **MOUNTS**			
MOUNT			
Fusion, Milan	06-09	6E5Z6038DF	96.81
Zephyr	06	6E5Z6038DF	96.81
STRUT			
Fusion, Milan	06	8E5Z6068D	63.48
FWD	07-09	8E5Z6068D	63.48
AWD	07-09	9E5Z6068D	60.94
Zephyr	06	8E5Z6068D	63.48
TRANS MOUNT			
Fusion, Milan	06-09	6H6Z6038BA	85.70
Zephyr	06	6H6Z6038BA	85.70

CYLINDER HEAD & VALVES

		Part No.	Price
...... **CYLINDER HEAD & VALVES**			
VALVE GRIND GASKET KIT			
Fusion, Milan	06-09	6E5Z6079AA	274.43
Zephyr	06	6E5Z6079AA	274.43
1-CYLINDER HEAD			
Fusion, Milan			
Right	06-09	6E5Z6049B	758.05
Left	06-09	6E5Z6049A	842.55
Zephyr			
Right	06	6E5Z6049B	758.05
Left	06	6E5Z6049A	842.55
2-HEAD GASKET			
Fusion, Milan			
Right	06-09	6E5Z6051A	35.52
Left	06-09	6E5Z6051B	36.32
Zephyr			
Right	06	6E5Z6051A	35.52
Left	06	6E5Z6051B	36.32
3-VALVE COVER			
Fusion, Milan			
Right	06-09	6E5Z6582AA	121.12
Left	06-09	6E5Z6582GA	160.52
Zephyr			
Right	06	6E5Z6582AA	121.12
Left	06	6E5Z6582GA	160.52
4-VALVE COVER GASKET			
Fusion, Milan			
Right	06-09	6E5Z6584AA	16.49
Left	06-09	6E5Z6584DA	21.08
Zephyr			
Right	06	6E5Z6584AA	16.49
Left	06	6E5Z6584DA	21.08
5-INTAKE VALVE			
Fusion, Milan	06-09	7L8Z6507A	11.37
Zephyr	06	7L8Z6507A	11.37
6-EXHAUST VALVE			
Fusion, Milan	06-09	7L8Z6505A	12.92
Zephyr	06	7L8Z6505A	12.92
7-VALVE SEALS			
Fusion, Milan	06-09	F5RZ6571B	1.94
Zephyr	06	F5RZ6571B	1.94
8-VALVE SPRINGS			
Fusion, Milan	06-09	2S7Z6513AC	5.40
Zephyr	06	2S7Z6513AC	5.40
9-VALVE SPRING RETAINERS			
Fusion, Milan	06-09	F5RZ6514B	1.27
Zephyr	06	F5RZ6514B	1.27
10-VALVE KEEPER			
Fusion, Milan	06-09	F5RZ6518B	1.01
Zephyr	06	F5RZ6518B	1.01

		Part No.	Price
11-VALVE LIFTERS			
Fusion, Milan	06-09	F6DZ6C501A	12.41
Zephyr	06	F6DZ6C501A	12.41
12-ROCKER ARMS			
Fusion, Milan	06-09	3F1Z6564AA	17.03
Zephyr	06	3F1Z6564AA	17.03

CAMSHAFT & TIMING

		Part No.	Price
.... **VARIABLE VALVE TIMING**			
CONTROL VALVE SOLENOID			
Fusion, Milan			
To 12/5/05	06	3M4Z6M280AE	
From 12/5/05	06-09	7T4Z6M280FA	126.48
Zephyr			
To 12/5/05	06	3M4Z6M280AE	
From 12/5/05	06	7T4Z6M280FA	126.48
13-ACTUATOR			
Fusion, Milan	06-09	3M4Z6256BA	266.51
Zephyr	06	3M4Z6256BA	266.51
.... **CAMSHAFT & TIMING**			
14-INTAKE CAMSHAFT			
Fusion, Milan			
Right	06-09	6E5Z6250DA	131.16
Left	06-09	6E5Z6250CA	168.79
Zephyr			
Right	06	6E5Z6250DA	131.16
Left	06	6E5Z6250CA	168.79
15-EXHAUST CAMSHAFT			
Fusion, Milan			
Right	06-09	3M4Z6250CAARH	312.46
Left	06-09	3M4Z6250CELH	181.38
Zephyr			
Right	06	3M4Z6250CAARH	312.46
Left	06	3M4Z6250CELH	181.38
16-CAMSHAFT SEAL			
Fusion, Milan	06-09	F5RZ6K292B	9.37
Zephyr	06	F5RZ6K292B	9.37
17-HOUSING			
Fusion, Milan	06-09	9L8Z6B293A	78.32
Zephyr	06	9L8Z6B293A	78.32
18-CRANKSHAFT GEAR			
Fusion, Milan	06-09	1S7Z6306CA	37.29
Zephyr	06	1S7Z6306CA	37.29
19-TIMING CHAIN			
Fusion, Milan	06-09	YF1Z6268AA	58.92
Zephyr	06	YF1Z6268AA	58.92
20-TENSIONER			
Fusion, Milan	06-09	YF1Z6L266BA	56.65
Zephyr	06	YF1Z6L266BA	56.65
21-ARM			
Fusion, Milan	06-09	4L8Z6K255AA	24.98
Zephyr	06	4L8Z6K255AA	24.98
22-HOUSING			
Fusion, Milan			
Right	06-09	6E5Z6C260AD	167.92
Left	06-09	6E5Z6C261AC	234.38
Zephyr			
Right	06	6E5Z6C260AD	167.92
Left	06	6E5Z6C261AC	234.38
23-FRONT COVER			
Fusion, Milan	06-09	8E5Z6019A	268.10
Zephyr	06	8E5Z6019A	268.10
24-GASKET			
Fusion, Milan			
Right	06-08	3M4Z6020AA	4.28
Left	06-08	3M4Z6020CA	4.28
Center	06-08	3M4Z6020BA	4.34
Zephyr			
Right	06	3M4Z6020AA	4.28
Left	06	3M4Z6020CA	4.28
Center	06	3M4Z6020BA	4.34

PISTONS, RINGS & BEARINGS

		Part No.	Price
.... **PISTONS, RINGS & BEARINGS**			
25-CONNECTING ROD			
Fusion, Milan			
Part Of Short Block. .	06-09		
Zephyr			
Part Of Short Block. .	06		
26-BEARING			
Fusion, Milan			
Part Of Short Block. .	06-09		
Zephyr			
Part Of Short Block. .	06		
27-BUSHING			
Fusion, Milan			
Part Of Short Block. .	06-09		
Zephyr			
Part Of Short Block. .	06		

CRANKSHAFT & BEARINGS

		Part No.	Price
.... **CRANKSHAFT & BEARINGS**			
28-FRONT CRANK SEAL			
Fusion, Milan	06-09	F5AZ6700A	10.45
Zephyr	06	F5AZ6700A	10.45
29-UPPER BEARINGS			
Fusion, Milan			
Part Of Short Block. .	06-09		
Zephyr			
Part Of Short Block. .	06		
30-LOWER BEARINGS			
Fusion, Milan			
Part Of Short Block. .	06-09		
Zephyr			
Part Of Short Block. .	06		
31-THRUST BEARING			
Fusion, Milan			
Part Of Short Block. .	06-09		
Zephyr			
Part Of Short Block. .	06		
32-CRANKSHAFT			
Fusion, Milan			
Part Of Short Block. .	06-09		
Zephyr			
Part Of Short Block. .	06		
33-REAR MAIN SEAL			
Fusion, Milan	06-09	F4AZ6701A	20.23
Zephyr	06	F4AZ6701A	20.23
34-PULLEY			
Fusion, Milan	06-09	6E5Z6312AA	58.07
Zephyr	06	6E5Z6312AA	58.07

LUBRICATION

		Part No.	Price
...... **OIL PAN**			
35-OIL PAN BAFFLE			
Fusion, Milan	06-09	YL8Z6687AA	27.48
Zephyr	06	YL8Z6687AA	27.48
36-OIL PAN			
Fusion, Milan	06-09	9L8Z6675A	208.98
Zephyr	06	9L8Z6675A	208.98
37-GASKET			
Fusion, Milan	06-09	3W4Z6710DA	42.12
Zephyr	06	3W4Z6710DA	42.12
...... **OIL PUMP**			
38-OIL PUMP			
Fusion, Milan	06-09	3W4Z6600AA	103.32
Zephyr	06	3W4Z6600AA	103.32
39-OIL PICK-UP			
Fusion, Milan	06-09	6E5Z6622EA	47.13
Zephyr	06	6E5Z6622EA	47.13

LABOR   10 ENGINE – 3.5L 10   LABOR

OPERATION INDEX

ENGINE

				(Factory) Time	Motor Time
1-ENGINE, R&I					
MKZ					
FWD		07-09 B		(8.2)	10.8
AWD		07-09 B		(9.3)	13.1
NOTES FOR: ENGINE, R&I					
Does Not Include: Transfer Of Any Part Of Engine Or Replacement Of Optional Equipment.					
2-ENGINE, R&R					
MKZ					
FWD		07-09 B			13.3

(CONTINUED)

FORD 1

FUSION, MILAN (06-09), ZEPHYR (06), MKZ (07-09), MILAN HYBRID (10)

ENGINE – 3.5L - Time Cont'd

		(Factory Time)	Motor Time
AWD	07-09 B		15.6

NOTES FOR: ENGINE, R&R
Includes: Transfer All Fuel & Electrical Units. Does Not Include Replacement Of Optional Equipment.

3-ENGINE, OVERHAUL
MKZ

FWD	07-09 A		30.3
AWD	07-09 A		32.6

NOTES FOR: ENGINE, OVERHAUL
Includes: Disassemble & Clean Engine, Ridge Ream & Hone Cylinders, Fit Pistons, Rings, Pins, Main & Rod Bearings, R&I Engine, Grind Valves & Tune-Up.

4-LONG BLOCK, R&R
MKZ

FWD	07-09 B		17.3
AWD	07-09 B		19.6

NOTES FOR: LONG BLOCK, R&R
Includes: R&I Engine And Transfer All Necessary Components Not Supplied With Long Block.

5-OIL LEAK, DIAGNOSIS

MKZ	07-09 B		1.0

6-SHORT BLOCK, R&R
MKZ

FWD	07-09 B		20.3
AWD	07-09 B		22.6

NOTES FOR: SHORT BLOCK, R&R
Includes: R&I Engine And Replacement Of All Necessary Components.

MOUNTS

MOUNTS

7-MOUNT, R&R
FRONT

		(Factory Time)	Motor Time
MKZ	07-09 B	(0.7)	1.1

UPPER

MKZ	07-09 B	(1.1)	1.5

8-TRANS MOUNT, R&R

MKZ	07-09 B	(1.8)	2.4

CYLINDER HEAD & VALVES

CYLINDER HEAD & VALVES

9-CYLINDER HEAD, R&R
MKZ

Right Bank	07-09 A		15.6
Left Bank	07-09 A		16.2
Both Banks	07-09 A		20.5

NOTES FOR: CYLINDER HEAD, R&R
Includes: R&I Cylinder Head, Grind All Valves & Make All Necessary Adjustments.

10-HEAD GASKET, R&R
MKZ

Right Bank	07-09 B		13.7
Left Bank	07-09 B		14.3
Both Banks	07-09 B		17.1

11-VALVE COVER GASKET, R&R
MKZ

Right Bank	07-09 B	(1.8)	2.4
Left Bank	07-09 B	(1.1)	1.5
Both Banks	07-09 B	(2.6)	3.6

12-VALVE LIFTERS, R&R
MKZ

One Bank	07-09 B	(10.2)	14.3
Both Banks	07-09 B	(10.8)	15.2

13-VALVE SEALS, R&R
MKZ
All Cyls

Right Bank	07-09 B		8.4
Left Bank	07-09 B		5.5
Both Banks	07-09 B		12.2

14-VALVE SPRINGS, R&R
MKZ
All Cyls

Right Bank	07-09 B		8.4
Left Bank	07-09 B		5.5
Both Banks	07-09 B		12.2

15-VALVES, GRIND
MKZ

Right Bank	07-09 A		15.6
Left Bank	07-09 A		16.2
Both Banks	07-09 A		20.5

NOTES FOR: VALVES, GRIND
Includes: R&I Cylinder Head, Timing Belt, Grind All Valves & Seats And Make All Necessary Adjustments.
To Ream Valve Guides, Add

Each	07-09		0.2

To R&R Valve Guides, Add

Each	07-09		0.3

To R&R Camshaft, Add

One	07-09		0.2
Each Additional	07-09		0.1

CAMSHAFT & TIMING

VARIABLE VALVE TIMING

16-SOLENOID, R&R
MKZ

			(Factory Time)	Motor Time
Right Bank	07-08 B		(2.6)	3.5
	09 B		(1.6)	2.2
Left Bank	07-08 B		(0.8)	1.1
	09 B		(0.8)	1.1
Both Banks	07-08 B		(3.2)	4.6
	09 B		(2.2)	3.0

CAMSHAFT & TIMING

17-CAMSHAFT, R&R
MZK

Right Bank	07-08 B			11.8
Left Bank	07-08 B			12.8
Both Banks	07-08 B			13.7

One Bank

FWD	09 B		(10.1)	13.8
AWD	09 B		(10.3)	14.0

Both Banks

FWD	09 B		(10.7)	14.5
AWD	09 B		(11.0)	14.9

NOTES FOR: CAMSHAFT, R&R
Includes: R&I Valve Cover.

18-FRONT COVER GASKET, R&R

MKZ	07-09 B		(9.2)	12.2

19-CAMSHAFT SEAL, R&R

MKZ	07-09 B			1.6

20-TENSIONER, R&R
MKZ
See TIMING CHAIN, R&R

21-TIMING CHAIN, R&R

MKZ	07-09 B		(10.0)	13.7

NOTES FOR: TIMING CHAIN, R&R
Includes: R&I Front Cover.
To R&R Tensioners, Add ... 07-09 ... 0.1

PISTONS, RINGS & BEARINGS

PISTONS, RINGS & BEARINGS

22-CONNECTING ROD BEARING, R&R

MKZ	07-09 A		4.7

NOTES FOR: CONNECTING ROD BEARING, R&R
Includes: R&I Oil Pan & Plastigage Bearings And Make All Necessary Adjustments.

23-PISTON RINGS, R&R
MKZ
FWD

		(Factory Time)	Motor Time
One Cylinder	07-09 A		20.7
Each Additional	07-09 A		0.3

AWD

One Cylinder	07-09 A		23.0
Each Additional	07-09 A		0.3

NOTES FOR: PISTON RINGS, R&R
Includes: Remove Cylinder Top Ridge, Deglaze Cylinder Walls And Clean Carbon.
To R&R Connecting Rod Bearings, Add

Each	07-09		0.1

To R&R Piston, Pin Or Connecting Rod, Add

One	07-09		0.2
Each Additional	07-09		0.1
All	07-09		0.7

CRANKSHAFT & BEARINGS

CRANKSHAFT & BEARINGS

24-CRANKSHAFT, R&R
MKZ

FWD	07-09 A		19.1
AWD	07-09 A		21.4

NOTES FOR: CRANKSHAFT, R&R
Includes: R&I Engine & Oil Pan, Renew All Bearings And Seals And Make All Necessary Adjustments.

25-FRONT CRANK SEAL, R&R

MKZ	07-09 B		(0.8)	1.1

26-MAIN & ROD BEARINGS, R&R
MKZ

FWD	07-09 A		19.1
AWD	07-09 A		21.4

NOTES FOR: MAIN & ROD BEARINGS, R&R
Includes: R&I Engine, Oil Pan And Plastigage Bearings.

27-MAIN BEARINGS, R&R
MKZ

FWD	07-09 A		18.5
AWD	07-09 A		20.8

NOTES FOR: MAIN BEARINGS, R&R
Includes: R&I Engine, Oil Pan And Plastigage Bearings.

28-PULLEY, R&R

MKZ	07-09 B		(0.7)	0.9

29-REAR MAIN SEAL, R&R

MKZ	07-09 B		0.5

NOTES FOR: REAR MAIN SEAL, R&R
After Trans Is Removed.

LUBRICATION

ADAPTER HOUSING

30-FILTER ADAPTER, R&R

MKZ	07-09 B		1.4

OIL PAN

31-OIL PAN, R&R

MKZ	07-09 B		(3.7)	5.1

32-OIL PAN GASKET, R&R

MKZ	07-09 B		(3.7)	5.1

OIL PUMP

33-OIL PUMP, R&R

MKZ	07-09 B		13.2

OIL COOLER

34-O-RING, R&R

MKZ	07-09 B		1.4

35-OIL COOLER, R&R

MKZ	07-09 B		1.4

PARTS 10 ENGINE – 3.5L 10 PARTS

ENGINE

ENGINE

			Part No.	Price
ENGINE				
MKZ				
To 1/16/09	07-09		9T4Z6006A	6026.60
From 1/16/09	09		9T4Z6006A	6026.60
SHORT BLOCK				
MKZ				
To 1/16/09	07-09		9T4Z6009A	2966.65
From 1/16/09	09		9T4Z6009A	2966.65
1-CYLINDER BLOCK				
MKZ	07-08		9T4Z6010AB	1594.70
1 Bolt Style Filter	09		9T4Z6010AB	1594.70
3 Bolt Style Filter	09		9T4Z6010A	1735.72

MOUNTS

MOUNTS

		Part No.	Price
MOUNT			
MKZ	07-09	7H6Z6038A	148.87
TRANS MOUNT			
MKZ	07-09	8H6Z6068A	70.11

CYLINDER HEAD & VALVES

CYLINDER HEAD & VALVES

		Part No.	Price
VALVE GRIND GASKET KIT			
MKZ	07-09	9T4Z6079A	254.11

		Part No.	Price
2-CYLINDER HEAD			
MKZ			
Right	07-09	7T4Z6049A	1283.32
Left	07-09	7T4Z6049B	1161.65
3-HEAD GASKET			
MKZ			
Right	07-09	AT4Z6051A	59.68
Left	07-09	AT4Z6051B	60.86
4-VALVE COVER			
MKZ			
Right	07-09	7T4Z6582F	90.28
Left	07-09	7T4Z6582E	141.65

(CONTINUED)

1 FORD
FUSION, MILAN (06-09), ZEPHYR (06), MKZ (07-09), MILAN HYBRID (10)

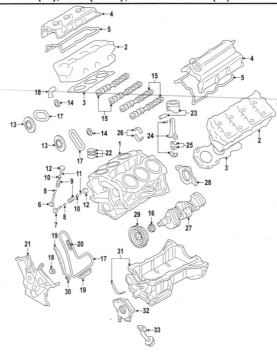

FPP075

ENGINE – 3.5L - Parts Cont'd	Part No.	Price
5-VALVE COVER GASKET		
MKZ		
Right 07-09	7T4Z6584B	14.51
Left 07-09	7T4Z6584A	14.58
6-INTAKE VALVE		
MKZ 07-09	7T4Z6507B	14.67
7-EXHAUST VALVE		
MKZ		
To 8/1/08 07-09	7T4Z6505A	19.17
From 8/1/08 09	AT4Z6505A	12.37
8-VALVE SEALS		
MKZ 07-09	7T4Z6571AA	1.81
9-VALVE SPRINGS		
MKZ 07-09	7T4Z6513AA	4.60
10-VALVE SPRING RETAINERS		
MKZ 07-09	7T4Z6514AA	4.10
11-VALVE KEEPER		
MKZ 07-09	1S7Z6518AA	4.41
NOTES FOR: VALVE KEEPER		
Price Is For One Half Valve Keeper.		
12-VALVE LIFTERS		
MKZ 07-09	AT4Z6500ALA	6.84
NOTES FOR: VALVE LIFTERS		
Order By Size.		

CAMSHAFT & TIMING

········ VARIABLE VALVE TIMING ········
CONTROL VALVE SOLENOID

	Part No.	Price
MKZ 07-09	7T4Z6M280B	98.32
HOUSING		
VALVE SOLENOID		
HOUSING		
MKZ		
Right 07-09	7T4Z6C260B	141.27
Left 07-09	7T4Z6C261B	141.27
13-ACTUATOR		
MKZ 07	7T4Z6A257D	245.45
08-09	8T4Z6A257B	271.65

	Part No.	Price
········· **CAMSHAFT & TIMING** ··········		
14-CAMSHAFT GEAR		
MKZ 07-09	7T4Z6256A	18.94
15-CAMSHAFT		
MKZ		
Intake		
Right 07-09	7T4Z6250A	88.57
Left 07-09	7T4Z6250B	100.16
Exhaust		
Right		
To 8/1/08 07-09	7T4Z6250C	109.35
From 8/1/08 09	9T4Z6250A	91.57
Left		
To 8/1/08 07-09	7T4Z6250D	94.17
From 8/1/08 09	9T4Z6250B	91.57
16-CRANKSHAFT GEAR		
MKZ 07-09	7T4Z6306A	19.68
17-TIMING CHAIN		
MKZ		
Primary 07-09	7T4Z6268A	55.46
Secondary 07-09	7T4Z6268CA	13.76
18-TENSIONER		
MKZ		
Primary 07-09	7T4Z6K254B	57.22
Secondary		
Right 07-09	7T4Z6K254AA	33.78
Left 07-09	7T4Z6K254BA	66.65
19-TENSIONER ARM		
MKZ		
Right 07-09	7T4Z6K255AA	24.50
Left 07-09	7T4Z6B274A	9.40
20-CHAIN GUIDE		
MKZ		
Right 07-09	7T4Z6M256A	11.29
Left 07-09	7T4Z6K297AA	16.05
21-FRONT COVER		
MKZ 07-09	7T4Z6019GD	151.55

PISTONS, RINGS & BEARINGS

···· PISTONS, RINGS & BEARINGS ····
22-PISTON RINGS
MKZ
  Part Of Short Block. . . 07-09

	Part No.	Price
23-PISTON		
MKZ		
Part Of Short Block. . . 07-09		
24-CONNECTING ROD		
MKZ 07-09	9T4Z6200A	48.32
25-BEARINGS		
MKZ		
Part Of Short Block. . . 07-09		

CRANKSHAFT & BEARINGS

······ CRANKSHAFT & BEARINGS ······
REAR MAIN SEAL

	Part No.	Price
MKZ 07-09	7T4Z6701AA	20.05
26-BEARINGS		
MKZ		
Grade 1 07-09	7T4Z6D309A	8.54
Grade 2 07-09	7T4Z6D309B	8.54
Grade 3 07-09	7T4Z6D309C	8.54
27-CRANKSHAFT		
MKZ		
Part Of Short Block. . . 07-09		
28-REAR MAIN SEAL RETAINER		
MKZ 07-08	7T4Z6K301A	57.31
09	AT4Z6K301AA	43.00
29-PULLEY		
MKZ 07-09	8T4Z6312A	88.43
30-FRONT CRANK SEAL		
MKZ 07-09	XW4Z6700AA	10.45

LUBRICATION

·········· OIL PAN ··········
31-OIL PAN

	Part No.	Price
MKZ 07-09	7T4Z6675A	169.98
·········· **OIL PUMP** ··········		
32-OIL PUMP		
MKZ		
To 4/4/08 07-08	7T4Z6600AA	115.27
From 4/4/08 08-09	AT4Z6600A	124.79
33-OIL PICK-UP		
MKZ		
To 9/20/07 07-08	7T4Z6622A	23.98
From 9/20/07 08-09	7T4Z6622B	22.10

FORD 1

FUSION, MILAN (06-09), ZEPHYR (06), MKZ (07-09), MILAN HYBRID (10)

LABOR — 11 CLUTCH 11 — LABOR

OPERATION INDEX

CLUTCH, R&R	2
MASTER CYLINDER, R&R	5
PEDAL ASSY, R&R	1
RELEASE BEARING, R&R	3
RELEASE FORK, R&R	4
SLAVE CYLINDER, R&R	6
STEEL LINE, R&R	8
SYSTEM, BLEED	7

CLUTCH & FLYWHEEL

.... CLUTCH & FLYWHEEL ....

		(Factory Time)	Motor Time
1-PEDAL ASSY, R&R			
Fusion, Milan	06-09 B		0.8
2-CLUTCH, R&R			
Fusion, Milan	06-09 B	(4.7)	6.3
NOTES FOR: CLUTCH, R&R			
Includes: R&I Transmission.			
To R&R Flywheel, Add	06-09		0.2
To R&R Pilot Bearing, Add	06-09		0.2
3-RELEASE BEARING, R&R			
Fusion, Milan	06-09 B		6.1
NOTES FOR: RELEASE BEARING, R&R			
Includes: R&I Transmission.			
4-RELEASE FORK, R&R			
Fusion, Milan	06-09 B		6.1

		(Factory Time)	Motor Time
NOTES FOR: RELEASE FORK, R&R			
Includes: R&I Transmission.			
HYDRAULIC SYSTEM			
.... HYDRAULIC SYSTEM			
5-MASTER CYLINDER, R&R			
Fusion, Milan	06-09 B	(1.0)	1.4
6-SLAVE CYLINDER, R&R			
Fusion, Milan	06-09 B	(0.6)	0.8
7-SYSTEM, BLEED			
Fusion, Milan	06-09 B	(0.4)	0.5
8-STEEL LINE, R&R			
Fusion, Milan	06-09 B		0.7

PARTS — 11 CLUTCH 11 — PARTS

PARTS INDEX

DISC	1
FLEX HOSE	8
FLYWHEEL	2
MASTER CYLINDER	9
PILOT BEARING	3
PRESSURE PLATE	4
RELEASE BEARING	5
RELEASE FORK	6
SLAVE CYLINDER	10
STEEL LINE	7

CLUTCH & FLYWHEEL

.... CLUTCH & FLYWHEEL ....

		Part No.	Price
1-DISC			
Fusion, Milan	06-09	7E5Z7B546A	420.00
NOTES FOR: DISC			
Includes Pressure Plate.			
2-FLYWHEEL			
Fusion, Milan	06-09	2L8Z6375A	123.67
3-PILOT BEARING			
Fusion, Milan	06-09	D4ZZ7600A	9.98
4-PRESSURE PLATE			
Fusion, Milan			
To 5/14/07	06-07	6E5Z7563A	137.32
From 5/14/07	07-09	7E5Z7B546A	420.00
5-RELEASE BEARING			
Fusion, Milan	06-09	6E5Z7548AA	57.00

		Part No.	Price
6-RELEASE FORK			
Fusion, Milan	06-09	1L8Z7515AA	34.82
HYDRAULIC SYSTEM			
.... HYDRAULIC SYSTEM			
7-STEEL LINE			
Fusion, Milan			
Master Cylinder To			
Hose	06-09	6E5Z7A512AA	7.65
Hose To Slave			
Cylinder	06-09	8M8Z7A512AA	19.19
8-FLEX HOSE			
Fusion, Milan	06-09	8M8Z7T504A	17.44
9-MASTER CYLINDER			
Fusion, Milan	06-09	6M8Z7A543A	54.94
10-SLAVE CYLINDER			
Fusion, Milan	06-09	6M8Z7A512A	100.11

LABOR — 12 MANUAL TRANSAXLE 12 — LABOR

OPERATION INDEX

SHIFT CONTROL CABLE, R&R	2
SHIFT LEVER, R&R	1
TRANSAXLE, OVERHAUL	5
TRANSAXLE, R&I	3
TRANSAXLE, R&R	4

MANUAL TRANSAXLE

.... MANUAL TRANSAXLE ....

		(Factory Time)	Motor Time
1-SHIFT LEVER, R&R			
Fusion, Milan	06-09 B		1.5
2-SHIFT CONTROL CABLE, R&R			
Fusion, Milan	06-09 B		3.5
3-TRANSAXLE, R&I			
Fusion, Milan	06-09 B	(4.0)	5.9

		(Factory Time)	Motor Time
4-TRANSAXLE, R&R			
Fusion, Milan	06-09 B		6.2
5-TRANSAXLE, OVERHAUL			
Fusion, Milan	06-09 A		13.9
NOTES FOR: TRANSAXLE, OVERHAUL			
Includes: R&I Transaxle.			
To Overhaul Differential,			
Add	06-09		1.5

PARTS — 12 MANUAL TRANSAXLE 12 — PARTS

PARTS INDEX

SHIFT CONTROL CABLE	1
TRANSAXLE	2

MANUAL TRANSAXLE

.... MANUAL TRANSAXLE ....

		Part No.	Price
1-SHIFT CONTROL CABLE			
Fusion, Milan			
To 2/8/06	06	6E5Z7E395D	99.98

		Part No.	Price
From 2/8/06	06-09	6E5Z7E395D	99.98
2-TRANSAXLE			
Fusion, Milan			
To 7/28/08	06-09	8E5Z7003AB	2430.16
From 7/28/08	09	8E5Z7003AB	2430.16

LABOR — 14 AUTOMATIC TRANSAXLE 14 — LABOR

OPERATION INDEX

HOSES, R&R	6
INTERLOCK CABLE, R&R	9
LINE, R&R	7
OIL COOLER, R&R	5
SHIFT CONTROL CABLE, R&R	8
TRANSAXLE, DIAGNOSIS	1
TRANSAXLE, OVERHAUL	4
TRANSAXLE, R&I	2
TRANSAXLE, R&R	3

AUTOMATIC TRANSAXLE

.... AUTOMATIC TRANSAXLE ....

		(Factory Time)	Motor Time
1-TRANSAXLE, DIAGNOSIS			
Fusion, Milan, Zephyr, MKZ	06-09 A		1.8
NOTES FOR: TRANSAXLE, DIAGNOSIS			
Includes: Time to Hook-Up And Disconnect Test Equipment. Checking System For Fault Codes, Road Test And Hydraulic Pressure Test Ad Required.			
2-TRANSAXLE, R&I			
Fusion, Milan, Zephyr			
FWD			
2.3L	06-09 B	(5.4)	7.4

		(Factory Time)	Motor Time
3.0L	06-09 B	(6.7)	8.8
AWD	07-09 B	(8.3)	10.4
MKZ			
FWD	07-09 B	(6.0)	8.0
AWD	07-09 B	(6.9)	9.4
NOTES FOR: TRANSAXLE, R&I			
To R&R Driveplate, Add	06-09		0.2
To R&R Torque Converter,			
Add	06-09		0.2
3-TRANSAXLE, R&R			
Fusion, Milan, Zephyr			
FWD			
2.3L	06-09 B		8.0
3.0L	06-09 B		9.4
AWD	07-09 B		11.0
MKZ			
FWD	07-09 B		8.6
AWD	07-09 B		10.0
NOTES FOR: TRANSAXLE, R&I			
To R&R Driveplate, Add	06-09		0.2
To R&R Torque Converter,			
Add	06-09		0.2
To Flush Cooler Lines. Add	06-09		0.6
4-TRANSAXLE, OVERHAUL			
Fusion, Milan, Zephyr			
FWD			
2.3L	06-09 B		20.5

		(Factory Time)	Motor Time
3.0L	06-09 B		15.4
AWD	07-09 B		16.5
MKZ			
FWD	07-09 B		15.4
AWD	07-09 B		16.5
NOTES FOR: TRANSAXLE, OVERHAUL			
Includes: R&I Transaxle.			
To Overhaul Valve Body, Add	06-09		1.5
To Overhaul Differential,			
Add	06-09		1.5
To Flush Cooler Lines. Add	06-09		0.6
.... OIL COOLER			
5-OIL COOLER, R&R			
Fusion, Milan, Zephyr, MKZ	06-09 B		4.6
6-HOSES, R&R			
Fusion, Milan, Zephyr, MKZ			
One	06-09 C		1.4
Both	06-09 C		2.1
7-LINE, R&R			
Fusion, Milan, Zephyr, MKZ			
One	06-09 C		0.9
Both	06-09 C		1.4
(CONTINUED)			

1 FORD
FUSION, MILAN (06-09), ZEPHYR (06), MKZ (07-09), MILAN HYBRID (10)

AUTOMATIC TRANSAXLE - Time Cont'd		
	(Factory Time)	Motor Time

GEAR SHIFT COMPONENTS

	(Factory Time)	Motor Time

·········· GEAR SHIFT CONTROL ··········

8-SHIFT CONTROL CABLE, R&R
Fusion, Milan, Zephyr

2.3L	06-09 B		3.5

·········· SHIFT INTERLOCK ··········

9-INTERLOCK CABLE, R&R
Fusion, Milan, Zephyr, MKZ ..

	06-09 B		1.8

	(Factory Time)	Motor Time
3.0L	06-09 B (4.1)	5.9
MKZ	07-09 B (4.1)	5.9

PARTS 14 AUTOMATIC TRANSAXLE 14 PARTS

PARTS INDEX
DRIVE PLATE	1
SHIFT CONTROL CABLE	4
TORQUE CONVERTER	3
TRANSAXLE	2

AUTOMATIC TRANSAXLE

········ AUTOMATIC TRANSAXLE ·········

	Part No.	Price

1-DRIVE PLATE
Fusion, Milan

2.3L	06-09	5M8Z6375BB	95.80
3.0L	06-09	5M8Z6375AA	105.42
Zephyr	06	5M8Z6375AA	105.42
MKZ	07-09	7U3Z6375AA	113.65

	Part No.	Price
2-TRANSAXLE		
Fusion, Milan		
5 Speed Auto	06	6E5Z7000ARM	3106.67
	07-09	8E5Z7000R	5421.60
6 Speed Auto	06	6E5Z7000AERM	3133.33
FWD			
To 9/8/06	07	6E5Z7000AERM	3133.33
From 9/8/06	07	7E5Z7000AERM	3133.33
To 6/25/08	08-09	8E5Z7000S	4534.51
From 6/25/08	09	8E5Z7000S	4534.51
AWD	07	7E5Z7000H	4534.51
To 1/18/08	08	8E5Z7000T	4534.51
From 1/18/08 To 12/23/08	08-09	8E5Z7000M	0.00
From 12/23/08	09	8E5Z7000T	4534.51
Zephyr	06	6E5Z7000AERM	3133.33
MKZ			
FWD			
To 12/4/06	07	7E5Z7000J	4534.51
From 12/4/06	07	7E5Z7000J	4534.51
To 1/18/08	08	8E5Z7000N	4318.57

	Part No.	Price
From 1/18/08	08-09	8E5Z7000V	4534.51
AWD	07	7E5Z7000K	4534.51
To 1/18/08	08	8E5Z7000D	4534.51
From 1/18/08	08-09	8E5Z7000U	4534.51

3-TORQUE CONVERTER
Fusion, Milan

2.3L	06-08	6E5Z7902A	1013.07
3.0L	06-09	6E5Z7902B	929.67
Zephyr	06	6E5Z7902B	929.67
MKZ	07-09	7E5Z7902A	848.98

GEAR SHIFT COMPONENTS

·········· GEAR SHIFT CONTROL ··········

4-SHIFT CONTROL CABLE
Fusion, Milan

5 Speed Auto	06-09	6E5Z7E395E	44.85
6 Speed Auto	06-09	AE5Z7E395G	66.65
Zephyr	06	AE5Z7E395G	66.65
MKZ	07-09	AE5Z7E395G	66.65

LABOR 16 TRANSFER CASE 16 LABOR

OPERATION INDEX
TRANSFER CASE, R&R	1

TRANSFER CASE

············ TRANSFER CASE ·············

	(Factory Time)	Motor Time

1-TRANSFER CASE, R&R
Fusion, Milan, MKZ ..

	07-09 B (3.6)		5.0

PARTS 16 TRANSFER CASE 16 PARTS

PARTS INDEX
CONTROL MODULE	1
TRANSFER CASE	2

TRANSFER CASE

············· TRANSFER CASE ··············

	Part No.	Price
1-CONTROL MODULE		
TRANSFER SHIFT MODULE		
Fusion, Milan, MKZ ... 07-09	7E5Z7E453B	299.27
2-TRANSFER CASE		
Fusion, Milan, MKZ ... 07-09	7E5Z7251E	940.05

LABOR 18 BRAKES 18 LABOR

OPERATION INDEX
ALL FLEX HOSES, R&R	4
BACKING PLATE, R&R	2
BACKING PLATE, R&R	3
BOOSTER CHECK VALVE, R&R	5
BRAKE BOOSTER, R&R	6
BRAKES, ADJUST	17
BRAKES, R&R	1
CONTROL MODULE, R&R	12
FLEX HOSE, R&R	10
FLEX HOSE, R&R	11
FRONT SPEED SENSOR, R&R	13
MASTER CYLINDER, R&R	7
PARKING BRAKE CONTROL, R&R	18
REAR CABLE, R&R	19
REAR SPEED SENSOR, R&R	15
SOLENOID VALVE BLOCK, R&R	14
SYSTEM, BLEED	8
SYSTEM, DIAGNOSIS	16
SYSTEM, OVERHAUL	9

BRAKE COMPONENTS

·········· BRAKE COMPONENTS ·········

	(Factory Time)	Motor Time

1-BRAKES, R&R
FRONT AXLE

Fusion,Milan,Zephyr,MKZ ...	06-09 B	(0.8)	1.1
REAR AXLE			
Fusion,Milan,Zephyr,MKZ ...	06-09 B	(0.7)	1.0
BOTH AXLES			
Fusion,Milan,Zephyr,MKZ ...	06-09 B	(1.3)	1.7

NOTES FOR: BRAKES, R&R
To R&I Or R&R Rotor, Add
Front

One Side	06-09	(0.2)	0.2
Both Sides	06-09	(0.4)	0.4
Rear			
One Side	06-09	(0.2)	0.2
Both Sides	06-09	(0.4)	0.4

	(Factory Time)	Motor Time
To Reface Rotor, Add		
Each	06-09	0.4
To R&R Caliper, Add		
Each	06-09	0.2
To Overhaul Caliper, Add		
Each	06-09	0.3
To R&R Brake Hose, Add		
Each	06-09	0.1
To R&R Rear Parking Brake Cable, Add		
One Side	06-09	0.7
Both Sides	06-09	0.9

··········· FRONT BRAKES ···········

2-BACKING PLATE, R&R
Fusion, Milan, Zephyr, MKZ

One Side	06-09 B	1.6
Both Sides	06-09 B	3.1

··········· REAR BRAKES ············

3-BACKING PLATE, R&R
Fusion, Milan, Zephyr, MKZ

One Side	06-09 B	0.6
Both Sides	06-09 B	1.0

HYDRAULIC SYSTEM

··········· HYDRAULIC SYSTEM ···········

4-ALL FLEX HOSES, R&R

Fusion, Milan, Zephyr, MKZ ..	06-09 B	(1.2)	1.6

5-BOOSTER CHECK VALVE, R&R

Fusion, Milan, Zephyr, MKZ ..	06-09 B	0.3

6-BRAKE BOOSTER, R&R
Fusion, Milan, Zephyr

2.3L	06-09 B		3.0
3.0L	06-09 B	(2.3)	3.0
MKZ	07-09 B		3.0

7-MASTER CYLINDER, R&R
Fusion, Milan, Zephyr

Manual Trans	06-09 B	(1.6)	2.0

	(Factory Time)	Motor Time
Auto Trans	06-09 B (1.5)	2.0
MKZ	07-09 B (1.5)	2.0

NOTES FOR: MASTER CYLINDER, R&R
Includes: Bleed Master Cylinder.

8-SYSTEM, BLEED

Fusion, Milan, Zephyr, MKZ ..	06-09 B	(0.6)	0.9

9-SYSTEM, OVERHAUL

Fusion, Milan, Zephyr, MKZ ..	06-09 B	5.5

NOTES FOR: SYSTEM, OVERHAUL
Includes: Overhaul Calipers, Wheel Cylinders & Master Cylinder. Drain, Flush, Refill & Bleed System.

··········· FRONT BRAKES ···········

10-FLEX HOSE, R&R
Fusion, Milan, Zephyr, MKZ

One Side	06-09 B	(0.6)	0.8
Both Sides	06-09 B	(0.9)	1.2

··········· REAR BRAKES ············

11-FLEX HOSE, R&R
Fusion, Milan, Zephyr, MKZ

One Side	06-09 B	(0.6)	0.8
Both Sides	06-09 B	(0.9)	1.2

ANTI-LOCK BRAKES

··········· ANTI-LOCK BRAKES ···········

12-CONTROL MODULE, R&R

Fusion, Milan, Zephyr, MKZ ..	06-09 B	(0.6)	0.8

NOTES FOR: CONTROL MODULE, R&R
Does Not Include: System, Diagnosis.

13-FRONT SPEED SENSOR, R&R
Fusion, Milan, Zephyr, MKZ

One Side	06-09 B	(0.3)	0.4
Both Sides	06-09 B	(0.5)	0.7

NOTES FOR: FRONT SPEED SENSOR, R&R
Does Not Include: System, Diagnosis.

(CONTINUED)

FORD 1

FUSION, MILAN (06-09), ZEPHYR (06), MKZ (07-09), MILAN HYBRID (10)

BRAKES - Time Cont'd

		(Factory) Time	Motor Time
14-SOLENOID VALVE BLOCK, R&R			
Fusion, Milan, Zephyr, MKZ ..	06-09 B	(2.0)	2.4
NOTES FOR: SOLENOID VALVE BLOCK, R&R			
Does Not Include: System, Diagnosis.			
15-REAR SPEED SENSOR, R&R			
Fusion, Milan, Zephyr, MKZ			
One Side	06-09 B	(0.5)	0.6
Both Sides	06-09 B	(0.8)	1.0

		(Factory) Time	Motor Time
NOTES FOR: REAR SPEED SENSOR, R&R			
Does Not Include: System, Diagnosis.			
16-SYSTEM, DIAGNOSIS			
Fusion, Milan, Zephyr, MKZ ..	06-09 B	(0.7)	0.9

PARKING BRAKE

.... PARKING BRAKE ....

17-BRAKES, ADJUST

		(Factory) Time	Motor Time
Fusion, Milan, Zephyr, MKZ ..	06-09 B	(0.2)	0.3

		(Factory) Time	Motor Time
18-PARKING BRAKE CONTROL, R&R			
Fusion, Milan	06-09 B	(0.8)	1.0
Zephyr, MKZ	06-09 B	(0.9)	1.3
19-REAR CABLE, R&R			
Fusion, Milan, Zephyr, MKZ			
Both Sides	06-09 B	(1.2)	1.8

18 BRAKES 18

PARTS

PARTS INDEX

BRAKE COMPONENTS

.... BRAKE COMPONENTS ....

		Part No.	Price
1-FRONT PADS			
Fusion, Milan	06-09	9E5Z2001A	100.97
Zephyr	06	9E5Z2001A	100.97
MKZ	07-09	9E5Z2001A	100.97
2-REAR PADS			
Fusion, Milan			
To 10/17/08	06-09	9E5Z2200A	104.97
From 10/17/08	09	9E5Z2200A	104.97
Zephyr	06	9E5Z2200A	104.97
MKZ			
To 10/17/08	07-09	9E5Z2200A	104.97
From 10/17/08	09	9E5Z2200A	104.97

.... FRONT BRAKES ....

		Part No.	Price
3-ROTOR			
Fusion, Milan			
To 6/25/08	06-09	9E5Z1125A	120.83
From 6/25/08	09	9E5Z1125A	120.83
Zephyr	06	9E5Z1125A	120.83
MKZ			
To 6/25/08	07-09	9E5Z1125A	120.83
From 6/25/08	09	9E5Z1125A	120.83

.... REAR BRAKES ....

		Part No.	Price
4-ROTOR			
Fusion, Milan	06-09	9E5Z2C026A	59.98
Zephyr	06	9E5Z2C026A	59.98
MKZ	07-09	9E5Z2C026A	59.98

HYDRAULIC SYSTEM

.... HYDRAULIC SYSTEM ....

		Part No.	Price
5-BRAKE BOOSTER			
Fusion, Milan	06-09	6E5Z2005B	157.31
Zephyr	06	6E5Z2005B	157.31
MKZ	07-09	6E5Z2005B	157.31
6-MASTER CYLINDER			
Fusion, Milan			
w/o ABS	06-07	6E5Z2140E	178.09
w/ABS	06-08	9E5Z2140C	194.96
w/o Stability Control .	09	9E5Z2140C	194.96
w/Stability Control ...	09	9E5Z2140A	177.36
Zephyr	06	9E5Z2140C	194.96
MKZ	07-08	9E5Z2140C	194.96

ANTI-LOCK BRAKES

.... ANTI-LOCK BRAKES ....

		Part No.	Price
11-FRONT SPEED SENSOR			
Fusion, Milan			
Right	06-09	AE5Z2C204A	43.68
Left	06-09	AE5Z2C205A	45.10

PARTS

		Part No.	Price
	09	9E5Z2140A	177.36

FRONT BRAKES

		Part No.	Price
7-CALIPER			
Fusion, Milan			
Right			
To 9/4/06	06-07	6E5Z2B120B	131.13
From 9/4/06	07-09	6E5Z2B120C	110.93
Left			
To 9/4/06	06-07	6E5Z2B121B	165.90
From 9/4/06	07-09	6E5Z2B121C	110.93
Zephyr			
Right	06	6E5Z2B120B	131.13
Left	06	6E5Z2B121B	165.90
MKZ			
Right			
To 9/4/06	07	6E5Z2B120B	131.13
From 9/4/06	07-09	6E5Z2B120C	110.93
Left			
To 9/4/06	07	6E5Z2B121B	165.90
From 9/4/06	07-09	6E5Z2B121C	110.93
8-FLEX HOSE			
Fusion, Milan			
Right	06-09	6E5Z2078BA	71.28
Left	06-09	6E5Z2078AA	71.28
Zephyr			
Right	07-08	6E5Z2078BA	71.28
Left	07-08	6E5Z2078AA	71.28
MKZ			
Right	07-09	6E5Z2078BA	71.28
Left	07-09	6E5Z2078AA	71.28

.... REAR BRAKES ....

		Part No.	Price
9-CALIPER			
Fusion, Milan			
Right	06-08	6E5Z2552C	295.28
Left	06-08	6E5Z2553C	242.40
Zephyr			
Right	06	6E5Z2552C	295.28
Left	06	6E5Z2553C	242.40
MKZ			
Right	07-08	6E5Z2552C	295.28
Left	07-08	6E5Z2553C	242.40
10-FLEX HOSE			
Fusion, Milan	06	6E5Z2282AA	29.67
FWD			
To 11/12/08	06-09	6E5Z2282AA	29.67
From 11/12/08			
Right	09	9E5Z2282A	28.50
Left	09	9E5Z2283A	28.50
AWD			
To 11/12/08	07-09	7E5Z2A442A	27.98
From 11/12/08	09	9E5Z2283B	31.75
Zephyr	06	6E5Z2282AA	29.67
MKZ			
FWD			
To 11/12/08	07-09	6E5Z2282AA	29.67
From 11/12/08			
Right	09	9E5Z2282A	28.50
Left	09	9E5Z2283A	28.50
AWD			
To 11/12/08	07-09	7E5Z2A442A	27.98
From 11/12/08	09	9E5Z2283B	31.75

PARTS

		Part No.	Price
Zephyr			
Right	06	AE5Z2C204A	43.68
Left	06	AE5Z2C205A	45.10
MKZ			
Right	07-09	AE5Z2C204A	43.68
Left	07-09	AE5Z2C205A	45.10
12-MOTOR & PUMP			
Fusion, Milan	06	6E5Z2C215AA	325.83
FWD	07-09	6E5Z2C215AA	325.83
AWD	07	7E5Z2C215A	786.65
	08	8E5Z2C215A	781.65
w/o Stability Control .	09	8E5Z2C215A	781.65
w/Stability Control ...	09	9E5Z2C215A	898.23
Zephyr	06	6E5Z2C215AA	325.83
MKZ			
FWD	07-09	6E5Z2C215AA	325.83
AWD	07	7E5Z2C215A	786.65
	08	8E5Z2C215A	781.65
	09	9E5Z2C215A	898.23
13-CONTROL MODULE			
Fusion, Milan	06	7E5Z2C219B	463.09
FWD	07-09	7E5Z2C219B	463.09
AWD	07	7E5Z2C219D	389.31
	08	8E5Z2C219A	610.64
w/o Stability Control .	09	8E5Z2C219A	610.64
w/Stability Control ...	09	9E5Z2C219A	604.60
Zephyr	06	7E5Z2C219B	463.09
MKZ			
FWD	07-09	7E5Z2C219B	463.09
AWD	07	7E5Z2C219D	389.31
	08	8E5Z2C219A	610.64
	09	9E5Z2C219A	604.60
14-REAR SPEED SENSOR			
Fusion, Milan			
Right	06	AE5Z2C182A	43.58
FWD	07-09	AE5Z2C182A	43.58
AWD	07-09	AE5Z2C182B	46.11
Left	06	AE5Z2C182C	45.31
FWD	07-09	AE5Z2C182C	45.31
AWD	07-09	AE5Z2C182D	46.11
Zephyr			
Right	06	AE5Z2C182A	43.58
Left	06	AE5Z2C182C	45.31
MKZ			
Right			
FWD	07-09	AE5Z2C182A	43.58
AWD	07-09	AE5Z2C182B	46.11
Left			
FWD	07-09	AE5Z2C182C	45.31
AWD	07-09	AE5Z2C182D	46.11

PARKING BRAKE

.... PARKING BRAKE ....

		Part No.	Price
15-PARKING BRAKE CONTROL			
Fusion, Milan	06-09	7M8Z2780BD	93.30
Zephyr	06	7M8Z2780BD	93.30
MKZ	07-09	7M8Z2780BD	93.30
NOTES FOR: PARKING BRAKE CONTROL			
Order By Color.			
16-REAR CABLE			
Fusion, Milan	06	AE5Z2A815B	56.63
FWD	07-09	AE5Z2A815A	66.63
AWD	07-09	AE5Z2A815A	56.63
Zephyr	06	AE5Z2A815B	56.63
MKZ			
FWD	07-09	AE5Z2A815A	66.63
AWD	07-09	AE5Z2A815B	56.63

19 FRONT SUSPENSION 19

LABOR

OPERATION INDEX

FRONT SUSPENSION

.... SUSPENSION SERVICE ....

(CONTINUED)

LABOR

1 FORD
FUSION, MILAN (06-09), ZEPHYR (06), MKZ (07-09), MILAN HYBRID (10)

FRONT SUSPENSION - Time Cont'd

		(Factory Time)	Motor Time
1-SUSPENSION, OVERHAUL			
Fusion, Milan, Zephyr, MKZ			
One Side	06-09 B		3.5
Both Sides	06-09 B		6.8
2-TOE-IN, ADJUST			
Fusion, Milan, Zephyr, MKZ ..	06-09 B		0.5
3-WHEELS, ALIGN			
Fusion, Milan, Zephyr, MKZ			
Front Wheel Alignment	06-09 B		1.5
Four Wheel Alignment	06-09 B		1.7
······ SUSPENSION COMPONENTS ·······			
4-COIL SPRING, R&R			
Fusion, Milan, Zephyr, MKZ			
One Side	06-09 B	(1.1)	1.2
Both Sides	06-09 B	(1.9)	2.0
5-FRONT HUB, R&R			
Fusion, Milan, Zephyr, MKZ			
One Side	06-09 B	(1.4)	1.6
Both Sides	06-09 B	(2.6)	2.8
6-FRONT WHEEL BEARING, R&R			
Fusion, Milan, Zephyr, MKZ			
One Side	06-09 B		1.6
Both Sides	06-09 B		2.8
7-KNUCKLE, R&R			
Fusion, Milan, Zephyr, MKZ			
One Side	06-09 B	(1.4)	1.6
Both Sides	06-09 B	(2.6)	2.8
8-LUG BOLTS, R&R			
Fusion, Milan, Zephyr, MKZ			
One Side	06-09 B		1.3

		(Factory Time)	Motor Time
9-STRUT, R&R			
Fusion, Milan, Zephyr, MKZ			
One Side	06-09 B	(1.0)	1.2
Both Sides	06-09 B	(1.7)	2.0
·········· UPPER CONTROL ARM ··········			
10-UPPER CONTROL ARM, R&R			
Fusion, Milan, Zephyr, MKZ			
One Side	06-09 B	(0.6)	0.7
Both Sides	06-09 B	(0.9)	1.1
·········· LOWER CONTROL ARM ··········			
11-BUSHINGS, R&R			
FRONT ARM			
Fusion, Milan, Zephyr			
2.3L			
One Side	06-09 B		0.9
Both Sides	06-09 B		1.4
3.0L			
Right Side	06-09 B		1.4
Left Side	06-09 B		1.0
Both Sides	06-09 B		2.0
MKZ			
Right Side	07-09 B		1.4
Left Side	07-09 B		1.0
Both Sides	07-09 B		2.0
REAR ARM			
Fusion, Milan, Zephyr, MKZ			
One Side	06-09 B		2.1
Both Sides	06-09 B		2.3
12-FRONT ARM, R&R			
Fusion, Milan, Zephyr			
2.3L			
One Side	06-09 B		0.7
Both Sides	06-09 B		1.0

		(Factory Time)	Motor Time
3.0L			
Right Side	06-09 B		1.2
Left Side	06-09 B		0.8
Both Sides	06-09 B		1.6
MKZ			
Right Side	07-09 B		1.2
Left Side	07-09 B		0.8
Both Sides	07-09 B		1.6
13-REAR ARM, R&R			
Fusion, Milan, Zephyr, MKZ			
One Side	06-09 B		2.0
Both Sides	06-09 B		2.1
············ STABILIZER BAR ············			
14-BUSHINGS, R&R			
Fusion, Milan, Zephyr, MKZ			
One Or Both	06-09 C	(1.7)	2.3
15-LINK, R&R			
Fusion, Milan, Zephyr, MKZ			
One Side	06-09 C	(0.3)	0.4
Both Sides	06-09 C	(0.5)	0.6
16-STABILIZER BAR, R&R			
Fusion, Milan			
FWD	06-09 C	(1.8)	2.3
AWD	07-08 C	(2.1)	2.8
Zephyr	06 C	(1.9)	2.3
MKZ			
FWD	07-09 C	(1.9)	2.3
AWD	07-09 C	(2.1)	2.8
············ FRONT SUSPENSION ············			
17-CROSSMEMBER, R&R			
Fusion, Milan, Zephyr, MKZ ..	06-09 B		3.5

PARTS   19   FRONT SUSPENSION   19   PARTS

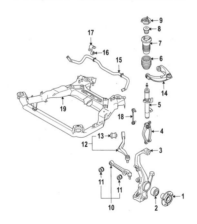

FPP080

FRONT SUSPENSION				Part No.	Price		Part No.	Price
······ SUSPENSION COMPONENTS ······			MKZ			MKZ		
	Part No.	Price	Right 07-09	4M8Z3462A	49.98	Right 07	7H6Z18124RL	61.05
1-FRONT HUB			Left 07-09	4M8Z3462B	51.65	 08-09	8H6Z18124A	58.25
Fusion, Milan 06-09	6E5Z1104AB	63.58	**5-STRUT**			Left 07	7H6Z18124LH	61.05
Zephyr 06	6E5Z1104AB	63.58	Fusion, Milan			 08-09	8H6Z18124B	58.25
MKZ 07-09	6E5Z1104AB	63.58	2.3L			**6-COIL SPRING**		
2-FRONT WHEEL BEARING			Right 06-07	6E5Z18124AR	60.97	Fusion, Milan		
Fusion, Milan 06-09	3M8Z1215A	71.48	Standard Suspension 08-09	6E5Z18124AR	60.97	2.3L 06-07	8E5Z5310G	58.30
Zephyr 06	3M8Z1215A	71.48	Sport Tuned			Standard Suspension 08-09	8E5Z5310G	58.30
MKZ 07-09	3M8Z1215A	71.48	Suspension 08-09	8E5Z18124R	65.57	Sport Tuned		
3-KNUCKLE			Left 06-07	6E5Z18124AL	60.97	Suspension 08-09	8E5Z5310E	60.17
Fusion, Milan			Standard Suspension 08-09	6E5Z18124AL	60.97	3.0L 06	8E5Z5310H	59.97
Right 06-09	8E5Z3K185A	166.82	Sport Tuned			FWD 07	8E5Z5310H	59.97
Left 06-09	8E5Z3K186A	166.82	Suspension 08-09	8E5Z18124L	61.82	Standard Suspension		
Zephyr			3.0L			sion 08-09	8E5Z5310H	59.97
Right 06	8E5Z3K185A	166.82	Right 06	6E5Z18124BR	60.97	Sport Tuned		
Left 06	8E5Z3K186A	166.82	FWD 07	6E5Z18124BR	60.97	Suspension 08-09	8E5Z5310F	61.22
MKZ			Standard Suspension 08-09	6E5Z18124BR	60.97	AWD 07-09	8E5Z5310J	58.22
Right 07-09	8E5Z3K185A	166.82	Sport Tuned			Zephyr 06	6H6Z5310A	182.40
Left 07-09	8E5Z3K186A	166.82	Suspension 08-09	8E5Z18124R	65.57	MKZ		
4-STRUT MOUNTING FORK			AWD 07-09	7E5Z18124RH	61.30	FWD 07-09	8H6Z5310A	61.63
Fusion, Milan			Left 06	6E5Z18124BL	60.97	AWD 07-09	8H6Z5310B	63.30
Right 06-09	4M8Z3462A	49.98	FWD 07	6E5Z18124BL	60.97	**7-DUST SHIELD**		
Left 06-09	4M8Z3462B	51.65	Standard Suspension 08-09	6E5Z18124BL	60.97	Fusion, Milan 06-09	4M8Z3C239A	12.82
Zephyr			Sport Tuned			Zephyr 06	4M8Z3C239A	12.82
Right 06	4M8Z3462A	49.98	Suspension 08-09	8E5Z18124L	61.82	MKZ 07-09	4M8Z3C239A	12.82
Left 06	4M8Z3462B	51.65	AWD 07-09	7E5Z18124LH	61.30	**8-STRUT BUMPER**		
			Left 06	6E5Z18124BL	60.97	Fusion, Milan 06-09	6E5Z18198A	14.98
			08	NOT LISTED		Zephyr 06	6E5Z18198A	14.98
			Zephyr			MKZ 07-09	6E5Z18198A	14.98
			Right 06	6H6Z18124AL	61.05			
			Left 06	6H6Z18124AR	60.92	**(CONTINUED)**		

FORD 1

FUSION, MILAN (06-09), ZEPHYR (06), MKZ (07-09), MILAN HYBRID (10)

FRONT SUSPENSION - Parts Cont'd

9-STRUT MOUNT

		Part No.	Price
Fusion, Milan			
To 12/3/07	06-08	6E5Z18183B	47.33
From 12/3/07	08-09	8E5Z18183A	50.30
Zephyr	06	6E5Z18183B	47.33
MKZ			
To 12/3/07	07-08	6E5Z18183B	47.33
From 12/3/07	08-09	8E5Z18183A	50.30

·········· UPPER CONTROL ARM ··········

14-UPPER CONTROL ARM

		Part No.	Price
Fusion, Milan			
Right			
To 9/4/06	06-07	6E5Z3084BA	111.40
From 9/4/06			
P205/60 R16 VAS			
Tires	07-09	6E5Z3084BA	111.40
P225/50VR 17 Tires	07-09	7E5Z3084R	132.43
Left			
To 9/4/06	06-07	6E5Z3085BA	111.33
From 9/4/06			
P205/60 R16 VAS			
Tires	07-09	6E5Z3085BA	111.33
P225/50VR 17 Tires	07-09	7E5Z3085L	132.43
Zephyr			
Right	06	6E5Z3084BA	111.40
Left	06	6E5Z3085BA	111.33
MKZ			
Right			
To 9/4/06	07	6E5Z3084BA	111.40
From 9/4/06			
P205/60 R16 VAS			
Tires	07-09	6E5Z3084BA	111.40
P225/50VR 17 Tires	07-09	7E5Z3084R	132.43
Left			
To 9/4/06	07	6E5Z3085BA	111.33
From 9/4/06			
P205/60 R16 VAS			
Tires	07-09	6E5Z3085BA	111.33
P225/50VR 17 Tires	07-09	7E5Z3085L	132.43

·········· LOWER CONTROL ARM ··········

10-FRONT ARM

		Part No.	Price
Fusion, Milan			
To 9/4/06	06-07	5M8Z3078S	382.50

		Part No.	Price
From 9/4/06	07-09	7E5Z3078R	227.23
Zephyr	06	5M8Z3078S	382.50
MKZ			
To 9/4/06	07	5M8Z3078S	382.50
From 9/4/06	07-09	7E5Z3078R	227.23

11-CONTROL ARM BUSHING

Fusion, Milan			
Part Of Front Arm.	06-08		
Zephyr			
Part Of Front Arm.	06		
MKZ			
Part Of Front Arm.	07-08		

12-REAR ARM

		Part No.	Price
Fusion, Milan			
Right	06	6M8Z3078R	309.05
FWD	07	6M8Z3078R	309.05
2.3L	08-09	6M8Z3078R	309.05
3.0L			
Standard Suspension	08-09	6M8Z3078R	309.05
Sport Tuned Suspension	08-09	7E5Z3078RA	218.73
AWD	07-09	7E5Z3078RA	218.73
Left	06	6M8Z3079L	324.95
FWD	07	6M8Z3079L	324.95
2.3L	08-09	6M8Z3079L	324.95
3.0L			
Standard Suspension	08-09	6M8Z3079L	324.95
Sport Tuned Suspension	08-09	7E5Z3079L	148.83
AWD	07-09	7E5Z3079L	148.83
Zephyr			
Right	06	6M8Z3078R	309.05
Left	06	6M8Z3079L	324.95
MKZ			
Right	07-09	7E5Z3078RA	218.73
Left	07-09	7E5Z3079L	148.83

13-BUSHING

Fusion, Milan			
Part Of Rear Arm.	06-09		
Zephyr			
Part Of Rear Arm.	06		
MKZ			
Part Of Rear Arm.	07-09		

·········· STABILIZER BAR ··········

15-STABILIZER BAR

		Part No.	Price
Fusion, Milan			
2.3L	06-09	8E5Z5482BA	80.65
3.0L			
To 9/4/06	06-07	6E5Z5482BA	67.82
From 9/4/06	07-09	7E5Z5482A	77.93
Zephyr	06	6H6Z5482AA	67.82
MKZ	07-09	7E5Z5482A	77.93

16-BUSHINGS

		Part No.	Price
Fusion, Milan			
To 9/4/07	06-07	6H6Z5484AA	7.32
From 9/4/07			
FWD	07-09	6H6Z5484AA	7.32
AWD	07-09	7E5Z5484A	4.63
Zephyr	06	6H6Z5484AA	7.32
MKZ			
To 9/4/07	07	6H6Z5484AA	7.32
From 9/4/07	07-09	7E5Z5484A	4.63

17-BRACKET

		Part No.	Price
Fusion, Milan	06-09	AE5Z5486A	22.57
Zephyr	06	AE5Z5486A	22.57
MKZ	07-09	AE5Z5486A	22.57

18-LINK

		Part No.	Price
Fusion, Milan			
Right	06-09	3M8Z5K483R	45.82
Left	06-09	3M8Z5K483L	45.82
Zephyr			
Right	06	3M8Z5K483R	45.82
Left	06	3M8Z5K483L	45.82
MKZ			
Right	07-09	3M8Z5K483R	45.82
Left	07-09	3M8Z5K483L	45.82

·········· FRONT SUSPENSION ··········

19-CROSSMEMBER

		Part No.	Price
Fusion, Milan			
2.3L	06	6E5Z5C145A	610.82
	07-09	8E5Z5C145F	589.43
3.0L			
To 9/4/06	06-07	6E5Z5C145BA	609.60
From 9/4/06			
FWD	07-09	8E5Z5C145E	619.02
AWD	07-09	8E5Z5C145D	628.27
Zephyr	06	6H6Z5C145AA	744.38
MKZ	07-09	AH6Z5C145A	590.58

LABOR 20 FRONT DRIVE AXLE 20 LABOR

OPERATION INDEX

DRIVE AXLES

·········· DRIVE AXLES ··········

1-AXLE ASSY, R&I

		(Factory Time)	Motor Time
Fusion, Milan, Zephyr, MKZ			
Right Side	06-09 B	(1.0)	1.3

		(Factory Time)	Motor Time
Left Side	06-09 B	(1.1)	1.3
Both Sides	06-09 B	(1.9)	2.3

NOTES FOR: AXLE ASSY, R&I
To R&R Outer Joint And/Or Boot, Add

Each	06-09		0.5

To R&R Inner Joint And/Or Boot, Add

Each	06-09		0.5

To R&R Axle Shaft Seals, Add

Each	06-09		0.2

2-AXLE ASSY, R&R

		(Factory Time)	Motor Time
Fusion, Milan, Zephyr, MKZ			
Right Side	06-09 B	(1.0)	1.3
Left Side	06-09 B	(1.1)	1.3
Both Sides	06-09 B	(1.9)	2.3

NOTES FOR: AXLE ASSY, R&R
To R&R Axle Shaft Seals, Add

Each	06-09		0.2

PARTS 20 FRONT DRIVE AXLE 20 PARTS

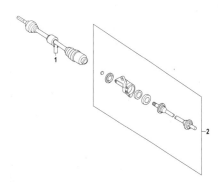

FPP090

DRIVE AXLES

·········· DRIVE AXLES ··········

1-AXLE ASSY

		Part No.	Price
Fusion, Milan			
Manual Trans			
Right	06-07	8E5Z3A428A	305.54
	08-09	8E5Z3A428A	305.54
Left	06-07	6E5Z3A427EA	347.58
	08-09	8E5Z3A427A	316.56

		Part No.	Price
Auto Trans			
5 Speed			
Right	06-09	8E5Z3A428B	336.58
Left	06-09	8E5Z3A427B	316.78
6 Speed			
Right	06-09	8E5Z3A428C	299.18
Left	06-09	8E5Z3A427C	314.58
Zephyr			
Right	06	8E5Z3A428C	299.18
Left	06	8E5Z3A427C	314.58
MKZ			
Right	07-09	8H6Z3A428A	343.18
Left	07-09	8H6Z3A427A	314.98

·········· AXLE SHAFTS & JOINTS ··········

2-INTERMEDIATE SHAFT

		Part No.	Price
Fusion, Milan			
Manual Trans	06-09	6E5Z3A329EA	99.25
Auto Trans			
5 Speed	06-09	6E5Z3A329B	113.95
6 Speed			
To 9/4/06	06-07	7E5Z3A329B	200.77
From 9/4/06			
FWD	07-09	7E5Z3A329B	200.77
AWD	07-09	7E5Z3A329A	168.38
Zephyr	06	7E5Z3A329B	200.77

(CONTINUED)

1 FORD
FUSION, MILAN (06-09), ZEPHYR (06), MKZ (07-09), MILAN HYBRID (10)

FRONT DRIVE AXLE - Parts Cont'd			Part No.	Price		Part No.	Price
	Part No.	Price	MKZ		AWD 07-09	7H6Z3A329B	165.20
			FWD 07-09 7H6Z3A329A	182.78			

LABOR 21 STEERING – POWER 21 LABOR

OPERATION INDEX

OUTER TIE ROD, R&R	6
P/S PRESSURE HOSE, R&R	1
P/S PUMP RESERVOIR, R&R	3
P/S PUMP, R&R	2
P/S RETURN HOSE, R&R	4
STEERING GEAR, R&I	7
STEERING GEAR, R&R	8
SYSTEM, DIAGNOSIS	5
TIE ROD BOOT, R&R	9

P/S PUMP & HOSES

········· P/S PUMP & HOSES ·········

		(Factory Time)	Motor Time
1-P/S PRESSURE HOSE, R&R			
Fusion, Milan, Zephyr			
2.3L .... 06-09 B			1.0
3.0L .... 06-09 B	(1.1)		1.3
MKZ .... 07-09 B			2.3
2-P/S PUMP, R&R			
Fusion, Milan, Zephyr, MKZ .. 06-09 B	(1.0)		1.4
3-P/S PUMP RESERVOIR, R&R			
Fusion, Milan, Zephyr			
2.3L .... 06-09 B	(0.9)		1.3
3.0L .... 06-09 B	(0.8)		1.0

		(Factory Time)	Motor Time
MKZ .... 07-09 B	(0.8)		1.0
4-P/S RETURN HOSE, R&R			
TO RESERVOIR			
Fusion, Milan, Zephyr			
2.3L .... 06-09 B			1.1
3.0L .... 06-09 B	(1.0)		1.3
MKZ .... 07-09 B			1.3
TO GEAR			
Fusion, Milan, Zephyr			
2.3L .... 06-09 B			0.9
3.0L .... 06-09 B	(0.7)		0.9
MKZ .... 07-09 B			0.9
TO PUMP			
Fusion, Milan, Zephyr			
2.3L .... 06-09 B			0.9
3.0L .... 06-09 B	(0.8)		1.0
MKZ .... 07-09 B			1.0
TO COOLER			
Fusion, Milan, Zephyr			
2.3L .... 06-09 B			1.1
3.0L .... 06-09 B	(1.0)		1.3
MKZ .... 07-09 B			1.3
5-SYSTEM, DIAGNOSIS			
Fusion, Milan, Zephyr, MKZ .. 06-09 B			0.5

NOTES FOR: SYSTEM, DIAGNOSIS
Includes: Test Pump And System Pressure. Check Pounds Pull On Steering Wheel And Check For Leaks.

		(Factory Time)	Motor Time
MKZ .... 07-09 B	(0.8)		1.0

STEERING GEAR & LINKAGE

····· STEERING GEAR & LINKAGE ·····

		(Factory Time)	Motor Time
6-OUTER TIE ROD, R&R			
Fusion, Milan, Zephyr, MKZ			
One Side .... 06-09 B	(0.3)		0.4
Both Sides .... 06-09 B	(0.4)		0.7

NOTES FOR: OUTER TIE ROD, R&R
Does Not Include: Adjust Toe-In.

7-STEERING GEAR, R&I			
Fusion, Milan, Zephyr, MKZ .. 06-09 B	(1.6)		2.0

NOTES FOR: STEERING GEAR, R&I
Does Not Include: Adjust Toe-In.

To R&R Short Rack, Add 06-09			0.6
To Overhaul Steering Gear, Add 06-09			1.5
To R&R Inner Tie Rods, Add Each 06-09			0.3
8-STEERING GEAR, R&R			
Fusion, Milan, Zephyr, MKZ .. 06-09 B			2.2

NOTES FOR: STEERING GEAR, R&R
Does Not Include: Adjust Toe-In.

9-TIE ROD BOOT, R&R			
Fusion, Milan, Zephyr, MKZ			
One Side .... 06-09 B			0.6
Both Sides .... 06-09 B			0.9

NOTES FOR: TIE ROD BOOT, R&R
Does Not Include: Adjust Toe-In.

PARTS 21 STEERING – POWER 21 PARTS

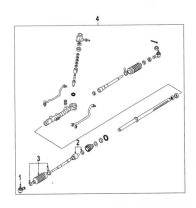

FPP100

P/S PUMP & HOSES

········· P/S PUMP & HOSES ·········

		Part No.	Price
P/S PUMP			
Fusion, Milan			
2.3L 06-09	6E5Z3A696AB	141.56
3.0L 06-09	6E5Z3A696B	145.28
Zephyr 06	6E5Z3A696B	145.28
MKZ 07-09	7H6Z3A696A	203.91
P/S PRESSURE SWITCH			
Fusion, Milan			
2.3L 06-09	6E5Z3N824A	71.98
3.0L 06-09	1W4Z3N824DA	33.30
Zephyr 06	1W4Z3N824DA	33.30
MKZ 07-09	7H6Z3N824A	35.58
P/S PUMP RESERVOIR			
Fusion, Milan			
2.3L 06-09	6E5Z3A697A	23.98
3.0L 06-09	6E5Z3A696B	145.28
Zephyr 06	6E5Z3A696B	145.28
MKZ 07-09	7H6Z3E764A	43.30
P/S PRESSURE HOSE			
Fusion, Milan			
2.3L 06-09	6E5Z3A719A	70.37
3.0L 06-09	7E5Z3A719B	80.12
Zephyr 06	7E5Z3A719B	80.12

		Part No.	Price
MKZ 07-09	AH6Z3A719A	114.54
P/S RETURN HOSE			
Fusion, Milan			
2.3L			
To Reservoir			
Front 06-09	6E5Z3A713AC	16.31
Rear 06-09	6E5Z3R807AA	24.37
To Cooler 06-09	6E5Z3A713AB	18.49
To Gear 06-09	7E5Z3A713A	27.40
3.0L			
To Cooler 06-09	6E5Z3A713AB	18.49
To Gear 06	6E5Z3A713BC	35.35
FWD 07-09	6E5Z3A713AB	18.49
AWD 07-09	7H6Z3A713BB	46.37
To Reservoir			
Front 06-09	6E5Z3A713BB	23.03
Rear 06-09	6E5Z3A713AB	18.49
Zephyr			
To Cooler 06	6E5Z3A713AB	18.49
To Gear 06	6E5Z3A713BC	35.35
To Reservoir			
Front 06	6E5Z3A713BB	23.03
Rear 06	6E5Z3A713AB	18.49
MKZ			
To Reservoir 07-09	7H6Z3A713BC	44.51
To Cooler 07-09	AH6Z3A713A	42.89
To Gear 07-09	7H6Z3A713BB	46.37
P/S SUCTION HOSE			
Fusion, Milan			
3.0L 06-09	7E5Z3691A	21.18
Zephyr 06	7E5Z3691A	21.18

		Part No.	Price
MKZ 07-09	7H6Z3691B	34.31
STEERING GEAR & LINKAGE			
····· STEERING GEAR & LINKAGE ·····			
1-OUTER TIE ROD			
Fusion, Milan			
Right 06-09	AE5Z3A130A	65.60
Left 06-09	AE5Z3A130B	64.20
Zephyr			
Right 06	AE5Z3A130A	65.60
Left 06	AE5Z3A130B	64.20
MKZ			
Right 07-09	AE5Z3A130A	65.60
Left 07-09	AE5Z3A130B	64.20
2-INNER TIE ROD			
Fusion, Milan 06-09	6E5Z3280AA	90.73
Zephyr 06	6E5Z3280AA	90.73
MKZ 07-09	6E5Z3280AA	90.73
3-TIE ROD BOOT			
Fusion, Milan 06-09	6E5Z3332A	35.40
Zephyr 06	6E5Z3332A	35.40
MKZ 07-09	6E5Z3332A	35.40
4-STEERING GEAR			
Fusion, Milan 06-09	7E5Z3504A	519.09
Zephyr 06	7H6Z3504A	672.48
MKZ 07-09	7H6Z3504A	672.48

FORD 1

FUSION, MILAN (06-09), ZEPHYR (06), MKZ (07-09), MILAN HYBRID (10)

LABOR 22 STEERING COLUMN 22 LABOR

OPERATION INDEX	
SHROUD, R&R	2
STEERING COLUMN, R&R	3
STEERING WHEEL, R&R	1

STEERING COLUMN

········· STEERING WHEEL ·············

	(Factory Time)	Motor Time
1-STEERING WHEEL, R&R		
Fusion, Milan, Zephyr, MKZ .. 06-09 B	(0.6)	0.9

NOTES FOR: STEERING WHEEL, R&R
WARNING: Before Repairing Any Air Restraint System, The Battery Cables And Any Back-up Power Supplies To The System Must Be Disconnected In Order To Prevent Accidental Deployment.

········· **STEERING COLUMN** ·············

	(Factory Time)	Motor Time
2-SHROUD, R&R		
Fusion, Milan, Zephyr, MKZ		
Upper/Lower ················ 06-09 B	(0.2)	0.3
3-STEERING COLUMN, R&R		
Fusion, Milan, Zephyr, MKZ .. 06-09 B	(1.0)	1.3

PARTS 22 STEERING COLUMN 22 PARTS

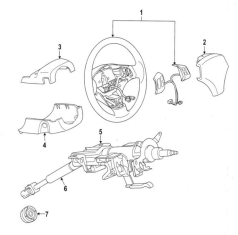

FPP110

STEERING COLUMN

········· STEERING WHEEL ·············

		Part No.	Price
1-STEERING WHEEL			
Fusion			
w/o Leather ········· 06-07		7E5Z3600AA	121.00
w/o Cell Phone			
System ············	08	8E5Z3600BC	248.87
2.3L	09	8E5Z3600BC	248.87
3.0L ·············	09	9E5Z3600EB	285.22
w/Mobile Phone			
System ············	08	8E5Z3600DC	245.35
2.3L	09	8E5Z3600DC	245.35
3.0L ·············	09	9E5Z3600GB	314.28
w/Leather			
w/o Audio Switch 06-08		7E5Z3600BA	249.18
w/Audio Switch ····· 06-08		7H6Z3600BD	694.00
2.3L			
w/o Cell Phone System			
w/o Navigation			
System ·········	09	8E5Z3600AA	300.92
w/Navigation System	09	8E5Z3600HA	97.98
w/Cell Phone System			
w/o Navigation			
System ··········	09	8E5Z3600CA	343.38
w/Navigation System	09	8E5Z3600NA	343.63
3.0L			
w/o Cell Phone System			
w/o Navigation			
System ··········	09	9E5Z3600JA	350.23
w/Navigation System	09	9E5Z3600KA	344.65
w/Cell Phone System			
w/o Navigation			
System ··········	09	9E5Z3600LA	332.62
w/Navigation System	09	9E5Z3600RA	366.90
Milan			
w/o Leather ········· 06-08		7E5Z3600AA	121.00
2.3L			
w/o Cell Phone System	09	8N7Z3600AC	292.07
w/Cell Phone System	09	8N7Z3600CC	326.02
3.0L			
w/o Cell Phone System	09	9N7Z3600DA	354.28
w/Cell Phone System em	09	9N7Z3600FA	361.38
w/Leather			
w/o Audio Switch ... 06-08		7E5Z3600BA	249.18
w/Audio Switch 06-08		7H6Z3600BD	694.00

		Part No.	Price
2.3L			
w/o Cell Phone System			
w/o Navigation			
System ··········	09	8N7Z3600AA	350.18
w/Navigation System	09	8N7Z3600BA	394.28
w/Cell Phone System			
w/o Navigation			
System ··········	09	8N7Z3600CA	360.95
w/Navigation System	09	8N7Z3600NA	368.05
3.0L			
w/o Cell Phone System			
w/o Navigation			
System ··········	09	9N7Z3600GA	392.28
w/Navigation System			
w/o Redundant			
Controls ········	09	8N7Z3600HA	382.20
w/Redundant			
Controls ··········	09	9N7Z3600HA	377.27
w/Cell Phone System			
w/o Navigation			
System ··········	09	9N7Z3600JA	332.98
w/Navigation System			
w/o Redundant			
Controls ········	09	8N7Z3600HA	382.20
w/Redundant			
Controls ·········	09	9N7Z3600RA	396.48
Zephyr			
Leather ············	06	7H6Z3600AC	462.77
Wood/Leather ········	06	7H6Z3600BC	696.98
MKZ			
Leather ············	07	7H6Z3600AC	462.77
w/o Cell Phone System			
w/o Navigation			
System ···········	08	8H6Z3600AA	514.17
	09	9H6Z3600JA	524.13
w/Navigation System	08	8H6Z3600AB	787.13
	09	9H6Z3600KA	503.33
w/Cell Phone System			
w/o Navigation			
System ···········	08	8H6Z3600CA	753.27
	09	9H6Z3600LA	526.72
w/Navigation System	09	9H6Z3600GA	527.53
Wood/Leather ········	07	7H6Z3600BC	696.98
w/o Cell Phone System			
w/o Navigation			
System ···········	08	8H6Z3600DA	698.00
	09	9H6Z3600MA	771.23
w/Navigation System	08	8H6Z3600EA	734.32
	09	9H6Z3600EA	718.33
w/Cell Phone System			
w/o Navigation			
System ··········	08	8H6Z3600FA	720.00
	09	9H6Z3600FA	702.73

		Part No.	Price
w/Navigation System	08	8H6Z3600HA	681.65
	09	9H6Z3600HA	742.63

NOTES FOR: STEERING WHEEL
Order By Color.

2-DRIVER INFLATOR MODULE

Fusion, Milan
See RESTRAINT
SYSTEM. ········· 06-09
Zephyr
See RESTRAINT
SYSTEM. ·········· 06
MKZ
See RESTRAINT
SYSTEM. ········· 07-09

········· **STEERING COLUMN** ·············

3-UPPER SHROUD			
Fusion, Milan			
Dark Stone ·········· 06-09		8E5Z3530AA	45.85
Charcoal Black ······ 06-09		AE5Z3530AF	57.07
Zephyr			
Charcoal Black ······	06	AE5Z3530AF	57.07
Medium Camel ······	06	8E5Z3530AB	45.85
MKZ			
Charcoal Black ······	07-09	AE5Z3530AF	57.07
Medium Camel ······	07-09	8E5Z3530AB	45.85

4-LOWER SHROUD			
Fusion, Milan			
Dark Stone ·········· 06-08		7E5Z3530AA	50.22
Charcoal Black ······ 06-08		AE5Z3530AB	126.00
Zephyr			
Charcoal Black ······	06	AE5Z3530AB	126.00
Medium Camel ······	06	7E5Z3530AB	61.47
MKZ			
Charcoal Black ······	07-09	AE5Z3530AB	126.00
Medium Camel ······	07-09	7E5Z3530AB	61.47

5-STEERING COLUMN			
Fusion, Milan			
To 12/3/07 ··········· 06-08		8E5Z3524C	262.12
From 12/3/07 ········ 08-09		8E5Z3524D	831.63
Zephyr ·············	06	8E5Z3524C	262.12
MKZ			
To 12/3/07 ··········· 07-08		8E5Z3524C	262.12
From 12/3/07 ········ 08-09		8E5Z3524D	831.63

6-SHAFT ASSY			
Fusion, Milan			
Part Of Steering			
Column. ··········· 06-09			
Zephyr			
Part Of Steering			
Column. ···········	06		
MZK			
Part Of Steering			
Column. ········· 07-09			

7-DUST SHIELD			
Fusion, Milan			
To 12/4/06 ·········· 06-07		3M8Z3C611B	34.97

(CONTINUED)

1 FORD
FUSION, MILAN (06-09), ZEPHYR (06), MKZ (07-09), MILAN HYBRID (10)

STEERING COLUMN - Parts Cont'd			

		Part No.	Price
Zephyr	06	3M8Z3C611B	34.97
MKZ			
To 12/4/06	07	3M8Z3C611B	34.97

		Part No.	Price
From 12/4/06	07-09	7E5Z3C611A	35.53

		Part No.	Price
From 12/4/06	07-09	7E5Z3C611A	35.53

LABOR 23 UNIVERSALS & REAR AXLE 23 LABOR

OPERATION INDEX

AXLE SHAFTS, R&I	2
AXLE SHAFTS, R&R	3
CARRIER, OVERHAUL	11
CARRIER, R&R	10
CLUTCH PLATES, R&R	4
COMPANION FLANGE, R&R	5
DIFFERENTIAL CASE, R&I	8
DIFFERENTIAL CASE, R&R	9
DRIVE SHAFT, R&I	1
PINION SEAL, R&R	7
VISCOUS COUPLER, R&R	6

REAR AXLE

·········· PROPELLER SHAFT ·········

		(Factory Time)	Motor Time
1-DRIVE SHAFT, R&I			
Fusion, Milan, MKZ .... 07-09 B	(1.0)		1.7
NOTES FOR: DRIVE SHAFT, R&I			
To R&R Universal Joints, Add			
Each 07-09			0.5

··········· DRIVE AXLES ···········

		(Factory Time)	Motor Time
2-AXLE SHAFTS, R&I			
Fusion, Milan, MKZ			
One Side	07-09 B		0.9
Both Sides	07-09 B		1.6
NOTES FOR: AXLE SHAFTS, R&I			
To R&R Outer Joint And/Or Boot, Add			
Each	07-09		0.5
To R&R Inner Joint And/Or Boot, Add			
Each	07-09		0.5
To R&R Axle Shaft Seals, Add			
Each	07-09		0.2
3-AXLE SHAFTS, R&R			
Fusion, Milan, MKZ			
One Side	07-09 B	(0.7)	0.9
Both Sides	07-09 B	(1.2)	1.6
NOTES FOR: AXLE SHAFTS, R&R			
To R&R Axle Shaft Seals, Add			
Each	07-09		0.2

··········· DIFFERENTIAL ···········

			Motor Time
4-CLUTCH PLATES, R&R			
Fusion, Milan, MKZ	07-09 B		8.8

		(Factory Time)	Motor Time
5-COMPANION FLANGE, R&R			
Fusion, Milan, MKZ .... 07-09 B			1.1
6-VISCOUS COUPLER, R&R			
Fusion, Milan, MKZ .... 07-09 B			4.3
7-PINION SEAL, R&R			
Fusion, Milan, MKZ .... 07-09 B			2.2
8-DIFFERENTIAL CASE, R&I			
Fusion, Milan, MKZ .... 07-09 B			8.8
NOTES FOR: DIFFERENTIAL CASE, R&I			
To R&R Ring & Pinion, Add	07-09		1.0
To R&R Side Bearings, Add	07-09		0.4
To R&R Gear Kit, Add	07-09		0.2
To R&R Pinion Cups, Add	07-09		0.2
To Check Run Out, Add	07-09		0.4
9-DIFFERENTIAL CASE, R&R			
Fusion, Milan, MKZ .... 07-09 B			9.3
NOTES FOR: DIFFERENTIAL CASE, R&R			
To Check Run Out, Add	07-09		0.4

··········· REAR AXLE ···········

		(Factory Time)	Motor Time
10-CARRIER, R&R			
Fusion, Milan, MKZ .... 07-09 B	(4.6)		6.8
11-CARRIER, OVERHAUL			
Fusion, Milan, MKZ .... 07-09 A			9.8

PARTS 23 UNIVERSALS & REAR AXLE 23 PARTS

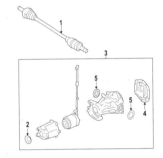

FPP115

REAR AXLE

·········· PROPELLER SHAFT ·········

		Part No.	Price
DRIVE SHAFT			
Fusion, Milan, MKZ ... 07-09		8E5Z4R602A	689.57

		Part No.	Price
·········· DRIVE AXLES ··········			
1-AXLE ASSY			
Fusion, Milan, MKZ			
Right 07-09		7E5Z4K138A	230.98
Left 07-09		AE5Z4K139A	193.62
·········· DIFFERENTIAL ··········			
2-PINION SEAL			
Fusion, Milan, MKZ ... 07-09		8G1Z4N046A	17.08

		Part No.	Price
·········· REAR AXLE ··········			
3-CARRIER			
Fusion, Milan, MKZ ...	07	7E534000CA	1258.33
	08-09	9E514000AA	1578.27
NOTES FOR: CARRIER			
Includes Viscous Coupler.			
4-COVER			
Fusion, Milan, MKZ ... 07-09		7E5Z4033A	44.97
5-AXLE SEALS			
Fusion, Milan, MKZ ... 07-09		5L8Z4B416AA	17.08

LABOR 24 REAR SUSPENSION 24 LABOR

OPERATION INDEX

BUSHINGS, R&R	13
COIL SPRING, R&R	4
CROSSMEMBER, R&R	16
HUB, R&R	6
KNUCKLE, R&R	5
LATERAL LINK, R&R	8
LINK, R&R	14
LOWER CONTROL ARM, R&R	12
LUG BOLTS, R&R	7
REAR WHEELS, ALIGN	1
SHOCK ABSORBER, R&R	9
STABILIZER BAR, R&R	15
SUSPENSION, OVERHAUL	2
TOE-IN, ADJUST	3
TRAILING LINK, R&R	10
UPPER CONTROL ARM, R&R	11

REAR SUSPENSION

·········· SUSPENSION SERVICE ·········

		(Factory Time)	Motor Time
1-REAR WHEELS, ALIGN			
Fusion, Milan, Zephyr, MKZ			
Rear Wheel Alignment	06-09 B		1.5
Four Wheel Alignment	06-09 B		1.7
2-SUSPENSION, OVERHAUL			
Fusion, Milan, Zephyr, MKZ			
One Side	06-09 B		3.5
Both Sides	06-09 B		6.8
3-TOE-IN, ADJUST			
Fusion, Milan, Zephyr, MKZ ..	06-09 B		0.7

····· SUSPENSION COMPONENTS ·····

		(Factory Time)	Motor Time
4-COIL SPRING, R&R			
Fusion, Milan, Zephyr, MKZ			
One Side	06-09 B	(0.6)	0.7
Both Sides	06-09 B	(0.9)	1.2

		(Factory Time)	Motor Time
5-KNUCKLE, R&R			
Fusion, Milan, Zephyr, MKZ			
FWD			
One Side	06-09 B		1.3
Both Sides	06-09 B		2.0
AWD			
One Side	06-09 B		1.8
Both Sides	06-09 B		3.2
6-HUB, R&R			
Fusion, Milan, Zephyr, MKZ			
FWD			
One Side	06-09 B	(0.6)	0.9
Both Sides	06-09 B	(0.9)	1.2
AWD			
One Side	06-09 B	(1.4)	2.0
Both Sides	06-09 B	(2.4)	3.3
7-LUG BOLTS, R&R			
Fusion, Milan, Zephyr, MKZ			
One Side	06-09 B		0.9
(CONTINUED)			

FORD 1

FUSION, MILAN (06-09), ZEPHYR (06), MKZ (07-09), MILAN HYBRID (10)

REAR SUSPENSION - Time Cont'd

	(Factory Time)	Motor Time
8-LATERAL LINK, R&R		
Fusion, Milan, Zephyr, MKZ		
One Side .... 06-09 B		1.2
Both Sides .... 06-09 B		1.6
9-SHOCK ABSORBER, R&R		
Fusion, Milan, Zephyr, MKZ		
One Side .... 06-09 B	(0.9)	1.2
Both Sides .... 06-09 B	(1.5)	1.9
10-TRAILING LINK, R&R		
Fusion, Milan, Zephyr, MKZ		
One Side .... 06-09 B		1.9
Both Sides .... 06-09 B		3.4

	(Factory Time)	Motor Time
········· **UPPER CONTROL ARM** ·········		
11-UPPER CONTROL ARM, R&R		
Fusion, Milan, Zephyr, MKZ		
One Side .... 06-09 B	(0.6)	0.7
Both Sides .... 06-09 B	(0.9)	1.1
········· **LOWER CONTROL ARM** ·········		
12-LOWER CONTROL ARM, R&R		
Fusion, Milan, Zephyr, MKZ		
One Side .... 06-09 B	(0.7)	0.8
Both Sides .... 06-09 B	(1.1)	1.3
············ **STABILIZER BAR** ············		
13-BUSHINGS, R&R		
Fusion, Milan, Zephyr, MKZ		
Both Sides .... 06-09 C	(0.4)	0.5

	(Factory Time)	Motor Time
14-LINK, R&R		
Fusion, Milan, Zephyr, MKZ		
One Side .... 06-09 C	(0.4)	0.5
Both Sides .... 06-09 C	(0.5)	0.6
15-STABILIZER BAR, R&R		
Fusion, Milan, Zephyr, MKZ		
FWD .... 06-09 C	(0.6)	0.7
AWD .... 07-09 C	(1.9)	2.5
············ **REAR SUSPENSION** ············		
16-CROSSMEMBER, R&R		
Fusion, Milan, Zephyr, MKZ .. 06-09 B		5.2

PARTS   24   REAR SUSPENSION   24   PARTS

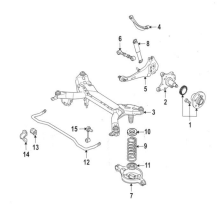

FPP120

REAR SUSPENSION		
······ **SUSPENSION COMPONENTS** ······		
	Part No.	Price
1-HUB & BEARING ASSY		
Fusion, Milan		
w/o ABS 06	6E5Z1104AA	114.00
w/ABS 06	6E5Z1104BA	116.87
FWD		
w/o ABS 07	6E5Z1104AA	114.00
w/ABS 07-09	6E5Z1104BA	116.87
AWD 07-09	6E5Z1104AB	63.58
Zephyr 06	6E5Z1104BA	116.87
MKZ		
FWD 07-09	6E5Z1104BA	116.87
AWD 07-09	6E5Z1104AB	63.58
2-SPINDLE		
Fusion, Milan		
Right 06	3M8Z4A013R	180.52
FWD 07-09	3M8Z4A013R	180.52
AWD 07-09	7E5Z4A013R	234.60
Left 06	3M8Z4A013L	122.33
FWD 07-09	3M8Z4A013L	122.33
AWD 07-09	7E5Z4A013L	234.60
Zephyr		
Right 06	3M8Z4A013R	180.52
Left 06	3M8Z4A013L	122.33
MKZ		
Right		
FWD 07-09	3M8Z4A013R	180.52
AWD 07-09	7E5Z4A013R	234.60
Left		
FWD 07-09	3M8Z4A013L	122.33
AWD 07-09	7E5Z4A013L	234.60
5-TRAILING LINK		
Fusion, Milan		
Right 06	4M8Z5500B	154.93
FWD 07-09	4M8Z5500B	154.93
AWD 07-09	7E5Z5A972R	95.33
Left 06	4M8Z5500A	154.93
FWD 07-09	4M8Z5500A	154.93
AWD 07-09	7E5Z5A972L	95.33
Zephyr		
Right 06	4M8Z5500B	154.93
Left 06	4M8Z5500A	154.93
MKZ		
Right		
FWD 07-09	4M8Z5500B	154.93
AWD 07-09	7E5Z5A972R	95.33
Left		
FWD 07-09	4M8Z5500A	154.93
AWD 07-09	7E5Z5A972L	95.33
6-LATERAL LINK		
Fusion, Milan 06	9E5Z5500A	33.60

		Part No.	Price
FWD 07-09		9E5Z5500A	33.60
AWD 07-09		9E5Z5500B	33.13
Zephyr 06		9E5Z5500A	33.60
MKZ			
FWD 07-09		9E5Z5500A	33.60
AWD 07-09		9E5Z5500B	33.13
8-SHOCK ABSORBER			
Fusion, Milan			
2.3L 06-07		6E5Z18125AA	60.98
Standard Suspension 08-09		6E5Z18125AA	60.98
Sport Tuned			
Suspension 08-09		8E5Z18125A	56.42
3.0L 06-07		6E5Z18125BA	60.98
FWD 08-09		8E5Z18125A	56.42
AWD 08-09		7E5Z18125A	42.42
Zephyr 06		6H6Z18125AA	54.88
MKZ			
FWD 07-09		7H6Z18125C	42.33
AWD 07-09		7H6Z18125D	42.33
9-COIL SPRING			
Fusion, Milan			
2.3L 06-07		8E5Z5560F	51.83
3.0L 07		8E5Z5560G	52.80
FWD			
Standard Suspen-			
sion 08-09		8E5Z5560G	52.80
Sport Tuned			
Suspension 08-09		8E5Z5560E	53.60
AWD 07-09		8E5Z5560H	49.49
Zephyr 06		8H6Z5560A	51.71
MKZ			
FWD 07-09		8H6Z5560A	51.71
AWD 07-09		8H6Z5560B	50.29
10-UPPER SPRING INSULATOR			
Fusion, Milan 06-09		3M8Z5536B	12.47
Zephyr 06		3M8Z5536B	12.47
MKZ 07-09		3M8Z5536B	12.47
11-LOWER SPRING INSULATOR			
Fusion, Milan 06-09		3M8Z5536A	7.98
Zephyr 06		3M8Z5536A	7.98
MKZ 07-09		3M8Z5536A	7.98
············ **UPPER CONTROL ARM** ············			
4-UPPER CONTROL ARM			
Fusion, Milan 06-08		6M8Z5500A	101.98
Zephyr 06		6M8Z5500A	101.98
MKZ 07-09		6M8Z5500A	101.98
············ **LOWER CONTROL ARM** ············			
7-LOWER CONTROL ARM			
Right 06-08		6E5Z5A649AA	97.60
	09	9E5Z5A649C	111.65
Left 06-08		6E5Z5A649AB	97.60

		Part No.	Price
	09	9E5Z5A649D	111.65
Zephyr			
Right 06		6E5Z5A649AA	97.60
Left 06		6E5Z5A649AB	97.60
MKZ			
Right 07-08		6E5Z5A649AA	97.60
	09	9E5Z5A649C	111.65
Left 07-08		6E5Z5A649AB	97.60
	09	9E5Z5A649D	111.65
············ **STABILIZER BAR** ············			
12-STABILIZER BAR			
Fusion, Milan 06		8E5Z5A772CA	47.65
FWD 07-09		8E5Z5A772CA	47.65
AWD			
To 9/4/06 07		8E5Z5A772CA	66.65
From 9/4/06 07-09		7E5Z5A772A	58.65
Zephyr 06		8E5Z5A772CA	47.65
MKZ			
FWD			
Standard Suspension 07-09		8E5Z5A772DA	66.65
Sport Tuned			
Suspension 07-09		8E5Z5A772A	86.15
AWD 07-09		8E5Z5A772DA	66.65
13-BUSHING			
Fusion, Milan 06		6E5Z5493AA	6.00
FWD 07-09		6E5Z5493AA	6.00
AWD 07-09		7E5Z5493A	5.55
Zephyr 06		6E5Z5493AA	6.00
MKZ 07-08		7H6Z5493A	5.55
14-MOUNTING BRACKET			
Fusion, Milan 06-09		3M8Z5B498A	12.82
Zephyr 06		3M8Z5B498A	12.82
MKZ 07-08		3M8Z5B498A	12.82
15-LINK			
Fusion, Milan 06-09		5M8Z5A486A	21.82
Zephyr 06		5M8Z5A486A	21.82
MKZ 07-08		5M8Z5A486A	21.82
············ **REAR SUSPENSION** ············			
3-CROSSMEMBER			
Fusion, Milan 06		6E5Z5035AA	571.78
FWD 07		6E5Z5035AA	571.78
To 12/3/07 08		8E5Z5035A	539.88
From 12/3/07 08		9E5Z5035A	636.33
	09	9E5Z5035A	636.33
AWD 07-09		8E5Z5035B	802.00
Zephyr 06		6E5Z5035A	571.78
MKZ			
FWD 07		6E5Z5035AA	571.78
To 12/3/07 08		8E5Z5035A	539.88
From 12/3/07 08		9E5Z5035A	636.33
	09	9E5Z5035A	636.33
AWD 07-09		8E5Z5035B	802.00

1 FORD
FUSION, MILAN (06-09), ZEPHYR (06), MKZ (07-09), MILAN HYBRID (10)

LABOR · 25 BODY HARDWARE 25 · LABOR

OPERATION INDEX

LIFT CYLINDER, R&R	10
LOCK CYLINDER, R&R	12
LOCK CYLINDER, R&R	3
LOCK, R&R	11
LOCK, R&R	1
LOCK, R&R	4
LOCK, R&R	7
RELEASE CABLE, R&R	2
STRIKER, R&R	13
STRIKER, R&R	5
STRIKER, R&R	8
WINDOW REGULATOR, R&R	6
WINDOW REGULATOR, R&R	9

HOOD

······· HOOD ·······

			(Factory) Time	Motor Time
1-LOCK, R&R				
Fusion, Milan, Zephyr, MKZ ..	06-09 B	(0.3)		0.4
2-RELEASE CABLE, R&R				
Fusion, Milan, Zephyr, MKZ ..	06-09 B	(0.7)		0.8

DOOR

		(Factory) Time	Motor Time
······· FRONT DOOR ·······			
3-LOCK CYLINDER, R&R			
Fusion, Milan, Zephyr, MKZ ..	06-09 B	(1.1)	1.2
4-LOCK, R&R			
Fusion, Milan, Zephyr, MKZ			
One Side	06-09 B	(0.8)	1.0
Both Sides	06-09 B	(1.6)	1.9
5-STRIKER, R&R			
Fusion, Milan, Zephyr, MKZ			
One	06-09 B	(0.2)	0.3
6-WINDOW REGULATOR, R&R			
Fusion, Milan, Zephyr			
One Side	06-09 B	(0.9)	1.2
Both Sides	06-09 B	(1.8)	2.2
MKZ			
One Side	07-09 B	(1.0)	1.3
Both Sides	07-09 B	(1.8)	2.3
······· REAR DOOR ·······			
7-LOCK, R&R			
Fusion, Milan, Zephyr, MKZ			
One Side	06-09 B	(1.1)	1.6

		(Factory) Time	Motor Time
Both Sides	06-09 B	(2.0)	2.8
8-STRIKER, R&R			
Fusion, Milan, Zephyr, MKZ			
One	06-09 B	(0.2)	0.3
9-WINDOW REGULATOR, R&R			
Fusion, Milan, Zephyr, MKZ			
One Side	06-09 B	(0.8)	1.0
Both Sides	06-09 B	(1.5)	1.9
TRUNK			
······· TRUNK ·······			
10-LIFT CYLINDER, R&R			
Fusion, Milan, Zephyr, MKZ			
One	06-09 C		0.2
Both	06-09 C		0.3
11-LOCK, R&R			
Fusion, Milan, Zephyr, MKZ ..	06-09 B	(0.3)	0.4
12-LOCK CYLINDER, R&R			
Fusion, Milan, Zephyr, MKZ ..	06-09 B	(0.6)	0.7
13-STRIKER, R&R			
Fusion, Milan, Zephyr, MKZ ..	06-09 B		0.4

PARTS · 25 BODY HARDWARE 25 · PARTS

PARTS INDEX

LIFT CYLINDER	10
LOCK	11
LOCK	1
LOCK	4
LOCK	7
LOCK CYLINDER	12
LOCK CYLINDER	3
RELEASE CABLE	2
STRIKER	13
STRIKER	5
STRIKER	8
WINDOW REGULATOR	6
WINDOW REGULATOR	9

HOOD

······· HOOD ·······

		Part No.	Price
1-LOCK			
Fusion, Milan	06-09	6E5Z16700AA	27.10
Zephyr	06	6H6Z16700AA	30.12
MKZ	07-09	6H6Z16700AA	30.12
2-RELEASE CABLE			
Fusion, Milan			
Camel Interior	06-09	6E5Z16916AG	30.82
Charcoal Black Interior			
To 9/29/05	06	6E5Z16916AE	36.30
From 9/29/05	06-09	6E5Z16916AH	25.68
Med. Light Stone			
Interior	06-09	6E5Z16916AF	25.73
Light Stone Interior ...	06-09	6E5Z16916AJ	34.72
Sand Interior	06-09	6E5Z16916AK	34.43
Zephyr			
Camel Interior	06	6E5Z16916AG	30.82
Charcoal Black Interior			
To 9/29/05	06	6E5Z16916AE	36.30
From 9/29/05	06	6E5Z16916AH	25.68
Med. Light Stone			
Interior	06	6E5Z16916AF	25.73
Light Stone Interior ...	06	6E5Z16916AJ	34.72
Sand Interior	06	6E5Z16916AK	34.43
MKZ			
Camel Interior	07-09	6E5Z16916AG	30.82
Charcoal Black Interior	07-09	6E5Z16916AH	25.68
Med. Light Stone			
Interior	07-09	6E5Z16916AF	25.73
Light Stone Interior ...	07-09	6E5Z16916AJ	34.72
Sand Interior	07-09	6E5Z16916AK	34.43

DOOR

		Part No.	Price
······· FRONT DOOR ·······			
3-LOCK CYLINDER			
Fusion, Milan	06-09	8E5Z5421991B	66.00
Zephyr	06	8E5Z5421991B	66.00
MKZ	07-09	8E5Z5421991B	66.00
4-LOCK			
Fusion, Milan			
Right	06-09	6E5Z5421812AA	113.80
Left			
w/o Alarm	06-09	6E5Z5421813AA	117.80
w/Alarm	06-09	6E5Z5421813BA	111.35
Zephyr			
Right	06	6E5Z5421812AA	113.80
Left	06	6E5Z5421813BA	111.35
MKZ			
Right	07-09	6E5Z5421812AA	113.80
Left	07-09	6E5Z5421813BA	111.35
5-STRIKER			
Fusion, Milan	06-09	8E5Z5422008A	12.87
Zephyr	06	8E5Z5422008A	12.87
MKZ	07-09	8E5Z5422008A	12.87
6-WINDOW REGULATOR			
Fusion, Milan			
Right	06-09	8E5Z5423200A	62.38
Left	06-09	8E5Z5423201A	62.38
Zephyr			
Right	06	8E5Z5423200A	62.38
Left	06	8E5Z5423201A	62.38
MKZ			
Right	07-09	8E5Z5423200A	62.38
Left	07-09	8E5Z5423201A	62.38
······· REAR DOOR ·······			
7-LOCK			
Fusion, Milan			
Right	06-09	6E5Z5426412AA	117.00
Left	06-09	6E5Z5426413AA	116.93
Zephyr			
Right	06	6E5Z5426412AA	117.00
Left	06	6E5Z5426413AA	116.93
MKZ			
Right	07-09	6E5Z5426412AA	117.00
Left	07-09	6E5Z5426413AA	116.93
8-STRIKER			
Fusion, Milan	06-09	8E5Z5422008A	12.87
Zephyr	06	8E5Z5422008A	12.87
MKZ	07-09	8E5Z5422008A	12.87
9-WINDOW REGULATOR			
Fusion, Milan			
Right	06-09	8E5Z5427000A	50.38

		Part No.	Price
Left	06-09	8E5Z5427001A	50.38
Zephyr			
Right	06	8E5Z5427000A	50.38
Left	06	8E5Z5427001A	50.38
MKZ			
Right	07-09	8E5Z5427000A	50.38
Left	07-09	8E5Z5427001A	50.38
TRUNK			
······· TRUNK ·······			
10-LIFT CYLINDER			
Fusion	06	6E5Z54406A10A	19.00
w/o Spoiler			
To 1/17/07	07	6E5Z54406A10A	19.00
From 1/17/07	07-09	7E5Z54406A10B	20.47
w/Spoiler			
To 12/4/06	07	7E5Z54406A10C	18.70
From 12/4/06	07-09	7E5Z54406A10C	24.00
Milan			
To 12/4/06	06-07	6N7Z54406A10A	20.80
From 12/4/06			
w/o Spoiler	07-09	7N7Z54406A10B	19.72
w/Spoiler	07-09	7N7Z54406A10A	23.28
Zephyr	06	6N7Z54406A10A	20.80
MKZ			
To 12/4/06	07	7N7Z54406A10A	20.80
From 12/4/06	07-09	7N7Z54406A10A	23.28
11-LOCK			
Fusion, Milan			
To 9/4/06	06-07	6E5Z5443200D	55.87
From 9/4/06	07-09	7E5Z5443200B	61.13
Zephyr	06	6E5Z5443200D	55.87
MKZ			
To 9/4/06	07	6E5Z5443200D	55.87
From 9/4/06	07-09	7E5Z5443200B	61.13
12-LOCK CYLINDER			
Fusion, Milan	06-09	6E5Z5443262B	66.73
Zephyr	06	6E5Z5443262B	66.73
MKZ	07-09	6E5Z5443262B	66.73
13-STRIKER			
Fusion			
To 10/14/08	06-09	9E5Z5443252A	30.05
From 10/14/08	09	9E5Z5443252A	30.05
Milan			
To 10/14/08	06-09	9E5Z5443252A	30.05
From 10/14/08	09	AH6Z5443252A	36.82
Zephyr	06	9E5Z5443252A	30.05
MKZ			
To 10/14/08	07-09	9E5Z5443252A	30.05
From 10/14/08	09	AH6Z5443252A	36.82
DP2FP0			

2 FORD
FUSION, MILAN HYBRID (2010)

BODY HARDWARE - Parts Cont'd

7-WINDOW REGULATOR

		Part No.	Price
Fusion, Milan Hybrid			
Right	10	8E5Z5427000A	50.38
Left	10	8E5Z5427001A	50.38

TRUNK

		Part No.	Price

TRUNK

8-LIFT CYLINDER

			Part No.	Price
Fusion, Milan Hybrid				
w/o Spoiler	10		AN7Z54406A10A	19.40
w/Spoiler	10		AH6Z54406A10A	17.88

9-LOCK

		Part No.	Price
Fusion, Milan Hybrid ..	10	AE5Z5443200A	64.13

10-LOCK CYLINDER

Fusion, Milan Hybrid ..	10	6E5Z5443262B	66.73

11-STRIKER

Fusion Hybrid	10	9E5Z5443252A	30.05
Milan Hybrid	10	AH6Z5443252A	30.68

LABOR 26 HYBRID COMPONENTS 26 LABOR

OPERATION INDEX

HYBRID COMPONENTS

BATTERY

		(Factory Time)	Motor Time
1-BATTERY, R&R			
Fusion, Milan Hybrid	10 B	(1.2)	1.6
2-CABLE, R&R			
HIGH VOLTAGE			
Fusion, Milan Hybrid	10 B	(2.8)	3.8

COOLING SYSTEM

3-RADIATOR, R&R			
Fusion, Milan Hybrid	10 B		1.6

		(Factory Time)	Motor Time
4-WATER PUMP, R&R			
Fusion, Milan Hybrid	10 B		1.4

HYBRID COMPONENTS

5-INVERTER ASSY, R&R			
DC TO AC			
Fusion, Milan Hybrid	10 B	(0.4)	0.5
DC TO DC			
Fusion, Milan Hybrid	10 B	(0.9)	1.3

PARTS 26 HYBRID COMPONENTS 26 PARTS

PARTS INDEX

HYBRID COMPONENTS

BATTERY

		Part No.	Price
1-BATTERY			
Fusion, Milan Hybrid	10	AE5Z10B759A	4686.81
2-POSITIVE CABLE			
Fusion, Milan Hybrid	10	AE5Z14300AA	1342.36
3-COOLING FAN			
Fusion, Milan Hybrid ..	10	AE5Z10C659A	209.84
4-CONTROL MODULE			
BATTERY MANAGEMENT SYSTEM			
Fusion, Milan Hybrid .	10	AE5Z10B830A	526.35
VOLTAGE CONTROL & BATTERY			
Fusion, Milan Hybrid	10	AE5Z10B687A	162.44
5-DISABLE SWITCH			
Fusion, Milan Hybrid ..	10	AE5Z10A757A	0.00

COOLING SYSTEM

		Part No.	Price
6-RADIATOR			
Fusion, Milan Hybrid ..	10	AE5Z8005D	404.26
7-UPPER HOSE			
Fusion, Milan Hybrid ..	10	9E5Z8260D	41.07
8-LOWER HOSE			
Fusion, Milan Hybrid ..	10	9E5Z8286D	15.53
9-WATER PUMP			
Fusion, Milan Hybrid			
To 3/28/09	10	9E5Z8C419A	146.68
From 3/28/09	10	9E5Z8C419B	146.68

HYBRID COMPONENTS

		Part No.	Price
10-INVERTER ASSY			
Fusion, Milan Hybrid			
Voltage Inverter DC To DC	10	AE5Z14B227A	1581.09
DP2FP1			

SUMMARY OF FORMULAS IN THIS TEXT-WORKBOOK

Area of a Circle

$$A = \pi r^2$$

or

$$A = 0.7854 \times d^2$$

where $\pi \approx 3.1416$, r is the radius of the circle, d is the diameter of the circle, and 0.7854 is a constant.

Carburetor Size

Carburetor size in cubic feet per minute (cfm) is determined by one of the following formulas:

$$\text{Street carburetor} = \frac{\text{rpm} \times \text{Displacement} \times 0.85}{3456}$$

$$\text{Racing carburetor} = \frac{\text{rpm} \times \text{Displacement} \times 1.10}{3456}$$

where displacement is in cubic inches.

Circumference of a Circle

$$C = \pi d$$
$$\approx 3.1416d$$

where C is the circumference of a circle and d is its diameter.

Cafe

The Corporate Average Fuel Economy (CAFE) figure for each manufacturer is calculated by using the following formula:

Cornering Force

Cornering force is usually expressed in terms of its relationship to gravity, or gs.

$$\text{Cornering Force} = \frac{\text{Velocity}^2}{32 \times \text{Radius}}$$

$$\text{Cornering Force} = \frac{\text{Velocity}^2}{16 \times \text{Diameter}}$$

where velocity is expressed in feet per second and the radius or diameter of the turn is given in feet.

If a skid pad is used, then cornering force is

$$\text{Cornering Force} = \frac{1.23 \times \text{Radius}}{\text{Time}^2}$$

where the radius is in feet, time is in seconds, and 1.23 is a constant.

$$\text{CAFE} = \frac{\text{Total Production Volume}}{\dfrac{\text{Production Vol. Vehicle A}}{\text{Fuel Economy Vehicle A}} + \dfrac{\text{Production Vol. Vehicle B}}{\text{Fuel Economy Vehicle B}} + \dfrac{\text{Production Vol. Vehicle C}}{\text{Fuel Economy Vehicle C}}}$$

Compression Ratio

To determine compression ratio (CR), the piston displacement (swept volume) is divided by the combustion volume. Thus, we have the two formulas shown below:

$$CR = \frac{\text{Piston Displacement} + \text{Combustion Volume}}{\text{Combustion Volume}}$$

or

$$CR = \frac{\text{Cylinder Vol.} + \text{Deck Height Vol.} + \text{Head Gasket Vol.} + \text{Combustion Chamber Vol.}}{\text{Deck Height Vol.} + \text{Head Gasket Vol.} + \text{Combustion Chamber Vol.}}$$

Drive Axle Ratio

To find the drive axle ratio needed to bring back original performance levels, use the following formula:

New Drive Ratio

$$= \frac{\text{New Tire Diameter}}{\text{Old Tire Diameter}} \times \text{Original Drive Ratio}$$

Engine Displacement

Displacement

$$= 0.7854 \times \text{Bore}^2 \times \text{Stroke} \times \text{Number of Cylinders}$$

where bore and stroke are both either in inches (in.) or millimeters (mm) and 0.7854 is a constant. If they are in inches, displacement is in in.^{3} and if the measurements are in mm, displacement is in mm^3.

Displacement in cubic centimeters (cc or cm^3)

$$= \frac{0.7854 \times \text{Bore}^2 \times \text{Stroke} \times \text{Number of Cylinders}}{1000}$$

if bore and stroke are in mm.

Displacement in liters (L)

$$= \frac{0.7854 \times \text{Bore}^2 \times \text{Stroke} \times \text{Number of Cylinders}}{1\ 000\ 000}$$

if bore and stroke are in mm.

Engine Revolutions per Minute

$$\text{rpm} = \frac{168 \times R \times \text{mph}}{r}$$

where mph is the vehicle ground speed in miles per hour, r is the rolling radius in inches of the loaded-driving tire, and R is the overall gear ratio.

Exhaust Header Tubing

Diameter

$$\text{Diameter} = \sqrt{\frac{\text{CID} \times 1900}{\text{Length} \times \text{rpm}}}$$

where CID is the cubic inch displacement of the engine, length is the exhaust header length in inches, and 1900 is a constant.

Length

$$\text{Length} = \frac{\text{CID} \times 1900}{\text{Diameter}^2 \times \text{rpm}}$$

where CID is the cubic inch displacement of the engine, diameter is the exhaust header diameter in inches, and 1900 is a constant.

Finance

Interest

To determine the amount of money accumulated with compound interest, use the formula

$$A = P\left(1 + \frac{r}{n}\right)^{ny}$$

where A is the accumulate amount, P is the principal, r is the annual interest rate, n is the number of payments in one year, and y is the number of years.

Payments

When money is borrowed, the amount of each payment, P, is

$$P = \frac{BR(1 + R)^{ny}}{(1 + R)^{ny} - 1}$$

where $R = \frac{r}{n}$, if r is the annual interest rate and n is the number of payments each year, B is the amount of money borrowed, and y is the length of the loan in years.

Profit Margin

$$\text{Profit Margin} = \frac{\text{Net Income After Taxes}}{\text{Revenue}} \times 100$$

$$= \frac{\text{Profit}}{\text{Revenue}} \times 100$$

Force, Pressure, and Area

$$F = PA, \qquad P = \frac{F}{A}, \qquad A = \frac{F}{P}$$

where F is the force in pounds or kilograms, P is the pressure in pounds per square inch (psi) or kilopascals (kPa), and A is the area in square inches (in.^{2}) or square centimeters (cm^2).

Gear Ratio

$$\text{Gear Ratio} = \frac{\text{Driven Gear}}{\text{Drive Gear}}$$

Gear Ratio for Planetary Gears

This gear set usually has an input member, an output member, and a reaction or locked member. These three members have different functions for different ratios. Table B–1 shows these different functions. The table also shows (on the right) the formulas used to determine the gear ratios for three operational modes where R equals the gear ratio, N equals the number of gear teeth of a particular gear, and the subscript identifies the particular gear in question. The bottom formula shows N_i as a negative number because the gear motion is in reverse.

Table B–1

	Input	Output	Locked	Ratio Formula
1st	S	P	I	$R = 1 + N_i/N_s$
2nd	I	P	S	$R = 1 + N_s/N_i$
Rev.	S	I	P	$R = -N_i/N_s$

Horsepower

$$1 \text{ hp} = 746 \text{ W}$$

or

$$\text{Horsepower} = \frac{\text{rpm} \times \text{Torque}}{5252}$$

Horsepower Loss

Horsepower Loss

$$= \frac{\text{Elevation in Feet}}{1000} \times 0.03 \times \text{Sea Level Horsepower}$$

Miles per Hour

$$\text{mph} = \frac{\text{rpm} \times r}{168 \times R}$$

where mph is the vehicle ground speed in miles per hour, r is the rolling radius in inches of the loaded-driving tire, R is the overall gear ratio, and 168 is a constant.

New Effective Drive Ratio

New Effective Drive Ratio

$$= \frac{\text{Old Tire Diameter}}{\text{New Tire Diameter}} \times \text{Drive Axle Ratio}$$

Ohm's Law

$$E = I \times R, \qquad I = \frac{E}{R}, \qquad R = \frac{E}{I}$$

where E is the electromotive force measured in volts, V; I is the current flow measured in amperes (or amps), A; and R is the resistance measured in ohms, Ω.

Overall Transmission Ratio (OTR)

$$\text{OTR} = \frac{\text{Front Counter Gear}}{\text{Clutch Shaft Gear}} \times \frac{\text{Last Gear}}{\text{Last Counter Gear in Use}}$$

$$\text{OTR} = \frac{\text{Front Driven Gear}}{\text{Front Drive Gear}} \times \frac{\text{Rear Driven Gear}}{\text{Rear Drive Gear}}$$

Resistance

Total Resistance in a Parallel Circuit

$$R_{\text{total}} = \frac{1}{\frac{1}{R_1} + \frac{1}{R_2} + \cdots + \frac{1}{R_n}}$$

where $R_1, R_2, \ldots, R_n$, and so on are the resistances of n resistors connect in parallel. Each resistance is measured in ohms, Ω.

Total Resistance in a Series Circuit

$$R_{\text{total}} = R_1 + R_2 + \cdots + R_n$$

where $R_1, R_2, \ldots, R_n$, and so on are the resistances of n resistors connect in series. Each resistance is measured in ohms, Ω.

Speed

True Speed

$$\text{True Speed} = \frac{\text{New Tire Diameter}}{\text{Old Tire Diameter}} \times \text{Indicated Speed}$$

or

$$\text{True Speed} = \frac{3600}{\text{Seconds per Mile}}$$

Equivalent Indicated Speed

$$\text{Equivalent Indicated Speed} = \frac{\text{Old Tire Diameter}}{\text{New Tire Diameter}} \times \text{True Speed}$$

Percent Error

$$\text{Percent Error} = \frac{\text{True Speed} - \text{Indicated Speed}}{\text{True Speed}} \times 100$$

Theoretical Air Capacity of an Engine

$$\text{Theoretical Air Capacity} = \frac{\text{rpm} \times \text{Displacement}}{2}$$

If engine displacement is given in cubic inches, the air capacity number becomes very large. To make it more manageable, refer to it in terms of cubic feet per minute (cfm). To convert cubic inches to cubic feet, divide the cubic inch number by 1728, as there are 1728 cubic inches per cubic foot. The above formula then becomes

$$\text{Theoretical Air Capacity (cfm)} = \frac{\text{rpm} \times \text{Displacement}}{2 \times 1728}$$

or

$$\text{Theoretical Air Capacity (cfm)} = \frac{\text{rpm} \times \text{Displacement}}{3456}$$

Tire Diameter

To determine the actual diameter in inches, use the following formula:

Tire Diameter

$$= \frac{\text{Section Width} \times \text{Aspect Ratio}}{2540} \times 2 + \text{Rim Diameter}$$

$$= \frac{\text{Section Width} \times \text{Aspect Ratio}}{1270} + \text{Rim Diameter}$$

where 2540 and 1270 are constants.

Torque

$$\text{Torque} = \text{Force} \times \text{Distance}$$

$$\text{Engine Torque} = \frac{5252 \times \text{Horsepower}}{\text{rpm}}$$

Torque is measured in lb-ft. The number 5252 is a constant.

Velocity

To convert velocity in miles per hour to velocity in feet per second (fps), use either of the following two formulas:

$$\text{Velocity (fps)} = \text{Velocity (mph)} \times \frac{22}{15}$$

$$= \text{Velocity (mph)} \times 1.467$$

where the constant $\frac{22}{15} \approx 1.467$.

Volume of a Cylinder

$$V = 0.7854 d^2 L$$

where d is the diameter of the cylinder, L is the length of the cylinder, and 0.7854 is a constant.

Volumetric Efficiency (VE)

Volumetric efficiency is given as a percent. It is a comparison of actual air flow to the theoretical air flow at a given speed. This comparison is usually made in either liters per minute (lpm) or cubic feet per minute (cfm). To determine volumetric efficiency, you must have access to actual air flow data while an engine is in operation. The formulas for determining volumetric efficiency are:

$$\text{VE} = \frac{\text{Actual Air Capacity (cfm)} \times 100}{\text{Theoretical Air Capacity (cfm)}}$$

$$\text{VE} = \frac{\text{Actual Air Capacity (lpm)} \times 100}{\text{Theoretical Air Capacity (lpm)}}$$

APPENDIX C

ANSWERS TO ODD-NUMBERED PRACTICE PROBLEMS

Practice Problems 1–1

Adding Whole Numbers

1. 86
3. 68
5. 177
7. 676
9. 86
11. 283
13. 611
15. 483
17. 2547
19. 7063
21. 41
23. 981
25. 14 quarts
27. 2421 miles
29. $258
31. 1661 kg
33. 27,441
35. 18,160
37. 88 cm^3
39. 482 g
41. $863

Practice Problems 1–2

Subtracting Whole Numbers

1. 32
3. 151
5. 4142
7. 2628

9. 15
11. 224
13. 6181
15. 954
17. 2124
19. 1029
21. 111
23. 178
25. 2111 miles
27. 17 mm
29. 47 minutes
31. $16,478
33. $9750
35. $9800
37. 251 mi
39. Horsepower: 40 hp
Torque: 40 lb-ft
41. $12,482

Practice Problems 1–3

Multiplying Whole Numbers

1. 328
3. 355
5. 288
7. 384
9. 825
11. 3256
13. 4410
15. 1008
17. 1242
19. 3392
21. 102,600

23. 543,000
25. 81,840
27. 280,016
29. 135,983
31. 465,975
33. 180 miles
35. $476
37. 8568 km
39. (a) 3276 miles; (b) $34.25;
    (c) $144.68; (d) $1796.26
41. 278 mi
43. (a) $676
    (b) $3380
45. $96

Practice Problems 1–4
Dividing Whole Numbers

1. 17
3. 83
5. 32
7. 76
9. 153
11. 3 R 5
13. 33 R 3
15. 58 R 9
17. 25 R 21
19. 69 R 5
21. 905 R 4
23. 203 R 13
25. 23 mpg
27. $14
29. 50 mph
31. $117
33. 123 lb
35. $67
37. (a) $231; (b) $2772; (c) $96

Practice Problems 2–1
Adding Decimals

1. 19
3. 46 or 47
5. 59
7. 49
    +5.7
9. 8.7
11. 9.99
13. 8.3
15. 22.81

17. 181.744
19. 29.257
21. 29.77
23. 271.38
25. 172.33
27. (a) $21; (b) $21.86
29. 3.7 hours
31. 174.5 gallons
33. (a) $1063.43; (b) $1107.30; (c) $1371.08
35. 4.0325 in.

Practice Problems 2–2
Subtracting Decimals

1. 35
3. 25
5. 18
7. 81.0
    −0.431
9. 1.2
11. 3.21
13. 8.3
15. 4.41
17. 4.6
19. 219.7
21. 31.1
23. 73.71
25. 131.63
27. 1128.9 km
29. 0.0059 in.
31. 0.029 in.
33. (a) 11.085 in.
    (b) Yes, the drum can be reused.
35. 3.6107 in.

Practice Problems 2–3
Multiplying Decimals

1. 18
3. 104
5. 2.3168
7. 100.1
9. 7.74
11. 8.925
13. 22.675
15. 0.1392
17. 0.15029
19. 0.3294961
21. 78.423375
23. 1108.4

25. 605.5 miles
27. 123.48 pounds
29. (a) 349.9488 in.^{3} or 350 in.^{3}
    (b) 459.844 in.^{3} or 460 in.^{3}
    (c) 181.194 in.^{3} or 181 in.^{3}
    (d) 150.8232 in.^{3} or 151 in.^{3}
31. 1960 rpm
33. 110,719,612.5 gal
35. $150

Practice Problems 2–4

Dividing Decimals

1. 214
3. 2.3
5. 5.4
7. 3.1
9. 300.71
11. 56,090
13. 22.53
15. 525.8
17. 9.6
19. 2000
21. 26.884058 mpg (about 26.9 mpg)
23. Every 1826 miles
25. 21.07 gallons
27. (a) 38.35 in.^{3}; (b) 10.7 : 1 or 10.7
29. $108.27
31. Total yearly cost: $7073; cost per mile driven: $0.47

Practice Problems 3–1

Fractions

1. $\dfrac{7}{3}$
3. $\dfrac{29}{8}$
5. $-\dfrac{25}{8}$
7. $-\dfrac{7}{4}$
9. Yes, $3 \times 8 = 6 \times 4$
11. Yes, $2 \times 18 = 3 \times 12$
13. No; $3 \times 7 = 21$; $5 \times 4 = 20$
15. No; $7 \times 16 = 112$; $8 \times 13 = 104$
17. 20
19. 24
21. -15
23. -60
25. $\dfrac{1}{16}$

27. $-\dfrac{3}{8}$
29. 0.25
31. -0.375
33. 0.875
35. -2.25
37. $\dfrac{1}{8}$ in.
39. 0.75 mm
41. 4.5 hr
43. (a) $\frac{3}{32}'' = 0.09375$ in.; (b) $\frac{1}{8}'' = 0.125$ in.
    (c) $\frac{3}{16}'' = 0.1875$ in.; (d) $\frac{29}{64}'' = 0.453125$ in.

Practice Problems 3–2

Addition of Fractions

1. 6
3. 16
5. 24
7. 12
9. $\dfrac{2}{3} = \dfrac{4}{6}, \dfrac{5}{6} = \dfrac{5}{6}$
11. $\dfrac{5}{8} = \dfrac{10}{16}, \dfrac{7}{16} = \dfrac{7}{16}$
13. $\dfrac{7}{8} = \dfrac{21}{24}, \dfrac{4}{3} = \dfrac{32}{24}$
15. $5\dfrac{2}{3} = \dfrac{17}{3} = \dfrac{68}{12}, \dfrac{3}{4} = \dfrac{9}{12}$
17. $\dfrac{9}{6} = \dfrac{3}{2} = 1\dfrac{1}{2}$
19. $\dfrac{17}{16} = 1\dfrac{1}{16}$
21. $\dfrac{53}{24} = 2\dfrac{5}{24}$
23. $\dfrac{77}{12} = 6\dfrac{5}{12}$
25. $\dfrac{11}{16}$
27. $\dfrac{9}{64}$
29. $\dfrac{11}{16}$
31. $9\dfrac{5}{32}$
33. $4\dfrac{5}{6}$
35. $9\dfrac{7}{12}$
37. $18\dfrac{1}{2}$ qt

39. $5\frac{19}{20}$

41. $4\frac{7}{8}$ hr

43. $32\frac{7}{16}''$ in.

Practice Problems 3–3

Subtraction of Fractions

1. 8

3. 8

5. 30

7. 30

9. $\frac{3}{4} = \frac{6}{8}$; $\frac{5}{8} = \frac{5}{8}$

11. $\frac{7}{8} = \frac{7}{8}$; $\frac{5}{8} = \frac{5}{8}$

13. $\frac{9}{10} = \frac{27}{30}$; $\frac{2}{3} = \frac{20}{30}$

15. $3\frac{4}{5} = \frac{19}{5} = \frac{114}{30}$; $\frac{5}{6} = \frac{25}{30}$

17. $\frac{1}{8}$

19. $\frac{2}{8} = \frac{1}{4}$

21. $\frac{7}{30}$

23. $\frac{89}{30} = 2\frac{29}{30}$

25. $\frac{3}{16}$

27. $\frac{11}{4} = 2\frac{3}{4}$

29. $\frac{1}{32}$

31. $2\frac{7}{8}$

33. $1\frac{1}{6}$

35. $2\frac{1}{16}$

37. $4\frac{3}{4}$ in.

39. 3 ft $7\frac{13}{16}$ in.

41. 5 ft $6\frac{3}{4}$ in.

43. $6\frac{11}{20}$ gallons

45. $4\frac{1}{4}$ L

Practice Problems 3–4

Multiplication of Fractions

1. $\frac{29}{1}$

3. $-\frac{8}{1}$

5. $-\frac{5}{2}$

7. $\frac{109}{8}$

9. $4\frac{2}{10} = \frac{42}{10}$ or $\frac{21}{5}$

11. $-19\frac{375}{1000} = -\frac{19375}{1000}$ or $-\frac{155}{8}$

13. $\frac{8}{15}$

15. $\frac{3}{32}$

17. $\frac{9}{8} = 1\frac{1}{8}$

19. $\frac{42}{48} = \frac{7}{8}$

21. $\frac{96}{40} = 2\frac{16}{40} = 2\frac{2}{5}$ or 2.4

23. $\frac{5075}{300} = 16\frac{275}{300} = 16\frac{11}{12}$

25. $-\frac{133}{40} = -3\frac{13}{40}$

27. $\frac{154}{10} = 15\frac{2}{5}$

29. $16\frac{1}{5}$ miles

31. $8\frac{2}{5}$ L

33. 210 miles

35. **(a)** $126\frac{1}{4}$ mi; **(b)** $151\frac{1}{2}$ mi; **(c)** $6186\frac{1}{4}$ mi

37. \$89.28

39. 22 in.

Practice Problems 3–5

Division of Fractions

1. $\frac{5}{8}$

3. $-3\frac{1}{6}$

5. $\frac{8}{5}$

7. $\dfrac{1}{2}$

9. 4

11. $-\dfrac{2}{3}$

13. $\dfrac{6}{5} = 1\dfrac{1}{5}$

15. $-\dfrac{8}{3} = -2\dfrac{2}{3}$

17. $\dfrac{3}{2} = 1\dfrac{1}{2}$

19. $\dfrac{34}{15} = 2\dfrac{4}{15}$

21. $-\dfrac{220}{375} = -\dfrac{44}{75}$

23. $\dfrac{43}{20} = 2\dfrac{3}{20}$

25. $-\dfrac{20}{17} = -1\dfrac{3}{17}$

27. $-\dfrac{3}{4}$

29. $25\dfrac{19}{69}$ or about 25.28 mpg

31. $\dfrac{133}{42}$ A $= 3\dfrac{1}{6}$ A

33. $50\dfrac{16}{22} = 50\dfrac{8}{11}$ miles

35. 166⅔ min = 2 hr 46⅔ min = 2 hr 46 min 40 s

37. 12 whole pieces with one 1½″ piece left over.

39. 19.3 mpg

Practice Problems 4–1
Angles

1. 30°

3. 75°

5. 105°

7. (b) 45°

9. (b) 135°

11. 120°

13. 0°45′

15. 2°30′

17. $\dfrac{15°}{60} = \dfrac{1°}{4}$

19. $7\dfrac{25°}{60} = 7\dfrac{5°}{12}$

21. 39°48′

23. 2°16′

25. 31°20′

27. 5°45′

29. 180°

31. The person turned completely around and faced the same way as before.

33. $1\dfrac{1°}{2}$

35. South

37. 1st gear: 56.7°; 2nd gear: 42.3°; 3rd gear: 30.6°; 4th gear: 18°; 5th gear: 10.8°

Practice Problems 4–2
Circular Measures

1. About 13.35 cm

3. About 27.10 in.

5. (a) $\dfrac{1}{15}$; (b) about 1.3 in.

6. About 22.56 cm²

9. About 35.13 in.²

11. About 228.26 in.³

13. (a) About 76.3 in.; (b) 76.3 in. $\approx$ 6.36 ft. $\approx 6'4\dfrac{1''}{4}$

15. (a) The P235/70R14 tire has the larger diameter; (b) 6.35 in.

17. (a) 29.06 ft; (b) 23.5 ft; (c) about 36.91 ft

19. (a) About 49.11 in.³; (b) 393 cu. in.

Practice Problems 4–3
The Metric System

1. 13.1232 ft

3. 52.92 lb

5. 54.6832 mph

7. 4.2588 L

9. 2.630·88 kg

11. 1.6764 mm

13. 0.00299212″–0.0090551″

15. 0.3720465″

17. 0.254 mm–0.508 mm

19. 2.0933029″

21. 50.546 mm

23. 3.4448875″

25. 0.5720461″

Practice Problems 5–1
Adding Integers and Signed Numbers

1. 9

3. 1

5. 39

7. -116
9. 9.9
11. -9.4
13. 6
15. -6
17. 5
19. -11
21. 3.1
23. -10.2
25. 0
27. -23
29. -6
31. -6
33. -16.6
35. 157
37. 831.07
39. 1018.95
41. 131.39
43. -48.164
45. $\dfrac{11}{24}$
47. $\dfrac{-23}{12} = -1\dfrac{11}{12}$
49. $\dfrac{5}{6}$
51. $\dfrac{-17}{5} = -3\dfrac{2}{5}$
53. $14°C$
55. $22°F$
57. $+5°$
59. $-0.04''$

Practice Problems 5–2
Subtracting Integers and Signed Numbers

1. 22
3. 53
5. 8
7. -4
9. -12
11. -35
13. -5
15. -11
17. 7.9
19. 35.8
21. -13.5
23. 15.73
25. 177

27. 80
29. -1055
31. 234
33. -148.13
35. 412.69
37. $-1\dfrac{3}{8}$
39. $\dfrac{4}{15}$
41. $7\dfrac{1}{5}$
43. $5\dfrac{2}{15}$
45. $29°F$
47. $7°$
49. $3.6°$ or P3.6

Practice Problems 6–1
Ratios

1. $6:1$
3. $\dfrac{5}{2}$
5. $\dfrac{14}{15}$ or $14:15$
7. $\dfrac{6}{1}$ or $6:1$
9. $2:1$
11. $9:5$
13. $\dfrac{3}{11}$
15. $1:3$
17. $4.5:1$
19. Approximately $3.286:1$
21. $1.38:16$
23. $37:1$
25. $\dfrac{3000}{1500}$ or $2:1$
27. $46:13 = 3.538:1$
29. $8:2304 = 1:288$

Practice Problems 6–2
Proportions

1. Yes, $2 \times 15 = 3 \times 10$
3. Yes, $145 \times 5 = 25 \times 29$
5. 3
7. 8
9. 1.4
11. 35.1

13. $4\frac{9}{14}$

15. $37\frac{1}{2}$

17. 4.267

19. 14

21. $5\frac{5}{6}$

23. $2\frac{242}{321}$

25. 39 teeth

27. 3.8 L

29. 712.5 rpm

31. 3437.5 rpm

33. 7 L

35. 121.5 mL conditioner

37. 72 teeth

39. 4

41. 751.5 rpm

Practice Problems 6–3

Percentages

1. 0.75

3. 0.0275

5. 1.26

7. 0.082

9. 24%

11. 97%

13. 259%

15. 4.5% or $4\frac{1}{2}\%$

17. $\frac{1}{2}$

19. $\frac{4}{5}$

21. $\frac{375}{1000} = \frac{15}{40} = \frac{3}{8}$

23. $\frac{7}{8}$

25. 75%

27. 70%

29. 125%

31. 240%

33. 9

35. 12.8

37. 150

39. 20%

41. 86%

43. 37.5%

45. 300

47. 6 L

49. 18.84%

51. $66.24

53. (a) 180 drivers

(b) 108 drivers were wearing seatbelts.

55. 4 quarts

57. (a) 367.5 mi

(b) 577.5 mi

(c) 183.75 mi

(d) 945 mi

(e) 735 mi

(f) 630 mi

(g) 488.25 mi

(h) 110.25 mi

(i) 425.25 mi

(j) 472.5 mi

(k) 315 mi

59. $241.66

61. (a) $72.38; (b) $181.09

63. 28.49 lb

65. 3.3%

67. Passenger vehicles: 63.4%;
Light trucks: 36.6%

69. Canadians: 128.9%;
Japanese: 360%;
British: 542.2%

Practice Problems 7–1

Discounts

1. $20.33

3. $121.48

5. (a) $1301.97; (b) $1236.87

7. $311.52

9. about 22%.

11. (a) $28.11; (b) 37.5%; (c) $52.33; (d) $22.61;
(e) 30%

13. (a) About 31.7%; (b) About 26.6%;
(c) About 18.6%; (d) About 13%; (e) Since the price
with next day shipping is $139.10, more than the price of
the spark plugs, there is no discount.

15. $2487.95

Practice Problems 7–2

Profit, Loss, and Commissions

1. About 8.6%

3. About 9.8%

5. (a) Yes, the expenses only add up to 90%;
  (b) $750; **(c)** 10%

7. (a) No, the expenses total 109.4%.

  (b) The company lost $587.50.

  (c) Since there was a loss, there is no profit margin.

9.

Junior's Auto Repair
Finances for September–December

	September	October	November	December
Net Sales	$21,178.00	$23,964.00	$32,451.00	$17,642.00
Variable Expenses: Salaries/Wages (42%)	8,894.76		13,629.42	
Parts (15%)	3,176.70		4,867.65	
Legal/Accounting ($250 + 1.5%)	567.67		736.77	
Advertising (1%)	211.78		324.51	
Office Supplies (0.5%)	105.89		162.26	
Utilities (6%)	1,270.68		1,947.06	
Payroll Taxes (19% of salaries)	1,690.00		2,589.59	
Total Variable Expenses	15,917.48		24,257.26	
Fixed Expenses: Rent	1,775.00	1,775.00	1,775.00	1,775.00
Insurance	215.00	215.00	215.00	215.00
Licenses/Permits	0	0	0	250.00
Loan Payments	2,135.00	2,135.00	2,135.00	2,135.00
Total Fixed Expenses	4,125.00		4,125.00	
Total Expenses	20,042.48		28,382.26	
Net Profit (loss) before Taxes	1,135.52		4,068.74	
Taxes (24% of net profit)	272.52		976.50	
Net Profit (loss) after Taxes	863.00		3,092.24	
Profit Margin	4.07		9.53	

Practice Problems 7–3

Interest and Payments

1. $42.75

3. (a) $961; **(b)** $13,361

5. $265.45

7. $4024.82

9. (a) 26; **(b)** $414.31 every two weeks;
  (c) $323,161.80; **(d)** $154.20

11. (a) $480

  (b) See table on the top of page 343.

A loan of $8,000 with payments made monthly at 9.9% compounded annually Each of the first 17 payments are $481.00. The last payment is $463.09.			
Payment	**Principal**	**Interest**	**Unpaid Balance**
1	415.00	66.00	7585.00
2	418.42	62.58	7166.58
3	421.88	59.12	6744.70
4	425.36	55.64	6319.34
5	428.87	52.13	5890.48
6	432.40	48.60	5458.08
7	435.97	45.03	5022.10
8	439.57	41.43	4582.54
9	443.19	37.81	4139.34
10	446.85	34.15	3692.49
11	450.54	30.46	3241.96
12	454.25	26.75	2787.70
13	458.00	23.00	2329.70
14	461.78	19.22	1867.92
15	465.59	15.41	1402.33
16	469.43	11.57	932.90
17	473.30	7.70	459.60
18	459.60	3.79	0.00
TOTAL	8000.00	640.39	

Practice Problems 7–4

Taxes

1. **(a)** $162.69, **(b)** $38.05, **(c)** $293.70
3. **(a)** $72.17, **(b)** $16.88, **(c)** $131.25
5. **(a)** $97.90, **(b)** $22.90, **(c)** $97.90
7. $12,065
9. $7,021.85

Practice Problems 7–5

1. Mean: 4.94;
   Median: 4.75;
   Mode: none

3. Mean: 88.35;
   Median: 90.5;
   Mode: 96.0
5. Mean: 113.675;
   Median: 107;
   Mode: none

Practice Problems 8–1

Completing Repair Orders

1. **(a)** 0.5
   (b) 0.2
   (c) 0.8
   (d) 0.1
   (e) 1.0
   (f) 1.6
   (g) 3.4
3. 0.9
5. $667.38
7. $43.75
9. $443.61
11. **(a)** $673.05 **(b)** $359.92
13. $855.32
15. $188.17

Practice Problems 9–1

The Automotive Engine

1. 124 kg
3. **(a)** 6.2 hours; **(b)** $492.90; **(c)** $221.81
5. $\frac{11}{32} = 0.34375$ inches
7. **(a)** 32.482 cu. in.; **(b)** 259.856 cu. in.
9. **(a)** 1.92 L; **(b)** 1920 cm^3
11. 0.0052″; remove
13. 0.13 mm; remove
15. 0.29 mm; remove

	Out-of-Round	Specifications	Within Specs?
17.	0.0045″	0.005″	Yes

19. 2000

	21. **Buick** **Exhaust Valve**	23. **Nissan** **Intake Valve**
Inside diameter of valve guide	7.948 mm	7.013 mm
Outside diameter of valve stem	7.912 mm	6.982 mm
Valve stem to guide clearance	0.036 mm	0.031 mm
Specifications	0.035–0.078 mm	0.1 mm Max tolerance
Within specs?	Yes	Yes

	25. **Ford 2.3 L**	27. **Nissan** **Intake Cam Lobe**
Dimension A	1.275″	
Dimension B	1.515″	1.7515″
Lobe lift	0.24″	
Specifications	0.2437″–0.005″ Max	1.7518–1.7520
Within specs?	Yes	No

29. **(a)** 2,880,000 revolutions; **(b)** 13; **(c)** 1,950,000; **(d)** 715

	31. **Buick** **1.8 L**	33. **Ford 2.3 L**
Taper	0.145 mm Max	0.0041″
Specifications	0.02 mm Max	0.010″
Within specs?	No	Yes

35. **(a)** 0.0037 in.; **(b)** Yes
37. **(a)** 0.0221 mm; **(b)** Yes, this exceeds the specs of 0.0065–0.0091 mm.
39. **(a)** 0.032 mm; **(b)** 0.025–0.079 mm; **(c)** Yes
41. 0.399668 inches
43. 360:45 or 8:1
45. **(a)** 6300 rpm; **(b)** 465 hp
47. 2300 rpm
49. The torque drop is 140 lb-ft.
51. **(a)** Yes, 128 is more than $156 \times 0.75 = 117$; **(b)** No, 130 is less than $180 \times 0.75 = 135$; **(c)** No; **(d)** Yes; **(e)** No; **(f)** No
53. 87° Valve duration; Intake 214°; Exhaust 242°; Valve overlap 36°
55. **(a)** 121.41 in.^{3}; **(c)** 137.49 in.^{3}
57. **(a)** 1.59 L; **(b)** 5.77 L; **(c)** 3 L; **(d)** 1.82 L
59. **(a)** 79.21 hp; **(b)** 153.47 hp; **(c)** 157.08 hp; **(d)** 233.05 hp

Practice Problems 10–1
Automobile Engine Systems

1. **(a)** 9 oil changes; **(b)** 45 quarts of oil; **(c)** $1^{21}/_{24}$ or 1.875 cases
3. **(a)** 210.5 pints; **(b)** 105.25 quarts; **(c)** 26.3 gallons
5. 62.3 L
7. **(a)** 29 gallons; **(b)** 26.4992 L; **(c)** 109.777 L
9. **(a)** You will save $4.68 on a case or **(b)** $0.20 (20¢) on each quart
11. 8 quarts
13. **(a)** $17.79; **(b)** $0.71; **(c)** $18.50; **(d)** $8.26 saved
15. −34°F
17. The protection would be 0°F.
19. **(a)** $4^{1}/_{2}$ quarts; **(b)** 5.4 quarts; **(c)** 46% antifreeze requires 4.14 quarts
21. 20 gallons

23. 97.5 pounds

25. 67.5 pounds

27. 375 miles

29. (a) 13.3 L ≈ 14.1 qt; (b) 4.3 L ≈ 4.5 qt;
(c) 70 L ≈ 18.4 gal; (d) 1.5 L ≈ 3.2 pt

31. $70.55

33. 286.5 L

35. Pressure must be 200 − 15 = 185 bars.
185 bars = 185 × 14.5 psi/bar = 2682.5 psi.

37. 6 GPM

39. (a) 46.2; (b) 431.2

41. (a) 2.48 in.; (b) 2.22 in.; (c) 2.06 in.; (d) 1.75 in.

43. (a) Theoretical air capacity ≈ 388.85995 cfm
VE ≈ 78.4%

(b) 89.4%

(c) Theoretical air capacity ≈ 253.81944 cfm
VE ≈ 87.07% or 87.1%

(d) 105.8%

45. 11.7 gal

47. 13%

Practice Problems 11–1
Automobile Electrical Systems

1. (a) 13.5 volts (13.5 V);

(b) 13.5 volts; (c) 10 V

3. (a) 0.78 ohms (0.78 Ω); (b) 7 ohms;
(c) 1.75 ohms

5. 84 watts (84 W)

7. 8.04 hp (or 8 hp) (6000 ÷ 746 = 8.04, which is about 8)

9. (a) 3 Ω; (b) 4 A

11. 3.5 Ω

13. 1

15. 750 ampere-turns (2 amps × 375 turns)

17. 204 A

19. 2 hours 30 minutes

21. 58.8%

23. (a) 1750 ohms per inch or 24 000 ohms
(2000 ohms/in. × 12 in.); (b) Yes

25. (a) 3600 W (360 A × 10 V); (b) 1080 W
(80 A × 13.5 V); (c) About 3.33 times as long (3600 ÷
1080) or approximately 50 seconds.

27. (a) $13,546.88; (b) $18,361.69; (c) $16,714.29;
(d) Diesel; (e) $1,647.40

29. (a) 2 h 15 min; (b) 101.25 miles; (c) 135 miles;
(d) Yes

31. (a) No; (b) Yes; (c) Yes; (d) No; (e) Yes;
(f) Yes; (g) No; (h) Yes; (i) No

33.

Wire Length	Current Flow (Amps)	Wire Gauge Size
Decrease	Increase	Same
Decrease	Same	Increase
Same	Increase	Decrease
Increase	Decrease	Same
Increase	Same	Decrease
Same	Decrease	Increase

35. The voltage drop at A is 2.7 Ω. The voltage drop at B is
4.4955 Ω. The voltage drop at C is 6.3045 Ω.

37. (a) The voltage drop across the starter relay is 3.4 V.

(b) The starter relay

(c) 3.4 V

(d) The neutral safety switch

39. $970.72

41. $720.99

Practice Problems 12–1
The Automobile Drive Train

1. (a) 116 lb-ft; (b) 1392 lb-in.

3. 18 N·m

5. 450 lb-ft.

7. 784.08 lb

9. $\frac{7°}{8} = 0.875°$

11. 870 lb-ft.

13. (a) 359.26 N·m; (b) 265.54 N·m; (c) 191.7 N·m;
(d) 142 N·m; (e) 408.96 N·m

15. 30%

17. (a) 2201.31 lb-ft; (b) 1339.21 lb-ft; (c) 837 lb-ft;
(d) 585.9 lb-ft

19. 15.895 : 1

21. A : B = 25 : 53 or 0.47 : 1; C : D = 28 : 56 or
0.5 : 1; Total = 0.236 : 1

23. 2848 lb

25. 0.060″

27. 4.25 turns

29. 96 psi

31. 2760 liters

33. (a) i. 1st gear, 3 : 1; ii. 2nd gear, 1.89 : 1;
iii. 3rd gear, 1.24 : 1; iv. 4th gear, 1 : 1

(b) i. 1st gear, 11.16 : 1; ii. 2nd gear, 7.03 : 1;
iii. 3rd gear, 4.61 : 1; iv. 4th gear, 3.72 : 1

35. (a) 72.41 mph; (b) 63.36 mph; (c) 81.47 mph

37. (a) 1st gear: 3952.36 lb-ft

2nd gear: 2754.67

3rd gear: 1902.99

4th gear: 1330.76

5th gear: 1064.61

6th gear: 825.07

(b) 1704.47 rpm

(c) 1st gear: 498.0 mph

2nd gear: 71.4

3rd gear: 103.3

4th gear: 147.8

5th gear: 155.2

6th gear: 162.1

39. (a) $1.45:1$; **(b)** $-2.21:1$

41. 31.9–32.5 mpg

Practice Problems 13–1

The Automobile Chassis

1. 2.40 sq. in.

3. 3.99 cu. in.

5. **(a)** 79.128 cm^3; **(b)** 79.128 mL; **(c)** 0.079 L (or 0.079 128 L)

7. **(a)** 17.785 cu. in.; **(b)** 18.111 cu. in.; **(c)** 0.326 cu. in.

9. 157 psi

11. 3500.02 lbf

13. **(a)** 1125 N; **(b)** 4.91 cm^2; **(c)** 229.12 N/cm^2 (2291.2 kPa); **(d)** 7.065 cm^2; **(e)** 1618.73 N; **(f)** 3.14 cm^2; **(g)** 719.44 N

15. 168.214 lb

17. 2.502 inches

19. Area of drum brake piston is 6.15 cm^2. Area of disc brake piston is 30.75 cm^2. Diameter of disc brake piston is 6.26 cm.

21. % Front: 51.9; % Rear: 48.1

23. 7659 kPa

25. **(a)** 0.038 in.; **(b)** 0.022 in.

27. **(a)** 0.0075 mm; **(b)** The maximum allowed is 0.1016 mm; **(c)** No, 0.0075 is less than the maximum allowed.

29. 12.636 cm

31. **(a)** $+\dfrac{3°}{4} - + 2\dfrac{1°}{4}$;

(b) $0°$. It is within specifications.

33. 6

35. **(a)** $180°$; **(b)** $60°$

37. Steering wheel turns $4\frac{2}{3}$ turns $\times 360°$/turn $= 1680°$. Steering ratio $= 1680:60 = 28:1$

39. 8 liters/min

41. 7592 kPa (9490×0.8)

43. **(a)** 3.65; **(b)** 3.96; **(c)** 3.01; **(d)** 3.36

45. **(a)** About 61.24 mph; **(b)** About 58.01 mph; **(c)** About 58.67 mph; **(d)** About 63.15 mph

47. **(a)** 65.45 mph; **(b)** 50.70 mph; **(c)** 200.00 mph; **(d)** 32.14 mph; **(e)** 58.06 mph

49. Front: 49%; Rear: 51%; L/side: 55%; R/side: 45%

(a) 0.51; **(b)** 0.77; **(c)** 1.01; **(d)** 0.98

51. **(a)** 0.51g; **(b)** 0.77g; **(c)** 1.01g; **(d)** 0.98g

53. In general, to get the overall gear ratio in first gear, multiply the rear axle ratio and the transmission ratio for that gear.

(a) $10.318:1$; **(b)** $6.13:1$; **(c)** $4.10:1$; **(d)** $3.08:1$; **(e)** $2.09:1$

55. **(a)** 1938 lb

(b) There is 1462 lb on the rear wheels.

57. **(a)** $167.35; **(b)** $459.86; **(c)** $30; **(d)** $197.35

59. **(a)** $246.70; **(b)** $986.80

Practice Problems 14–1

Automobile Heating, Ventilation, and Air Conditioning

1. 51 Btu

3. **(a)** $88°F$; **(b)** 1144 Btu

5. **(a)** $59°C$; **(b)** 295 000 cal; **(c)** 295 kcal ($5 \text{ kg} \times 59°C$)

7. 140 lb

9. 998.960 cal

11. 80 000 000 cal

13. $\dfrac{2}{3}$ ton

15. $51.7°C$

17. 103.9 psi

19. **(a)** 40.0; **(b)** 113.5 psi; **(c)** 73.5 psi

21. **(a)** $-20.6°C$; 4.1 psi; **(c)** 28.3 kPa

23. At $70°F$, the pressure/temperature is $70°/70.9$. As one moves from $70°$, the relationships between pressure and temperature become more distant.

25. $10°C$

27. $197.6°F$

29. 11 oz

31. **(a)** 5 oz; **(b)** 45.5%

33. **(a)** 60 oz; **(b)** 5 cans

35. 1.15 kg

37. **(a)** 4.6 hours; **(b)** $303.60

39. **(a)** $436.47; **(b)** $135.20; **(c)** $237.75

41. **(a)** 350.61 or 351 horsepower; **(b)** 16 horsepower; **(c)** 11.93 kW $\approx$ 12 kW; **(d)** 4.8%

43. **(a)** $212°F$; **(b)** $236°F$; **(c)** $254°F$; **(d)** $230°F$; **(e)** $257°F$; **(f)** $269°F$; **(g)** $246°F$; **(h)** $267°F$; **(i)** $291°F$

Practice Problems 15–1

Digital Meters

1. 1850 revolutions per minute
3. 13.28 volts of direct current
5. 18 milliseconds
7. 12 milliamps of direct current

Practice Problems 15–3

Dial Indicators

1. (a) 0.562 mm; (b) 0.062 mm; (c) 0.262 mm
3. (a) 0.492 mm; (b) 0.092 mm; (c) 0.292 mm; (d) 0.592 mm

Practice Problems 15–2

Scale Measurement

1. (a) $^{13}/_{32}$ in.; (b) $^{31}/_{32}$ in.; (c) $1^{7}/_{16}$ in.; (d) $2^{1}/_{32}$ in.; (e) $2^{3}/_{4}$ in.; (f) $3^{1}/_{8}$ in.; (g) $3^{15}/_{32}$ in.; (h) $3^{13}/_{16}$ in.
3. (a) $1^{12}/_{16} = 1^{3}/_{4}$ in.; (b) $2^{3}/_{8}$ in.; (c) $1^{5}/_{16}$ in.; (d) $1^{7}/_{8}$ in.

5.

Car: Component:	Nissan Valve Deflection	Buick Connecting Rod Side Clearance	Ford Camshaft End Play
Specification	0.008″	0.10–0.61 mm	0.009 service limit
Reading	0.0040″	0.022 mm	0.006″
Pass/Fail	Pass	Fail	Pass

Car: Component:	Dodge Crankshaft End Play	Ford Lobe Lift	Buick Crankshaft End Play
Specification	0.002″–0.007″	0.2437″	0.05–0.18 mm
Reading	0.0064″	0.206″	0.0068 mm
Pass/Fail	Pass	Fail	Fail

Car: Component:	Ford Camshaft Runout	Dodge Camshaft End Play	Buick Valve Seat Runout
Specification	0.005 T.I.R.	0.004/0.008″	0.05 mm
Reading	0.0021″	0.0089″	0.0068 mm
Pass/Fail	Pass	Fail	Pass

Car: Component:	Buick Camshaft Lift	Nissan Camshaft End Play	Dodge Front Brake Disc Runout
Specification	6.65 mm ± 0.05 mm	0.1016 mm	0.08 mm max
Reading	0.214″	0.0085″	0.011 mm
Pass/Fail	Fail	Fail	Pass

Practice Problems 15–4

Micrometers

1. 0.240″
3. 1.800″
5. 2.599″
7. 3.434″
9. 4.898″
11. 21.05 mm
13. 5.78 mm
15. 6.82 mm
17. 7.54 mm
19. 2.374″

21. 3.245″
23. 3.837″

Practice Problems 15–5
Drill Sizes

1. (a) Q; **(b)** U; **(c)** $^{11}/_{16}$ in.; **(d)** 14; **(e)** 7;
(f) $^{27}/_{64}$ in.; **(g)** 3; **(h)** F; **(i)** $^{13}/_{16}$ in.; **(j)** $^{7}/_{8}$ in.

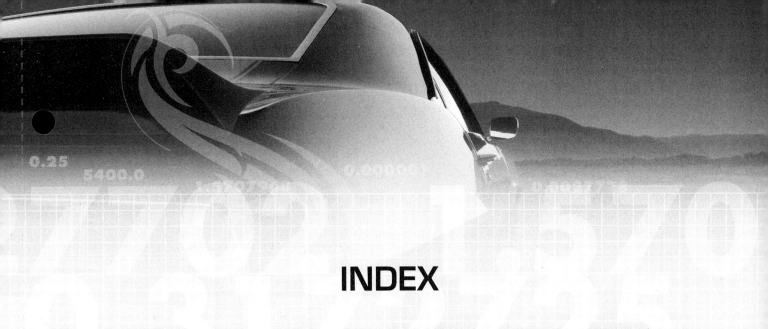

INDEX